MASTER ELECTRICIAN'S Exam Workbook

based on the
2017 NEC® National Electrical Code

atp AMERICAN TECHNICAL PUBLISHERS
Orland Park, Illinois 60467-5756

Lowell Reith

Master Electrician's Exam Workbook Based on the 2017 NEC® contains procedures commonly practiced in industry and the trade. Specific procedures vary with each task and must be performed by a qualified person. For maximum safety, always refer to specific manufacturer recommendations, insurance regulations, specific job site and plant procedures, applicable federal, state, and local regulations, and any authority having jurisdiction. The material contained herein is intended to be an educational resource for the user. American Technical Publishers, Inc. assumes no responsibility or liability in connection with this material or its use by any individual or organization.

American Technical Publishers, Inc., Editorial Staff

Editor in Chief:
    Jonathan F. Gosse
Vice President—Editorial:
    Peter A. Zurlis
Assistant Production Manager:
    Nicole D. Bigos
Technical Editor:
    Greg A. Gasior
Supervising Copy Editor:
    Catherine A. Mini
Copy Editor:
    Talia J. Lambarki

Cover Design:
    Steven E. Gibbs
Art Supervisor:
    Sarah E. Kaducak
Illustration/Layout:
    Steven E. Gibbs
Digital Media Manager:
    Adam T. Schuldt
Digital Resources:
    Lauren M. Lenoir
    Cory S. Butler

Acknowledgments

The author and publisher are grateful to the following organizations and individuals for providing images.

ABB, Inc.
Blodgett Oven Company
Ductsox
Entourage
Fluke Corporation

Ideal Industries, Inc.
MayTag
Roger A. Brooks, Architect
Ruud Lighting, Inc.
UE Systems, Inc.

Technical review provided by:
Roger Zieg
Instructor
Lewellyn Technology

National Electrical Code and NEC are registered trademarks of National Fire Protection Association, Inc. Underwriters Laboratories Inc. is a registered trademark of Underwriters Laboratories Inc. Quick Quiz, Quick Quizzes, and QuickLinks are either registered trademarks or trademarks of American Technical Publishers, Inc.

© 2017 by American Technical Publishers, Inc.
All rights reserved

1 2 3 4 5 6 7 8 9 – 17 – 9 8 7 6 5 4 3 2 1

Printed in the United States of America

    ISBN 978-0-8269-2044-7

 This book is printed on recycled paper.

Contents

CHAPTER 1   **The Master Electrician's Exam**   1

Becoming a Master Electrician • Apprentice • Journeyman Electrician • Master Electrician • Master Electrician's Exam • Preparing the NEC® • Tabbing the NEC® • Studying for the Exam • Exam Preparation • Taking the Exam • Master Electrician's Exam Workbook

CHAPTER 2   **Direct Current Principles**   7

Atoms • Current • Resistance • Voltage • Batteries • Ohm's Law • Voltage, Current, and Resistance Relationship • Determining Voltage, Current, and Resistance Using Ohm's Law • Determining Power, Voltage, and Current Using the Power Formula • Formula Wheel • Voltmeters • Ammeters • Ohmmeters • Megohmmeters • Watt-Hour Meters

Review Questions 19

CHAPTER 3   **Alternating Current Principles**   21

Generators • 1ϕ Generators • 3ϕ Generators • Sine Waves • Frequency • Effective Voltage • Inductance • Inductive Reactance • Capacitance • Checking Capacitors • Discharging Capacitors • Uses for Capacitors • Capacitive Reactance • Impedance • Power Factor • Lagging/Leading Power Factor • In-Phase/Out-of-Phase • Reactive Power

Review Questions 33

CHAPTER 4   **Circuits**   35

Circuit Requirements • Determining Current, Voltage, Resistance, and Power in Series Circuits • Determining Current, Voltage, Resistance, and Power in Parallel Circuits • Combination Circuits • Resistance in Combination Circuits • Multiwire Circuits

Review Questions 45

CHAPTER 5   **Transformers**   47

Transformer Ratios • Turns Ratios • Voltage Ratios • Current Ratios • Rating Transformers • Transformer Efficiency • Sizing Transformers • Sizing 1ϕ Transformers • Sizing 3ϕ Transformers • Autotransformers • Current Transformers • Potential Transformers • Transformer Protection Devices • Transformer Protection—Over 1000 Volts • Transformer Protection—Under 1000 Volts • Delta-Wye Calculations • Transformer Connections • 1ϕ Transformer Connections • 3ϕ Transformer Connections

Review Questions 63

CHAPTER 6   **Grounding, Bonding, and Neutrals**   65

Grounding • Bonding • Sizing the Grounded Conductor • Sizing the Main Bonding Jumper • Sizing the Grounding Electrode Conductor • Sizing Equipment Grounding Conductors • Sizing System Bonding Jumpers • Sizing System Equipment Bonding Jumpers • Sizing System Grounding Electrode Conductors • Sizing System-Grounded Conductors • Neutral Conductors for 1ϕ and 3ϕ Systems • 3ϕ Delta Neutrals • 3ϕ Wye Neutrals • Derating Neutrals • Ungrounded Systems

Review Questions 81

Contents

CHAPTER 7 — Conductor Ampacity and Protective Devices — 85

Current-Carrying Conductors • Parallel Conductors • Minimum Size of Conductors • Ampacity • Ambient Temperature • Conduit Fill • Neutral as a Current-Carrying Conductor • Temperature and Conduit-Fill Derating Factors • Auxiliary Gutters and Wireways • Terminal Temperature Rating • Sizing Conductors for Branch Circuits • Sizing Branch Circuit Conductors for Continuous Loads • Busbar Ampacity • Overcurrent Protection for Devices Rated 800 A or Less • Overcurrent Protection for Devices Rated over 800 A • Overcurrent Protection for Small Conductors • Types of Overcurrent Protective Devices

Review Questions 103

CHAPTER 8 — Voltage Drop — 105

Voltage Drop Recommendations from the NEC® • Factors that Affect Voltage Drop • Reducing Voltage Drop • Table 8 — Conductor Properties • Single-Phase (1ϕ) Branch Circuit Calculations • Using Ohm's Law to Determine Voltage Drop • Using the Voltage Drop Formula to Determine Voltage Drop • Three-Phase (3ϕ) Branch Circuit Calculations • Determining Power Loss in Circuits • Voltage Drop for Feeder and Branch Circuits Combined

Review Questions 119

CHAPTER 9 — Motors — 121

Motor Calculations • Determining Branch-Circuit Conductor Size • Overload Calculations for Motors • Branch/Short-Circuit Ground Fault Protection Calculation • Feeder Conductor Sizing Calculation • Motor Input Volt-Amp Rating • Motor Speeds • Motor Torque • Motor Horsepower • Motor Nameplate Current • Motor Connections • Motor Voltage Drop • Motor Control Circuit Conductors • Locked-Rotor Current Ratings • Variable-Frequency Drives

Review Questions 143

CHAPTER 10 — Conductor-Fill, Box-Fill, and Pull and Junction Box Sizing — 147

Conductor Fill • Sizing Raceways for Different-Size Conductors • Nipple Conduits • Wireways • Maximum Number of Same-Size Conductors in Wireways • Sizing Wireways • Installing Conductors into Existing Raceways • Box Fill • Plaster Rings and Raised Covers • Pull Boxes • Sizing Pull Boxes — 600 Volts and Under • Distance between Conduits in Pull Boxes that Contain the Same Conductors — 1000 Volts and Under • Sizing Pull Boxes — Over 1000 Volts

Review Questions 167

CHAPTER 11 — Dwellings — 171

Standard Calculations for One-Family Dwellings • Sizing Service Conductors for One-Family Dwellings • Optional Calculation for One-Family Dwellings • Standard Calculation for Multifamily Dwellings • Optional Calculation for Multifamily Dwellings • Farm Loads • Standard Calculation for Mobile Homes and Manufactured Homes • Calculating Total Demand Load for Mobile and Manufactured Homes • Sizing Service Load for Mobile and Manufactured Homes • Calculations for Mobile Home Parks

Review Questions 197

CHAPTER 12

Nondwelling Occupancies — 201

Continuous Lighting Loads • Demand Factors • Required Number of Lighting Circuits • Nondwelling Lighting • Show Window Lighting • Track Lighting • Signs • Receptacle Loads • Multioutlet Assemblies • Heating or Air Conditioning • Commercial Cooking Equipment • Fixed Appliances • Optional Service Calculations • Optional School and Restaurant Calculations • Welding Machines • Single Non-Motor-Operated Appliances • Capacitors • Refrigeration Equipment • Phase Converters • Generators • Other Equipment Used in Nondwelling Occupancies

Review Questions 213

CHAPTER 13

Printreading — 215

Prints • Floor Plans • Plot Plans • Elevation View • One-Line Riser Drawing • Specifications • Support Information • Title Blocks • Drawing Scale • Drawing Notes • Symbols • Schedules • Panelboard Schedules • Luminaire Schedule • Mechanical Equipment Circuit Schedule

Review Questions 225

CHAPTER 14

Practice Exams — 227

Practice Exam 14-1 229
Practice Exam 14-2 243
Practice Exam 14-3 257
Practice Exam 14-4 271

Practice Question Answers — 287

Appendix — 293

Glossary — 317

Index — 321

Learner Resources

- Quick Quizzes®
- Illustrated Glossary
- Flash Cards
- Voltage Drop Calculator
- NEC® Sample Exams
- Media Library
- ATPeResources.com

Introduction

Master Electrician's Exam Workbook Based on the 2017 NEC® is designed to help prepare for a master electrician's exam. This comprehensive workbook covers information ranging from basic electrical theory to complex load calculations. Key topics include transformers, grounding, overcurrent protection, voltage drop, motors, box-fill calculations, and load calculations. Step-by-step examples are used to simplify complex calculations. Practice questions and review questions reinforce comprehension of material throughout the workbook. The four 100 question practice exams in the back of the book simulate actual testing conditions and help the learner become efficient at looking up code information.

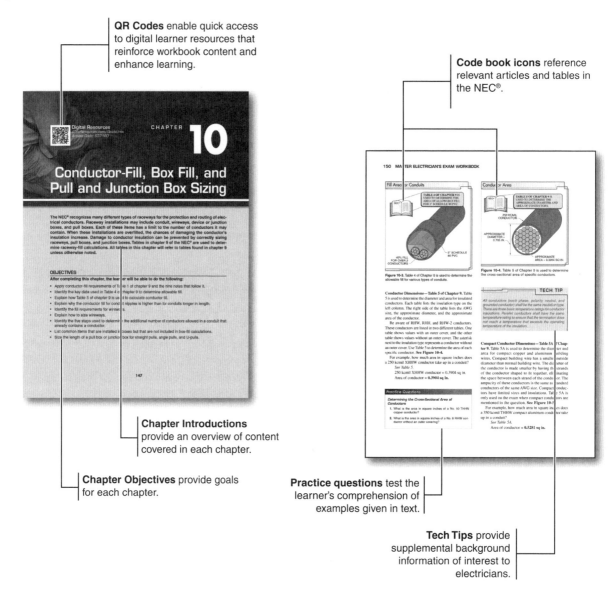

QR Codes enable quick access to digital learner resources that reinforce workbook content and enhance learning.

Code book icons reference relevant articles and tables in the NEC®.

Chapter Introductions provide an overview of content covered in each chapter.

Chapter Objectives provide goals for each chapter.

Practice questions test the learner's comprehension of examples given in text.

Tech Tips provide supplemental background information of interest to electricians.

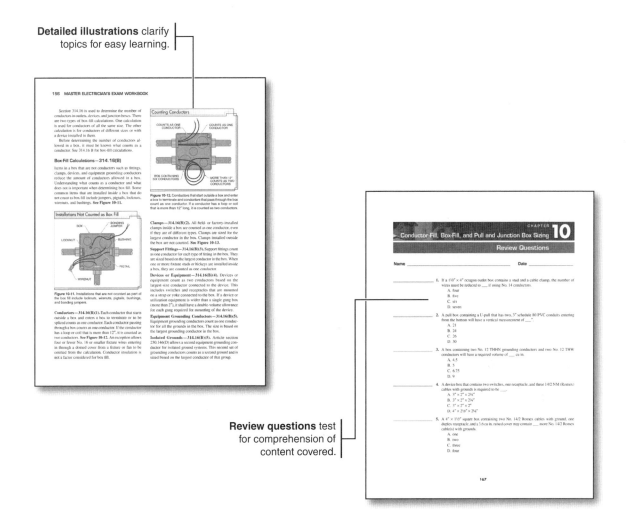

About the NEC®

- Calculations resulting in a fraction less than 0.5 are permitted to be dropped per Informative Annex D of the NEC®, Examples.
- Calculations for range loads using the Standard Calculation or the Optional Calculation that result in a fraction less than 0.5 kW are permitted to be dropped per Informative Annex D of the NEC®, Examples.
- Calculations for conductors having the same cross-sectional area and the same insulation resulting in 0.8 or larger are rounded to the next higher number per Note 7 of Notes to Tables, Chapter 9.
- The word "shall" indicates a mandatory rule.
- Informational Notes (Notes) are explanatory or provide additional information. Notes are not mandatory and do not contain any mandatory provisions.

Learner Resources

Master Electrician's Exam Workbook Based on the 2017 NEC® includes access to online Learner Resources that reinforce textbook content and enhance learning. These online resources can be accessed using either of the following methods:

- Key ATPeResources.com/QuickLinks into a web browser and enter QuickLinks™ Access code 637160.
- Use a Quick Response (QR) reader app to scan the QR Code with a mobile device.

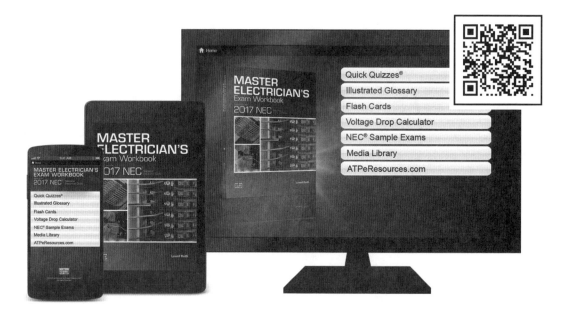

The learner resources include the following:

- **Quick Quizzes®** that provide an interactive review of fundamental concepts covered in the chapters
- **Illustrated Glossary** that provides a helpful reference for key terms included in the text; selected terms are linked to illustrations that enhance the provided definitions
- **Flash Cards** that enable a review of electrical terms and definitions
- **Voltage Drop Calculator** that determines voltage drop for various types of circuits
- **NEC® Sample Exams** that provide practice in preparing for a master electrician's exam; each exam includes a timer to simulate an examination setting
- **Media Library** that contains video clips and animated graphics that depict related electrical/electronic concepts
- **ATPeResources.com,** which provides a comprehensive array of instructional resources, including Internet links to manufacturers, associations, and ATP resources

Digital Resources
ATPeResources.com/QuickLinks
Access Code: 637160

CHAPTER 1

The Master Electrician's Exam

The highest level of achievement in an electrician's career is master electrician. As with any type of profession, electricians may spend years trying to master their craft. It may take an electrician up to six years to be fully prepared to test for the level of master electrician. The path to becoming a master electrician will require years of formal education and on-the-job training, which all culminates with passing of the master electrician's exam.

OBJECTIVES

After completing this chapter, the learner will be able to do the following:

- List the steps involved in becoming a master electrician.
- Describe the types of questions on the exam.
- Apply tips for studying, preparing, and taking the exam.
- Understand how to use the *Master Electrician's Exam Workbook*.

BECOMING A MASTER ELECTRICIAN

An individual training to become a master electrician typically begins as an apprentice. After four or five years as an apprentice, the apprentice becomes a journeyman by passing a journeyman electrician's exam. After journeyman, an electrician must pass the master electrician's exam to become a master electrician. Other career opportunities for electricians include foreman, project manager, supervisor, estimator, and owner of an electrical contracting company.

Apprentice

Most electricians generally start their careers by enrolling in an electrical apprenticeship program in which a learner typically works as an electrical apprentice during the day and takes classes at night. There are also technical/vocational schools that offer certificate or degree programs for electricians. The required education varies by state. Most electrical apprenticeship programs last about four years. During that time, an apprentice works with an experienced electrician on various types of projects to gain knowledge and experience. The responsibilities of the apprentice increase as they move through the different levels of an apprenticeship program. Apprentices will usually attend classes that cover subjects related to the electrical field, such as safety and general electrical theory.

Journeyman Electrician

The next major step in an electrician's career is becoming a journeyman electrician. A journeyman electrician is viewed in the field as a competent and qualified person. Journeymen will often train apprentices. Most states require at least 8000 hours of job experience to be recognized as a journeyman electrician. Many states will accept up to 2000 hours of schooling from recognized technical colleges or other electrical training facilities. Once apprentices reach the required working hours, they may apply to the appropriate office in the state in which they are working in order to take the journeyman electrician's exam.

Master Electrician

The highest level to achieve in an electrician's career is master electrician. A master electrician oversees all of the work being done by the journeyman and apprentices. While some states require one master electrician for each company, other states may require a master electrician for each job site. Most states require between 10,000 and 12,000 hours of job experience to be qualified to take the master electrician's exam. A journeyman, after reaching the required working hours, may apply to the appropriate office in the state in which they are working in order to take the master electrician's exam. Some states also require that the electrician hold a journeyman license for at least one year before taking the master electrician's exam.

Along with the work experience, most states require an electrician to pass a trade knowledge test. The trade knowledge test is usually a multiple choice test of 80 to 100 questions. Applicants who pass the master electrician's exam are issued a master electrician's card. The card contains the name of the licensing board, the name of the cardholder, license number, expiration date, and issuance date. **See Figure 1-1.**

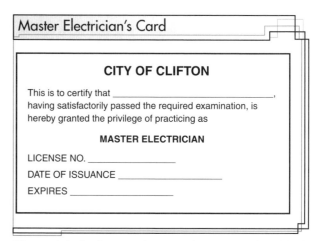

Figure 1-1. Applicants who pass the master electrician's exam are issued a master electrician card by the authority having jurisdiction in the area.

MASTER ELECTRICIAN'S EXAM

The time required to prepare for the exam depends on the individual. Some individuals study for up to three to six months before taking the exam. An electrician preparing to take the exam should start by reviewing the outline of the exam. The exam outline typically lists the topics that will be on the exam along with the number of questions on each topic. **See Figure 1-2.** The outline of each state's exam can usually be found on the Internet. The major difference between individual state exams is the number of questions and how the questions are answered. For example, some exams are taken with a paper and pencil while others may be taken on a computer.

Exam Outline

Master Electrician's Exam

Exam Outline
100 multiple choice questions
5 hr time limit
75% needed to pass exam

Major Content Area	No. of Questions	Percent of Exam
Service Transformers and Equipment	19	19%
Wiring Methods and Installation	16	16%
Cabinets, Panelboards, Switchboards, Boxes, and Conduit Bodies	4	4%
Conductors	14	14%
Control Devices	5	5%
Motors and Generators	10	10%
Utilization Equipment and Devices	7	7%
Special Occupancies and Uses	10	10%
Miscellaneous	5	5%
Plan Reading and Analysis	10	10%

The following references are permitted during the exam:

NFPA 70-National Electrical Code®, 2017 Edition
National Fire Protection Association
1 Batterymarch Park
P. O. Box 9101
Quincy, MA 02169-9101
www.nfpa.org

Printreading Based on the 2017 National Electrical Code® (NEC®), 2017 Edition
American Technical Publishers, Inc.
10100 Orland Parkway, Suite 200
Orland Park, IL 60467
www.atplearning.com

Figure 1-2. The exam outline provides a list of the topics that will be on the exam along with the number of questions for each topic.

The exam is a broad-based test containing questions on calculations, trade knowledge, and finding answers to questions in the NEC®. When taking the exam, time is critical. Looking up questions in a timely manner is essential. All questions not answered are marked wrong. Estimates on how long it will take to answer each question can be determined by answering a set of timed questions. For example, a test with 100 questions that is to be taken in 4 hr will require the test taker to complete 25 questions in an hour. This breaks down to about 2.5 min per question.

Most of the exams are produced by national testing organizations such as the International Code Council (ICC) or PSI Examination Services. Some states may create their own exam. States that use a testing service will have the exam graded by the testing service. The states that create their own exams will be responsible for grading them. Most of these tests will require a passing grade of 70% or 75%. Some exams are a combination of a closed book session, which covers basic NEC® rules and theory, and an open book session, which would require the electrician to reference material in the NEC®. Some exams require the NEC® article be given along with each answer.

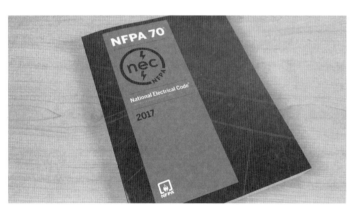

Some master electrician's exams require the most current edition of the NEC® to answer questions.

Preparing the NEC®

Each exam has its own rules and requirements when it comes to allowing supplemental material for the exam. It is necessary to check the requirements for the test ahead of time. Most exams require individuals to bring their own copy of the NEC®. A few states will provide a copy of the NEC®, but everyone should bring their own. The NEC® is essential for passing the exam.

The National Electrical Code® is sponsored and controlled by the National Fire Protection Association, Inc. (NFPA). The primary function of the NEC® is to safeguard people and property against electrical hazards. Copies of the current NEC® (NFPA 70) may be ordered directly from its publisher:
National Fire Protection Association, Inc.
1 Batterymarch Park
Quincy, MA 02169
www.nfpa.org

Most tests will allow the NEC® book to be highlighted and tabbed. When highlighting the NEC®, it is important to choose a method that makes sense and is easy to read. It is not recommended to highlight the whole section and tables of an article. Words or numbers that stand out should be chosen. Individuals must remember that the whole NEC® may be used for the test.

Tabbing the NEC®

Tabbing the NEC® will make the process of referencing the NEC® more efficient. Tabs specific to the NEC® may be purchased on the Internet. Blank tabs may be bought at any office supply store. Post-it® notes are not allowed to be used as tabs. Tabs should be easy to read and large sections should not be left untabbed. Using the index or turning page-by-page through a large section to find a particular article can take up a lot of time. A list of NEC® sections and tables that are recommended to tab can be found in the appendix of this text. **See Appendix.**

The index of the book should be tabbed as well. Too often, an individual will be in a hurry and turn pages in the wrong direction when looking for a specific section in the index. The index of the NEC® is in alphabetical order. Each page has a section listed at the top corner just like a phone book. The page on the left will have the name of the section that begins on that page. The page on the right will list the section that is at the end of that page. **See Figure 1-3.**

STUDYING FOR THE EXAM

Studying involves obtaining the information required to answer the questions on the exam. There are several tips that can be used to make studying for the exam easier and more enjoyable. These tips include the following:

- Have the right attitude. Having a positive attitude is a must when preparing for the test.
- Let others know when you are studying. We all have people around us. Letting them know when you need some time alone to study will eliminate interruptions.
- Take breaks from studying. Taking a break once in a while will allow you a chance to recharge your mind.
- Have your eyes checked. The NEC® is printed in small letters; if your eyesight is impaired, you may get tired faster.

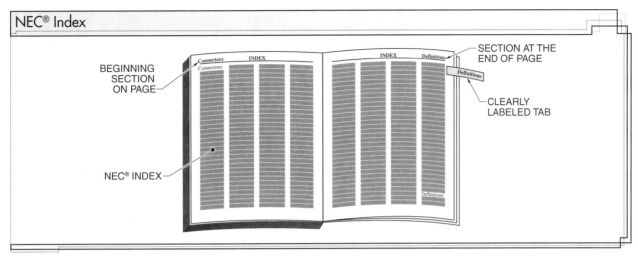

Figure 1-3. The index of the NEC® is in alphabetical order and the appropriate section is printed in the top corner of each page just like a phone book.

- Find a quiet place to study.
- Turn off cell phones and other electronic gadgets. This will eliminate interruptions and keep you focused on the study material.
- Divide practice tests into smaller sections to concentrate learning time. For example, instead of spending four hours to answer 100 questions, spend one hour answering 25 questions.

EXAM PREPARATION TIPS

The day before the exam can be stressful, so it is important to have everything prepared in the days leading up to the exam. There are several tips to follow to prepare for the exam, including the following:
- Have all materials such as books, pencils, and calculators ready the night before the exam. Trying to get things ready in the morning will only make you hurry and may cause you to forget something.
- Get plenty of rest the night before the exam. If you are well prepared for the exam, there is no sense in cramming more information the night before. Relax and go to bed at a reasonable time.
- Do not go to the test hungry. The test can be several hours long and being hungry will only make you lose focus. Eat breakfast if the test is in the morning, and have a light lunch if the exam is in the afternoon. However, a larger meal in the afternoon may make you tired.
- Get to the test site at least 30 min to 60 min ahead of the exam time. This will allow time to find the location and any extra time can be used to relax. If you are traveling from out of town, go to the exam site a day or two before the exam to eliminate any difficulty in finding the location.

TAKING THE EXAM

The day of the exam can be stressful. There are several tips to follow to relieve stress and help to efficiently complete the exam in the allotted time. These tips include the following:
- Read each question completely and read every possible answer of the multiple choice questions. Make sure you understand what the question is asking.
- First answer all questions that do not require referencing the NEC® or that you already know the answer to. This allows more time for questions that require you to look up information.
- Do not spend a lot of time on one question. If the answer cannot be found in a reasonable time, then move on to the next one.
- Leave some time to finalize the test. If time permits, go back and answer any unanswered questions.
- If you do not know an answer, then guess. Do not leave a question unanswered. Unanswered questions will be marked wrong.
- Be cautious when changing answers. Usually your first answer is the correct one. If an answer does need to be changed, be sure to change the correct answer on the answer sheet.

MASTER ELECTRICIAN'S EXAM WORKBOOK

The *Master Electrician's Exam Workbook* will cover the most common material on the master electrician's exam. While the test for each state might be different, they all cover the same information. This information includes the NEC® as well as most of the information in this text.

Each chapter in the *Master Electrician's Exam Workbook* contains example questions. Example questions are step-by-step procedures for formulas that may be on the exam. Each example question is followed by at least one practice question. Practice questions are designed to test the learner's knowledge of the material covered in the example question. **See Figure 1-4.** The answers to the practice questions are in the back of the text. Review questions at the end of each chapter are similar to the questions that will be on the exam. Review questions cover the material that pertains to the corresponding chapter.

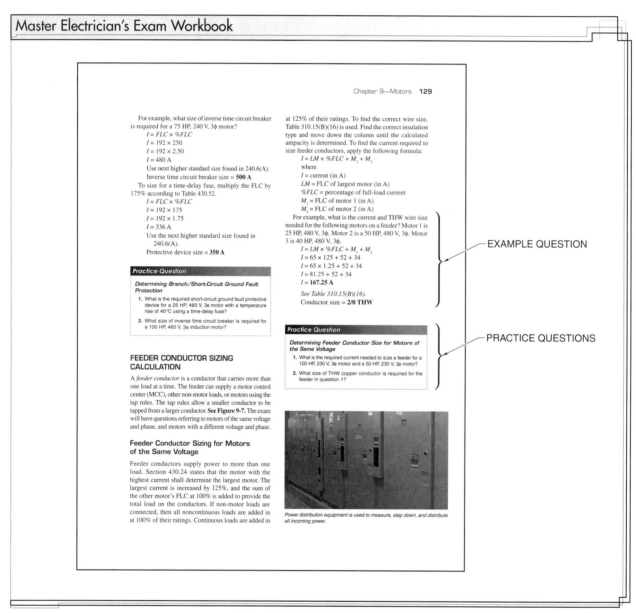

Figure 1-4. Practice questions follow the example questions and are designed to test the learner's knowledge of the material covered.

Digital Resources
ATPeResources.com/QuickLinks
Access Code: 637160

CHAPTER 2

Direct Current Principles

Electricity is the most widely used form of energy. Electricity is used to provide energy for traditional electrical applications such as lighting, heating, cooling, cooking, communication, and transportation. Direct current (DC) is a form of electricity in which current moves in one direction in a circuit. To understand the basic concepts of electricity and DC systems, the building blocks of matter and its subatomic structure must be understood.

OBJECTIVES

After completing this chapter, the learner will be able to do the following:

- Describe the basic theory of electricity.
- Explain the relationship between voltage, current, and resistance using Ohm's law.
- Demonstrate the relationship between power, voltage, and current using the power formula.
- Solve problems using equations from the formula wheel.
- Describe the functionality of various electrical meters.

MATTER

Matter is anything that has mass and occupies space. All objects are made up of matter. Matter can exist in a solid, liquid, or gaseous state. All matter has some electrical properties. The electrical behavior of matter varies according to the physical makeup of the matter. Some matter, such as copper, allows electricity to easily move through it and can act as a conductor. A *conductor* is a material that has low electrical resistance and permits electrons to move through it easily.

Atoms

An *atom* is the smallest particle to which an element can be reduced and still maintain the properties of the element. All materials are made of matter and all matter contains atoms. These atoms contain particles. Positive-charged particles are called protons, negative-charged particles are called electrons, and particles with no charge are called neutrons. Particles of the same charges repel each other, and particles with opposite charges attract each other.

The neutrons and protons are located in the nucleus, which is in the center of the atom. The electrons circle the nucleus in orbits called shells. Each shell contains a certain number of electrons. Shell one, which is the closest shell to the nucleus, can have up to 2 electrons. Shell two can have up to 8 electrons; shell three can have up to 18 electrons; shell four can have up to 32 electrons; shell five can have up to 50 electrons; shell six can also have up to 72 electrons; and shell seven can have 98 electrons. **See Figure 2-1.** Not all atoms use all seven shells.

The outer shell of an atom is called the valance shell. Atoms that have 3 or fewer electrons in the valance shell make good conductors. Copper makes a good conductor because it has one electron in the valance shell. The atoms that make up materials that are used to insulate conductors will have 7 or 8 electrons in the outer shell. Atoms with 4 to 6 electrons are not good conductors or good insulators. They are called semiconductors.

In a copper atom, there is a free electron in the outer shell that can be transferred from one atom to another when there is a potential difference placed on it. This potential difference is called voltage. Voltage is the pressure that makes electrons move. This movement of electrons is called current. **See Figure 2-2.**

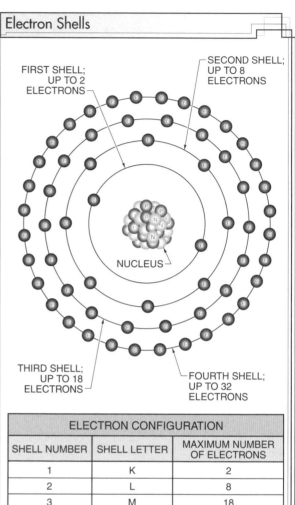

Figure 2-1. The shells of an atom are numbered from innermost to outermost (1, 2, 3, 4, 5, 6, and 7) or lettered (K, L, M, N, O, P, and Q).

Practice Questions

Atoms

1. Which part of an atom has a negative charge? **electron**
2. Which type of charges will attract each other? **opposite**
3. What is the outside shell of an atom called? **valance shell**
4. A material used as a conductor has atoms that have **3** or fewer electrons in the outer shell.
5. An insulator has atoms with **4** or more electrons in the outer shell. **4-6**

Chapter 2—Direct Current Principles

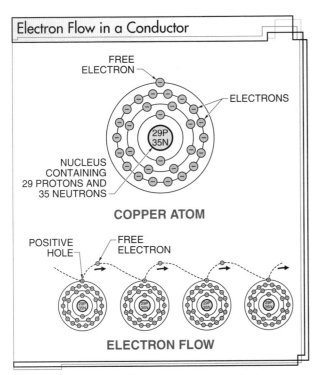

Figure 2-2. Electron flow in a copper conductor occurs as free electrons jump from the outer orbit of one atom to another.

Current

Current (I) is the flow of electrons moving past a point in a circuit. Current is measured in amperes, or amps (A). Current will flow as long as there is an imbalance between the positive and negative charges.

Current flows from the negative side of a battery to the positive side. The negative charged electrons will leave the negative side of a battery, flow through a conductor to the load, and then return to the battery on the positive side. **See Figure 2-3.** Over time, as the circuit is operating, the electrons moving from the negative pole to the positive pole will eventually equalize when the battery has discharged. With no electron flow, the load will not function.

Practice Questions

Current
1. The movement of electrons is called ___. *Current*
2. Current flows from − to + in a battery.

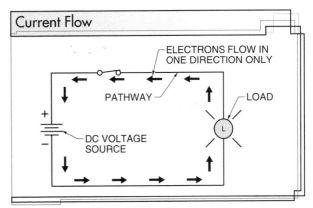

Figure 2-3. Current flows from the negative side of a battery to the positive side in a DC circuit.

Resistance

Resistance (R) is the opposition to current flow. All conductive materials have a certain amount of resistance in them. This resistance is measured in ohms (Ω). Some materials have less resistance than others. Of the four types of materials normally used for conductors, silver has the least amount of resistance, followed by copper, gold, and aluminum. Some contacts use gold or silver to help make better connections.

Wire is usually used to carry the current that is flowing through a circuit. The amount of current a wire can safely carry is determined by wire size. Wire is sized by using the American wire gauge (AWG) numbering system. The smaller the AWG number, the larger the diameter of the wire. Generally, the larger the diameter of the wire, the less resistance and the more current it can safely carry. **See Figure 2-4.**

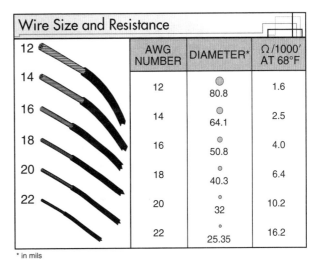

AWG NUMBER	DIAMETER*	Ω/1000′ AT 68°F
12	80.8	1.6
14	64.1	2.5
16	50.8	4.0
18	40.3	6.4
20	32	10.2
22	25.35	16.2

*in mils

Figure 2-4. Wire is sized using the American wire gauge (AWG) numbering system.

Practice Questions

Resistance

1. The opposition to current flow in DC circuits is called ___. *Resistance*
2. Of copper, silver, gold, or aluminum, which has the lowest resistance? *Gold*

Voltage

Voltage (E) is the amount of electrical pressure in a circuit. Voltage is measured in volts (V). Voltage is produced any time there is an excess of electrons at one terminal of a voltage source and a shortage of electrons at the other terminal. The six common methods used to produce a voltage are:
- mechanical (induction generators)
- chemical (batteries)
- thermoelectric (heat)
- photovoltaic (light)
- piezoelectric (pressure)
- friction (static electricity)

Voltages can be produced as alternating current (AC) or direct current (DC). With AC voltages, the flow of electrons alternate direction. With DC voltages, the electrons flow in one direction from a power source. One of the most common power sources that produce DC voltages is a battery.

Practice Questions

Voltage

1. ___ is electrical pressure that moves electrons in a circuit.
2. What method uses pressure to produce voltage?
3. AC voltage has electrons that move in ___ directions.

Batteries and Battery Cells

A *battery* is an electrical energy storage device made up of cell plates placed together in a series or parallel configuration to produce a voltage. A battery produces electricity through the chemical reaction between battery plates and acid. A cell is a unit that produces electricity at a fixed voltage or current level. A standard dry cell will produce 1.5 V. A 9 V battery is made up of six 1.5 V dry cells connected in series. A wet cell will produce around 2 V per cell. The amount of voltage required from a battery depends on the device it is used in. **See Figure 2-5.**

Standard DC Voltages	
DEVICE	VOLTAGE LEVEL
Flashlights; watches; etc.	1.5, 3
Toys; automobiles; trucks	6, 9, 12, 24, 36
Printing presses; small electric railway systems	115, 230, 460
Large electric railway systems	1200, 1500, 3000

Figure 2-5. The type of power used depends on the application and the amount of power required.

Batteries are rated in voltage, amperage, and amp-hours (A/hr) as well. An amp-hour is the amount of time that a battery produces a usable amount of current. It is found by multiplying the amperage rating of the battery by the length of time the battery will last. When cells are combined in series, the voltage is added together. The current will be the same as for one cell. For example, if three 2 V cells at 20 A each are placed in series, they will produce 6 V at 20 A. When cells are connected in parallel, the voltage will be the same as for one cell, but the current of each cell is added together. For example, if three 2 V cells at 20 A each are placed in parallel, they will produce 2 V at 60 A.

Practice Questions

Batteries

1. A dry cell battery has a voltage of _1.5_ V per cell.
2. A wet cell battery has a voltage of _2_ V per cell.

OHM'S LAW

Ohm's law is fundamental to electrical formulas. In 1827, Georg Ohm showed through experiments that a relationship existed between voltage, current, and resistance. Ohm's law states that current is directly proportional to voltage and indirectly proportional to resistance. Ohm's law is beneficial when designing and troubleshooting electrical circuits. Electricians use Ohm's law to calculate the size of conductors required to carry current to a load. It is also used to determine if a load is drawing the correct amperage. The Ohm's law formula chart is designed to show the relationship of current, voltage, and resistance. Three formulas are created from this relationship. **See Figure 2-6.** Any variable can be found when the other two are known.

Chapter 2—Direct Current Principles

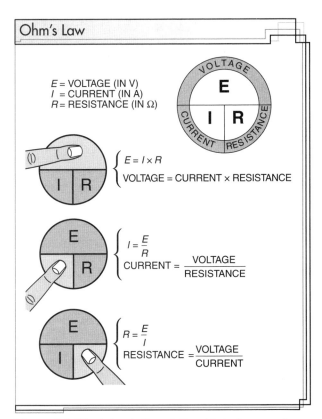

Figure 2-6. Ohm's law is the relationship between voltage, current, and resistance in an electrical circuit.

Voltage, Current, and Resistance Relationship

Voltage is directly proportional to current. This means that if the voltage increases, then the current increases at the same rate, while the resistance stays the same. This applies to both DC circuits and AC circuits when resistance is the only type of load.

Since current is inversely proportional to resistance, the opposite happens. As the resistance increases, the current will decrease and the voltage remains the same. If the resistance decreases, the current increases and the voltage still remains the same.

TECH TIP

Ohm's law is used during troubleshooting to determine circuit conditions. Voltage and current measurements are taken because resistance cannot be measured on a circuit that is energized. After voltage and current measurements are taken, Ohm's law is applied to determine the resistance of the circuit.

Determining Voltage Using Ohm's Law

According to Ohm's law, the voltage (E) in a circuit is equal to resistance (R) times current (I). **See Figure 2-7.** To determine voltage using Ohm's law, apply the following formula:

$$E = I \times R$$

where
E = voltage (in V)
I = current (in A)
R = resistance (in Ω)

For example, what is the voltage of a circuit that includes a 24 Ω resistor that draws 5 A?

$$E = I \times R$$
$$E = 5 \times 24$$
$$E = \mathbf{120\ V}$$

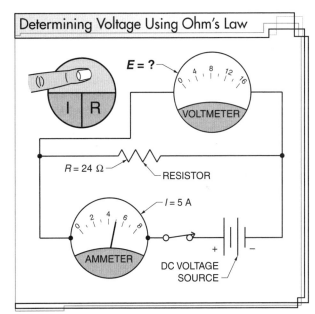

Figure 2-7. To determine voltage using Ohm's law, multiply current by resistance.

Practice Questions

Determining Voltage Using Ohm's Law

1. What is the voltage of a circuit that includes a 120 Ω resistor that draws 1 A?

2. What is the voltage across a 16 A load with a resistance of 0.2 Ω?

Determining Current Using Ohm's Law

According to Ohm's law, the current in a circuit is equal to voltage divided by resistance. **See Figure 2-8.** To determine current using Ohm's law, apply the following formula:

$$I = \frac{E}{R}$$

where
I = current (in A)
E = voltage (in V)
R = resistance (in Ω)

For example, what is the current in a circuit that has a 3 Ω resistor connected to a 12 V power source?

$$I = \frac{E}{R}$$
$$I = \frac{12}{3}$$
$$I = 4\,\text{A}$$

Determining Resistance Using Ohm's Law

According to Ohm's law, the resistance in a circuit is equal to voltage divided by current. **See Figure 2-9.** To determine resistance using Ohm's law, apply the following formula:

$$R = \frac{E}{I}$$

where
R = resistance (in Ω)
E = voltage (in V)
I = current (in A)

For example, what is the resistance of a circuit when a load that draws 5 A is connected to a 50 V power source?

$$R = \frac{E}{I}$$
$$R = \frac{50}{5}$$
$$R = 10\,\Omega$$

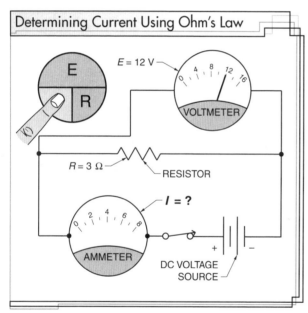

Figure 2-8. To determine current using Ohm's law, divide voltage by resistance.

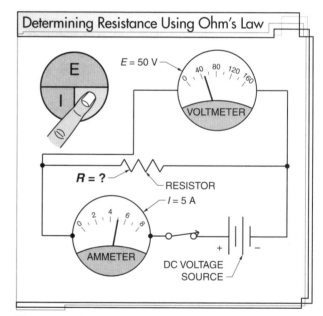

Figure 2-9. To determine resistance using Ohm's law, divide voltage by current.

Practice Questions

Determining Current Using Ohm's Law

1. What is the current flow of a circuit with a 162 Ω lamp connected to a 120 V power source?

2. What is the current flow of a circuit if the resistance is 20 Ω and the power source is 60 V?

Practice Questions

Determining Resistance Using Ohm's Law

1. What is the resistance of a conductor that has a voltage of 25 V and a current of 15 A?

2. What is the resistance of a lamp that is rated at 120 V and uses 0.625 A to produce 75 W of power?

POWER FORMULA

The power formula shows the relationships between power, voltage, and current in an electrical circuit. **See Figure 2-10.** Any variable in this relationship can be determined when the other two are known. The power formula demonstrates that power will increase at the same rate that the current or voltage increases on the circuit. Power is measured in watts (W).

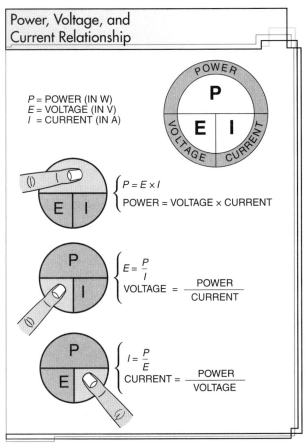

Figure 2-10. The power formula is the relationship between power, voltage, and current in an electrical circuit.

TECH TIP

Like Ohm's law, the power formula can be used when troubleshooting and to predict circuit characteristics before power is applied. Any electrical value (P, E, or I) can be calculated when the other two values are known or measured. The power formula is useful when determining expected current values because most electrical equipment lists a voltage and power rating. The listed power rating is given in watts for most appliances and heating elements or in horsepower for motors.

Determining Power Using the Power Formula

According to the power formula, power (P) in a circuit is equal to voltage (E) times current (I). **See Figure 2-11.** Power is determined using the power formula as follows:

$P = E \times I$

where

P = power (in W)

E = voltage (in V)

I = current (in A)

For example, what is the power of a load that draws 0.5 A when connected to a 120 V power source?

$P = E \times I$

$P = 120 \times 0.5$

$P = \mathbf{60\ W}$

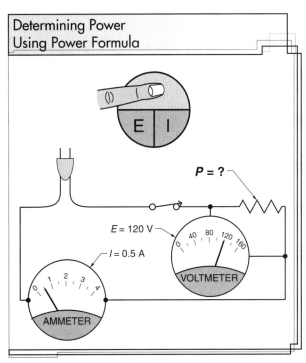

Figure 2-11. To determine power using the power formula, multiply voltage times current.

Practice Questions

Determining Power Using the Power Formula

1. What is the power of a load that draws 4.5 A when connected to a 30 V power source?

2. What is the power of a lamp that uses 0.833 A on a 120 V circuit?

3. What is the power of a heating element rated at 115 V and using 10 A?

Determining Voltage Using the Power Formula

According to the power formula, voltage in a circuit is equal to power divided by current. **See Figure 2-12.** Voltage is determined using the power formula as follows:

$$E = \frac{P}{I}$$

where
E = voltage (in V)
P = power (in W)
I = current (in A)

For example, what is the voltage of a lamp that has a rating of 500 W and uses 2.08 A?

$$E = \frac{P}{I}$$
$$E = \frac{500}{2.08}$$
$$E = \mathbf{240\ V}$$

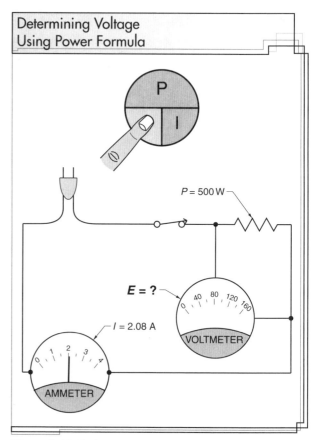

Figure 2-12. To determine voltage using the power formula, divide power by current.

Fluke Corporation
When troubleshooting electrical problems, electricians take power and current readings in order to perform power formula calculations.

Practice Questions

Determining Voltage Using the Power Formula

1. What is the voltage of a 150 W lamp that uses 1.3 A?
2. What is the voltage of a circuit with a solenoid that draws 2.3 A and produces 550 W of power?

TECH TIP

Heat is produced any time electricity passes through a conductor that has resistance. This method is used to produce heat in the heating element included in toasters, portable space heaters, hair dryers, and electric water heaters. The heating capacity of the heating element is usually stated in watts. The size selected is based on the application. The greater the required heat output, the higher the required wattage. Once a wattage size is selected, the current draw is determined by using the power formula.

Determining Current Using the Power Formula

According to the power formula, current in a circuit is equal to power divided by voltage. **See Figure 2-13.** Current is determined using the power formula as follows:

$$I = \frac{P}{E}$$

where
I = current (in A)
P = power (in W)
E = voltage (in V)

For example, what is the amount of current in a circuit that has 330 W of power and is operating on 50 V?

$$I = \frac{P}{E}$$
$$I = \frac{330}{50}$$
$$I = \mathbf{6.6\ A}$$

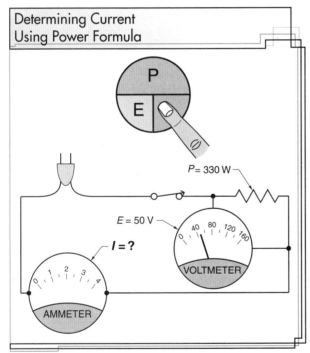

Figure 2-13. To determine current using the power formula, divide power by voltage.

Practice Questions

Determining Current Using the Power Formula

1. What is the current in a circuit for a 5 kW heater that operates on 240 V?
2. A hot tub has a 115 V heater that uses 7500 W to heat the water. How much current is the heater drawing?

FORMULA WHEEL

Ohm's law and the power formula can be combined to produce a formula wheel, which includes a total of 12 formulas that are broken down into four sections. **See Figure 2-14.** The four sections include formulas for determining power, resistance, amps, and voltage. Six of the formulas are taken directly from Ohm's law and the power formula. The other six formulas are derived from Ohm's law and the power formula. Electricians must be able to work these basic formulas in order to pass the exam.

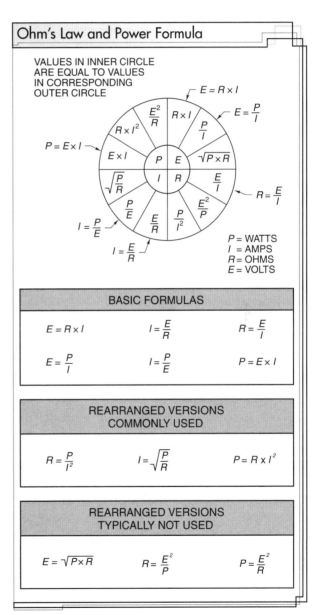

Figure 2-14. The formula wheel is made up of equations derived from Ohm's law and the power formula.

> **Practice Questions**
>
> *Using the Formula Wheel*
> 1. What is the resistance of a load that has a voltage rating of 120 V and a power rating of 500 W?
> 2. What is the power of a circuit that has 23 Ω and a load that draws 12 A?
> 3. What is the power of a 240 V circuit that has a resistance of 10 Ω?

METERS

Meters measure quantities such as voltage, current, speed, pressure, counts, etc. There are different types of meters used in the electrical field. Some are used every day, while others are used for special applications.

Voltmeters

Voltmeters are used to measure the voltage of a circuit. Voltage is measured by placing leads in parallel across the power source. **See Figure 2-15.** If using an analog meter on DC voltages, an electrician must take polarity into consideration. If the meter is connected in reverse polarity, the needle of the meter will move in the wrong direction and the meter could be damaged.

To measure higher voltages, the meter has multiplier resistors installed. Multiplier resistors are connected in series with a moving coil meter movement to extend the voltage range. A voltmeter will have a very high amount of resistance. Having a high resistance means the meter will have little effect on the circuit when connected.

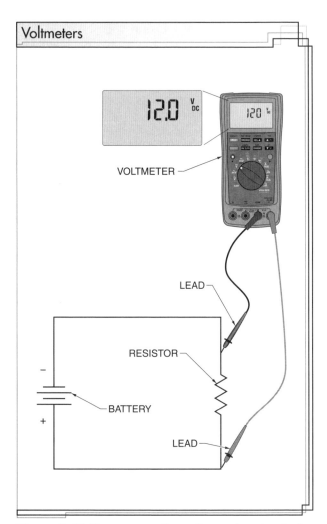

Figure 2-15. To measure the voltage of a circuit, the leads are placed from the voltmeter across (parallel to) the power source.

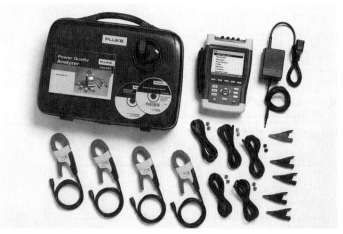

Fluke Corporation
Power quality meters are used to maintain power systems, troubleshoot power problems, and diagnose equipment failures.

Ammeters

Ammeters are used to measure the current flow through a circuit. The meter must be connected in series with the load. The ammeter has a very low resistance. If it is connected in parallel across the conductors, it will cause the circuit to draw a high current, which could damage the meter. If an electrician is using an analog ammeter on DC amperage, polarity must be taken into consideration. Like a voltmeter, if an ammeter is connected in reverse polarity, the needle of the meter will move in the wrong direction and the meter may be damaged. For an ammeter to read higher currents, shunt resistors are added in parallel with the meter.

Clamp-On Ammeters

Clamp-on ammeters measure the current in a circuit by measuring the strength of the magnetic field around the conductor. The clamp of the meter is placed around one conductor of the circuit to read the magnetic field around the conductor. **See Figure 2-16.** If the meter is clamped around two different conductors in the circuit, the magnetic fields of the two conductors will cancel each other out and the meter will read zero. If the same conductor is wrapped around the clamp twice, then the amperage will double. This method may be used on very low amperage amounts. The reading can be divided by the number of turns to determine the correct amperage of the circuit.

A clamp-on ammeter works best when placed at a 90° angle to the conductor. Surrounding conductors should be separated from the tested conductor to ensure an accurate reading. Some clamp-on ammeters will read AC only, while others can read AC and DC amperages.

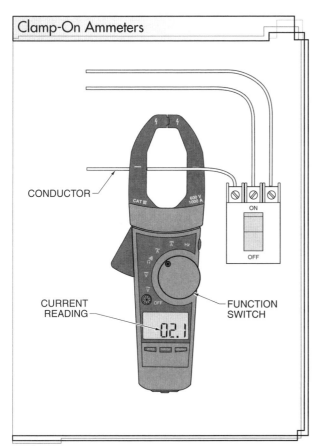

Figure 2-16. Whenever possible, a conductor being tested by a clamp-on ammeter should be separated from other surrounding conductors by a few inches.

Ohmmeters

Ohmmeters are used to measure the resistance of a circuit. **See Figure 2-17.** They can also be used to check the continuity of a circuit. If the circuit is complete, the resistance reading will be low. If the circuit is open, the meter will read an infinite amount. The ohmmeter supplies its own power source and should not be used on a live circuit.

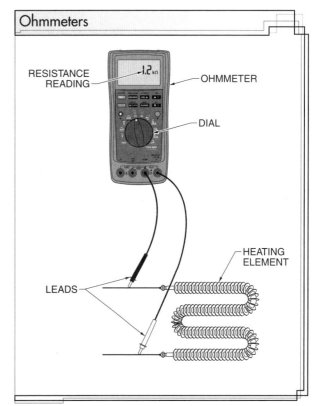

Figure 2-17. Ohmmeters may be used to measure the resistance in a circuit or to test the continuity of the circuit.

TECH TIP

A multimeter is a test tool used to measure two or more electrical values. Multimeters can be used to measure voltage, current, continuity, resistance, capacitance, frequency, duty cycle, etc. They can be either analog or digital devices. Most electricians use a multimeter because it combines the functions of a voltmeter, ammeter, and ohmmeter.

Megohmmeters

Megohmmeters measure high resistance. **See Figure 2-18.** They are mainly used to detect insulation failure or potential insulation failure caused by excessive moisture, dirt, extreme temperatures, corrosive substances, vibration, and aging. Megohmmeters produce high voltages, so caution must be used to prevent shock. Once testing is done, the conductors should be shorted to ground to discharge any voltage remaining on the conductor. The testing voltage is usually two times the circuit voltage.

Watt-Hour Meters

A watt-hour meter is a combination of a voltmeter and an ammeter. It is used by a utility company on a dwelling and other buildings or structures to show how much electricity is being used. A watt-hour meter is connected in series-parallel to measure both the voltage and the current flow of circuits.

Ideal Industries, Inc.
Some clamp-on ammeters are capable of determining power by measuring current and voltage.

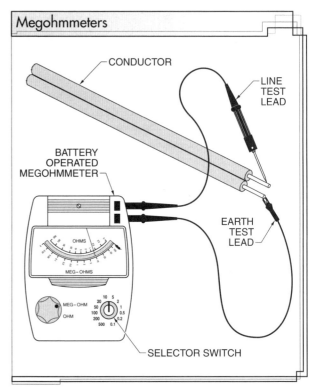

Figure 2-18. Megohmmeters are used to detect insulation failure or potential insulation failure.

Practice Questions

Meters

1. A voltmeter is connected in ___ with the load.
2. An ammeter is connected in ___ with the load.
3. A(n) ___ is connected in a series-parallel fashion.
4. A clamp-on ammeter works best when placed at a(n) ___ angle to the conductor.

CHAPTER 2
Direct Current Principles
Review Questions

Name _____   Date _____

_____ 1. A(n) ___ is a material that has low electrical resistance and permits electrons to move through it easily.
   A. nucleus
   B. atom
   C. cell
   D. conductor

_____ 2. Copper makes a good conductor because it has ___ electron(s) in the outer shell.
   A. one
   B. four
   C. eight
   D. twelve

_____ 3. ___ is the flow of electrons moving past a point in a circuit.
   A. Voltage
   B. Resistance
   C. Current
   D. Ampacity

_____ 4. The material with the least amount of resistance is ___.
   A. silver
   B. copper
   C. gold
   D. aluminum

_____ 5. A circuit that has a 10 Ω resistor connected to 4 A will have a voltage of ___ V.
   A. 0.4
   B. 2.5
   C. 25
   D. 40

_____ 6. A circuit that has a 5 Ω resistor connected to a 25 V power source will have a current of ___ A.
   A. 0.2
   B. 5
   C. 10
   D. 125

I   E

_____ 7. A load that draws 20 A connected to a 100 V power source will have a resistance of ___ Ω.
   A. 0.2 $E = IR$
   B. 5
   C. 200 $R = E/I$
   D. 2000

_____ 8. The power of a circuit with a 12 A load connected to a 240 V power source is ___ W.
   A. 0.05 $P = IE$
   B. 20
   C. 200
   D. 2880

_____ 9. ___ are used to measure the current flow through a circuit.
   A. Voltmeters
   B. Ohmmeters
   C. Ammeters
   D. Megohmmeters

_____ 10. A clamp-on ammeter is used to read the magnetic field of a circuit by placing it around ___ conductor(s).
   A. one
   B. two
   C. three
   D. four

Digital Resources
ATPeResources.com/QuickLinks
Access Code: 637160

CHAPTER 3

Alternating Current Principles

Alternating current (AC) is the voltage used in residential, commercial, and industrial locations. It is more commonly used because it is easily and efficiently distributed at higher voltages and lower current. Alternating current is generated at power plants and distributed to homes and businesses through a distribution network of step-up and step-down transformers, substations, and distribution transformers. The current in an AC circuit flows in two directions. Knowledge of AC terminology and formulas is critical when taking the master electrician's exam.

OBJECTIVES

After completing this chapter, the learner will be able to do the following:

- Explain the basic operations of alternating current (AC).
- List the different types of generators and describe how they work.
- Describe the formation of a sine wave as it relates to an AC generator.
- Explain how capacitance, inductance, and impedance affect AC circuits.
- Describe how to calculate power factor.
- Explain the relationship between reactive power, true power, and apparent power.

GENERATORS

A *generator* is a machine that converts mechanical energy into electrical energy by means of electromagnetic induction. AC generators consist of field windings, an armature, slip rings, and brushes. **See Figure 3-1.** A *field winding* is a magnet used to produce the magnetic field in a generator. An *armature* is the movable coil of wire in a generator that rotates through a magnetic field. The ends of the coil are connected to slip rings. A *slip ring* is a metallic ring connected to the ends of an armature and is used to conduct the induced voltage to the brushes. A *brush* is a sliding contact that rides against the slip rings and is used to connect the armature to the external circuit.

As the armature rotates, each half cuts across the magnetic lines of force at the same speed. The strength of the voltage induced on one side of the armature is always the same as the strength of the voltage induced on the opposite side of the armature. Each half of the armature cuts the magnetic lines of force in a different direction. For example, as the armature rotates in the clockwise direction, the lower half of the armature cuts the magnetic lines of force from the bottom up to the left, while the top half of the coil cuts the magnetic lines of force from the top down to the right. The voltage induced on one side of the armature is opposite to the voltage induced on the other side. The voltage in the lower half of the coil produces current flow in one direction, and the voltage on the upper half enables current flow in the opposite direction.

Since the two halves of the coil are connected in a closed loop, the voltages are added together. The total voltage of a full rotation of the armature is twice the voltage of each coil half. This total voltage is obtained where the brushes connect to the slip rings. Since a generator produces voltage with movement in two directions, the voltage and current will have a varying rate.

Single-Phase (1ϕ) Generators

A 1ϕ generator uses one armature to produce voltage and current. Each complete rotation of the armature in a 1ϕ AC generator produces one complete AC cycle. The rotation speed of the armature determines the output frequency of an AC generator. Before the armature begins to rotate in a clockwise direction, there is no voltage or current in the external circuit because the armature is not cutting across any magnetic lines of force. As the armature rotates, each half of the armature cuts across the magnetic lines of force, producing current in the external circuit. One complete rotation of the armature equals one complete 360° cycle of the sine wave. **See Figure 3-2.**

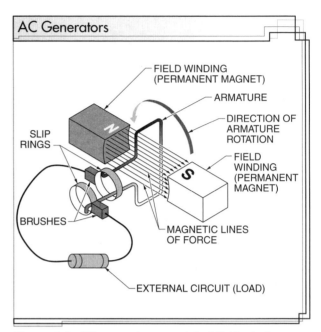

Figure 3-1. AC generators consist of field windings, an armature, slip rings, and brushes.

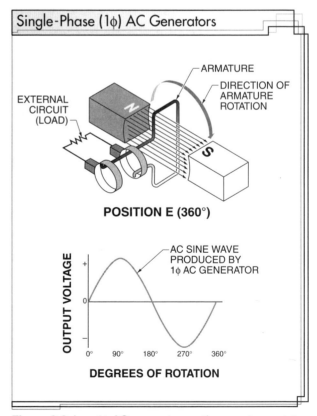

Figure 3-2. In a 1ϕ AC generator, as the armature rotates through 360° of motion, the voltage generated is a continuously changing AC sine wave.

Three-Phase (3ϕ) Generators

The principles of a 3ϕ generator are the same as a 1ϕ generator, except that there are three equally spaced armature windings 120° apart. The output of a 3ϕ generator results in three output voltages that are 120° out-of-phase with each other. **See Figure 3-3.**

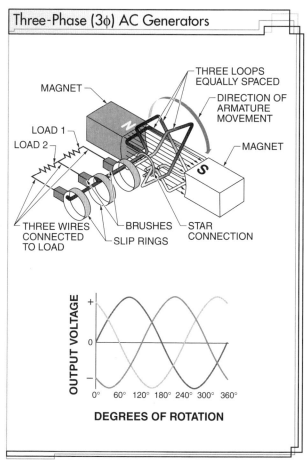

Figure 3-3. A 3ϕ generator has three equally spaced armature windings 120° out-of-phase with each other.

TECH TIP

Most electric power is produced by alternating current (AC) generators that are driven by rotating prime movers. Most prime movers are steam turbines whose thermal energy comes from steam generators that use either fossil fuel or nuclear energy. Combustion turbines are often used for smaller units and in applications where gas, diesel, or fuel oil is the only available fuel.

SINE WAVES

A *sine wave* is a symmetrical waveform that shows how AC varies over time. A sine wave is generated as the armature spins inside a magnetic field. The armature starts parallel with the magnetic field. When the armature is parallel, it is not crossing any magnetic lines of force. When no magnetic lines of force are being crossed, no voltage is being produced.

As the armature starts to spin, it starts to cut across the magnetic lines of force and voltage is generated. The voltage will continue to increase until it reaches its peak voltage. This is when the armature crosses the maximum number of magnetic lines of force at 90°. As the armature continues, it will cross fewer magnetic lines of force and the voltage being produced will decrease until it is again parallel with the force lines at 180°. This is called an alternation. One complete revolution of the armature produces two alternations, a positive alternation and a negative alternation. **See Figure 3-4.**

Generators are available in numerous sizes and configurations and can be built to exact specifications. Generators are commonly used as standby or back-up power sources but can also be used as primary power sources.

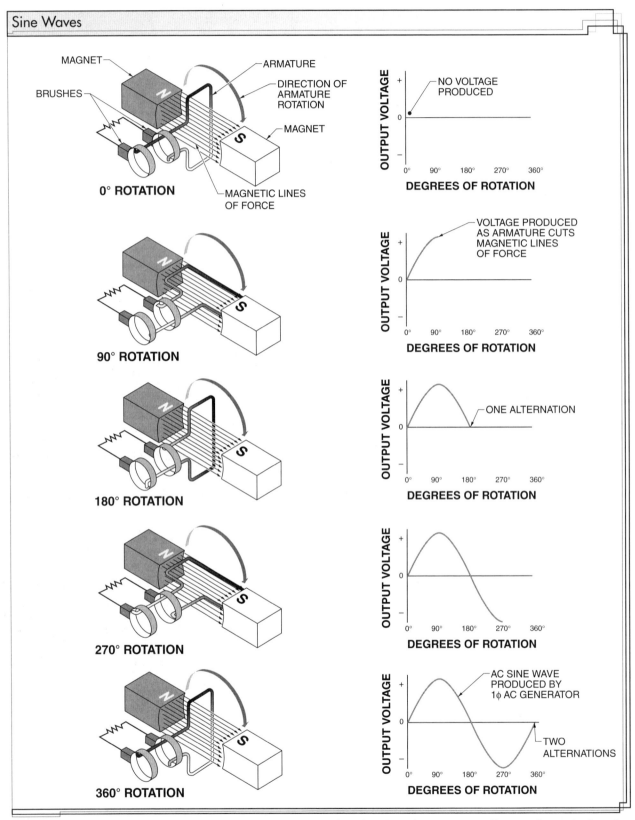

Figure 3-4. As the armature in a generator rotates through 360° of motion, the voltage generated is a continuously changing AC sine wave.

Frequency

Frequency (f) is the speed (cycles per second) at which an armature spins in a magnetic field. Frequency is measured in hertz (Hz). If the armature spins 60 complete revolutions in 1 sec, then the frequency is 60 Hz. The standard frequency in the United States is 60 Hz. The number of magnetic poles in a generator will affect the speed of the armature. A 2-pole generator will require the motor to turn at 3600 revolutions per minute (rpm), whereas a 6-pole generator will only require it to spin at 1200 rpm. To determine the frequency of a sine wave, apply the following formula:

$$f = \frac{1}{T}$$

where
f = frequency (in Hz)
1 = constant
T = period (in sec)

For example, what is the frequency of a sine wave that requires 0.5 sec to complete one cycle?

$$f = \frac{1}{T}$$
$$f = \frac{1}{0.5}$$
$$f = \mathbf{2\,Hz}$$

Practice Question

Determining the Frequency of a Sine Wave

1. What is the frequency of a sine wave that requires 0.2 sec to complete one cycle?

One complete revolution of the armature is called a cycle or period. A period is measured in time (T). To determine the period of the sine wave, apply the following formula:

$$T = \frac{1}{f}$$

where
T = period (in sec)
1 = constant
f = frequency (in Hz)

For example, what is the period of a 50 Hz sine wave?

$$T = \frac{1}{f}$$
$$T = \frac{1}{50}$$
$$T = \mathbf{0.02\,sec}$$

Practice Question

Determining the Period of a Sine Wave

1. What is the period of a 200 Hz sine wave?

The frequency of a sine wave can be determined if the period of the sine wave is given. To determine the frequency of a sine wave when the period is known, apply the following formula:

$$f = \frac{1}{T}$$

where
f = frequency (in Hz)
1 = constant
T = period (in sec)

For example, what is the frequency of a sine wave that has a period of 0.1667 sec?

$$f = \frac{1}{T}$$
$$f = \frac{1}{0.1667}$$
$$f = \mathbf{6\,Hz}$$

Practice Question

Determining the Frequency of a Sine Wave When the Period Is Known

1. What is the frequency of a sine wave if the period is 0.01 sec?

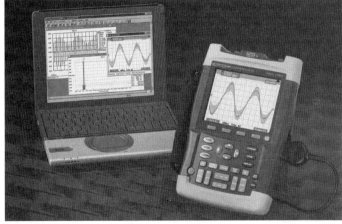

Fluke Corporation
Graphic-display test instruments provide a visual display of the frequency in a circuit.

Effective Voltage

The *effective voltage* (V_{EFF}) is the voltage value of a sine wave that produces the same amount of heat in a pure resistive circuit as DC of the same value. It is also referred to as the root-mean-square (RMS) voltage. When an electrician measures the voltage of a circuit, the effective voltage is displayed on the meter. Effective voltage occurs on the sine wave at 45° and 135° on the positive alternation, and at 225° and 315° on the negative alternation. **See Figure 3-5.** To determine the effective voltage, apply the following formula:

$$V_{EFF} = 0.707 \times V_P$$

where

V_{EFF} = effective voltage (in V)
0.707 = constant
V_P = peak voltage (in V)

For example, what is the effective voltage if the peak voltage is equal to 230 V?

$$V_{EFF} = 0.707 \times V_P$$
$$V_{EFF} = 0.707 \times 230$$
$$V_{EFF} = \mathbf{162.6\ V}$$

The *peak voltage* (V_P) is the maximum voltage of either a positive or negative alternation. The peak voltage can be found by rearranging the effective voltage equation. To determine the peak voltage, apply the following formula:

$$V_P = \frac{V_{EFF}}{0.707}$$

where

V_P = peak voltage (in V)
V_{EFF} = effective voltage (in V)
0.707 = constant

For example, what is the peak voltage if the effective voltage is equal to 162.6 V?

$$V_P = \frac{V_{EFF}}{0.707}$$
$$V_P = \frac{162.6}{0.707}$$
$$V_P = \mathbf{230\ V}$$

TECH TIP

Georg S. Ohm discovered that all electrical quantities are proportional to each other and can therefore be expressed as mathematical formulas.

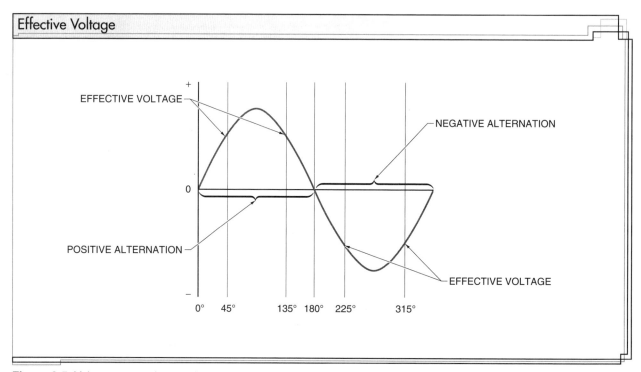

Figure 3-5. Voltmeters can be used to determine effective voltage. Effective voltage occurs on a sine wave at 45°, 135°, 225°, and 315°.

INDUCTANCE

Inductance (L) is a property that opposes the change of current in an AC circuit. Inductance is used in generators, transformers, and standard induction electric motors to generate, transform, or use AC. As the magnetic field around a conductor expands and collapses, this field cuts across the conductor and produces current. The current produced is opposite the normal current flow. The expanding and collapsing magnetic field around a conductor tends to cause electrons to travel on the outer edges of the conductor. This is known as the skin effect. Skin effect increases as the frequency throughout the conductor increases. **See Figure 3-6.**

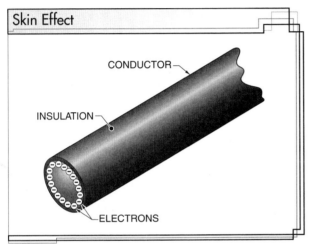

Figure 3-6. Electrons tend to move to the surface of the conductor at high frequencies.

All electrical equipment that has a conductor has some inductance. When the conductor is looped, the inductance is increased. Devices such as motors, transformers, coils in starters, and ballasts in HID lighting all have a certain amount of inductance. The more devices that are connected to a circuit, the greater the inductance will be. Inductance is measured in henrys (H).

An *inductor* is an electrical device designed to store electrical energy by means of a magnetic field. It is made up of a coil of wire that may be wound around a core of metal or paper. It may also be self-supporting. One of the factors that can affect the amount of inductance is the amount of current flow. The higher the current flow, the higher the inductance will be. Other factors that will affect the amount of inductance include the number of coil turns, coil spacing, length of the coil wire, and type of core material. **See Figure 3-7.**

Figure 3-7. Factors that affect inductance include the number of coil turns, coil spacing, length of the coil, and the type of core material.

Inductive Reactance

Inductive reactance (X_L) is the opposition of an inductor to alternating current. When inductors are used in AC circuits, the induction causes an opposition to current flow. Inductive reactance opposes the flow of current just like resistance. Inductive reactance is measured in ohms (Ω). As the frequency or inductance goes up or down, the inductive reactance will do the same. Inductive reactance is determined by applying the following formula:

$$X_L = 2\pi f L$$

where
X_L = inductive reactance (in Ω)
2π = 6.28 (indicates circular motion that produces the AC sine wave)
f = applied frequency (in Hz)
L = inductance (in H)

For example, what is the inductive reactance of an inductor that has an inductance value of 100 H and operates on a 60 Hz circuit?

$$X_L = 2\pi f L$$
$$X_L = 6.28 \times 60 \times 100$$
$$X_L = \mathbf{37{,}680 \ \Omega}$$

Practice Question

Determining Inductive Reactance

1. What is the inductive reactance of a 50 Hz circuit with an inductance value of 50 H?

CAPACITANCE

Capacitance (*C*) is the ability of a component or circuit to store energy in the form of an electrical charge. Capacitance is measured in farads (F). A *capacitor* is an electrical device designed to store electrical energy by means of an electrostatic field. It is made of two conductors separated by an insulator (dielectric), which allows an electrostatic charge to develop. **See Figure 3-8.** The amount of capacitance in a capacitor depends on the following:
- size of the plates
- type of insulation between the plates
- spacing between the plates
- type of connection (series or parallel)

When capacitors are connected in parallel, the values are added together as when finding resistance in series. When capacitors are connected in series, the total capacitance is determined by using the reciprocal method as when finding parallel resistance.

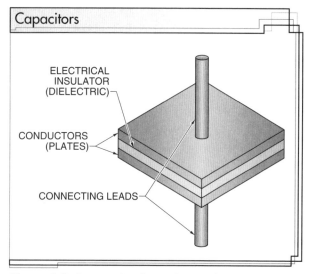

Figure 3-8. A capacitor is made up of two conductors separated by an electrical insulator.

Checking Capacitors

Capacitors should be checked to see if they are in good working condition. A capacitor can be checked with one of the following methods:
- Use a capacitance meter.
- Use an analog meter set to measure resistance. If the capacitor is good, the needle will move to the far right of the meter face and then slowly move back to the left.
- Use an incandescent lamp in series with the capacitor. If the lamp blinks and stays off, the capacitor is good. If it does not light at all, the capacitor is open. If it stays lit, the capacitor has shorted out.

Discharging Capacitors

Safety is a high priority when working around capacitors because they can hold a high charge of electricity. Article 460 in Section 460.6(A) states that a capacitor shall be discharged down to a voltage level of 50 V or less within 1 min after being disconnected from power. Per 460.6(B), this discharge shall be done by a permanent means.

Uses for Capacitors

Since capacitors store electrical energy, they are used to help start 1ϕ electric motors or in hard start kits for AC units. Capacitors are also used to correct a lagging power factor caused by inductive loads.

Capacitive Reactance

Capacitors oppose current flow in AC circuits just like inductors. *Capacitive reactance* (X_C) is the opposition to current flow by a capacitor. Capacitive reactance is measured in ohms (Ω). As the capacitance goes up or the frequency goes up, the reactance goes down. As the capacitance goes down or the frequency goes down, the reactance goes up. To determine capacitive reactance, apply the following formula:

$$X_C = \frac{1}{2\pi f C}$$

where
X_C = capacitive reactance (in Ω)
2π = 6.28 (indicates circular motion that produces the AC sine wave)
f = applied frequency (in Hz)
C = capacitance (in F)

For example, what is the capacitive reactance of a 250 μF capacitor connected to a 60 Hz supply?

$$X_C = \frac{1}{2\pi f C}$$

$$X_C = \frac{1}{6.28 \times 60 \times 0.000250}$$

$$X_C = \frac{1}{0.0942}$$

$$X_C = \mathbf{10.62 \ \Omega}$$

> **Practice Question**
>
> *Determining Capacitive Reactance*
> 1. What is the capacitive reactance of a circuit that has a 500 μF capacitor and operates at 50 Hz?

IMPEDANCE

In AC circuits, resistance, inductive reactance, and capacitive reactance all limit current flow. Most AC circuits contain all three oppositions to current flow. The behavior of current in an AC circuit depends on the amount of resistance, inductive reactance, and capacitive reactance.

Impedance (Z) is the total opposition of any combination of resistance, inductive reactance, and capacitive reactance offered to the flow of current in an AC circuit. Like reactance and resistance, impedance is measured in ohms (Ω). To determine total impedance, apply the following formula:

$$Z = \sqrt{R_T^2 + (X_L - X_C)^2}$$

or

$$Z = \sqrt{R_T^2 + (X_C - X_L)^2}$$

where
Z = total impedance (in Ω)
R_T = total resistance (in Ω)
X_L = inductive reactance (in Ω)
X_C = capacitive reactance (in Ω)

For example, what is the impedance of a circuit that has a resistance of 20 Ω, a capacitive reactance of 50 Ω, and an inductive reactance of 100 Ω?

$$Z = \sqrt{R_T^2 + (X_L - X_C)^2}$$
$$Z = \sqrt{20^2 + (100 - 50)^2}$$
$$Z = \sqrt{20^2 + (50)^2}$$
$$Z = \sqrt{400 + 2500}$$
$$Z = \sqrt{2900}$$
$$Z = \mathbf{53.85 \ \Omega}$$

> **Practice Question**
>
> *Determining Impedance*
> 1. What is the impedance of a circuit that has a resistance of 10 Ω, a capacitive reactance of 20 Ω, and an inductive reactance of 30 Ω?

POWER FACTOR

When capacitance and inductance are added to an AC circuit, the electrical efficiency of the circuit goes down. *Power factor (PF)* is the electrical efficiency of a circuit. The power factor is the ratio of true power to apparent power. *True power* (in W) is the actual work being done by the load using electricity. Apparent power (in VA) is the amount of power that is required to operate a load. Power factor is usually expressed as a percentage. For example, a power factor of 0.8 would be expressed as 80%. The power factor can be found by dividing the true power by the apparent power. To determine the power factor of a circuit, apply the following formula:

$$PF = \frac{P_T}{P_A}$$

where
PF = power factor (in %)
P_T = true power (in W)
P_A = apparent power (in VA)

For example, what is the power factor of a circuit that has a load of 5200 W and a rating of 6456 VA?

$$PF = \frac{P_T}{P_A}$$
$$PF = \frac{5200}{6456}$$
$$PF = \mathbf{80.5\%}$$

By changing the power factor formula around, true power or apparent power can be found by dividing or multiplying by the power factor. To determine the true power or apparent power when the power factor is known, apply one of the following formulas:

$$P_T = P_A \times PF$$

where
P_T = true power (in W)
P_A = apparent power (in VA)
PF = power factor (in %)

or

$$P_A = \frac{P_T}{PF}$$

where
P_A = apparent power (in VA)
P_T = true power (in W)
PF = power factor (in %)

For example, what is the wattage of a load that has a voltage of 240 V, draws 15 A, and has a power factor of 85%?

$$VA = V \times A$$
$$VA = 240 \times 15$$
$$VA = 3600 \text{ VA}$$

$$P_T = P_A \times PF$$
$$P_T = 3600 \times 0.85$$
$$P_T = \mathbf{3060 \text{ W}}$$

Practice Questions

Power Factor

1. What is the power factor of a circuit that has a 6400 W load, operates on 240 V, and draws 30 A?
2. What is the true power of a 3φ load that operates at 208 V, draws 20 A, and has a power factor of 90%?
3. What is the apparent power in volt-amps of a circuit with a load of 3056 W and a power factor of 80%?

Lagging/Leading Power Factor

An *inductive circuit* is a circuit in which current lags voltage. The greater the inductance is in a circuit, the larger the phase shift will be. Inductance creates a lagging power factor. A capacitive circuit is a circuit in which current leads voltage. The greater the capacitance is in a circuit, the larger the phase shift will be. Capacitance creates a leading power factor.

A convenient way to remember this is to use the phrase "ELI the ICE man." In the word "ELI," the L stands for inductance; the I (current) comes after the E (voltage). In an inductive circuit, the current lags the voltage. In the word "ICE," the C stands for capacitance; the I (current) comes before the E (voltage). In a capacitive circuit, the current leads the voltage. **See Figure 3-9.**

Most industrial plants have more inductive equipment than capacitive equipment, so the power factor is usually lagging. A suitable power factor is 98% or better. Power supply companies will place a surcharge on industrial plants with a low power factor. To fix this problem, plants will add power factor correction capacitors to raise the power factor.

Circuits that contain only resistive elements have a power factor of 1 or 100%. These circuits are called unity circuits. A *unity circuit* is a pure inductive circuit that is 90° out-of-phase lagging or a pure capacitive circuit that is 90° out-of-phase leading.

Since inductance has a lagging power factor and capacitance has a leading power factor, the two oppose each other. If the inductive reactance and capacitive reactance equal each other, then they will cancel each other out. This is known as resonance. *Resonance* is a condition where the inductive reactance equals capacitive reactance.

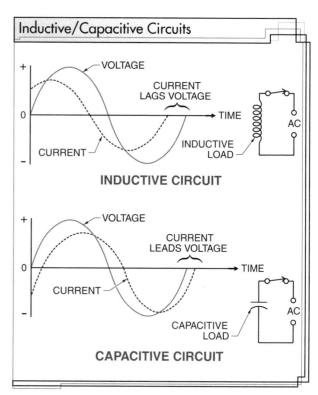

Figure 3-9. In an inductive circuit, the current lags voltage. In a capacitive circuit, the current leads voltage.

In-Phase/Out-of-Phase

A circuit with a 100% power factor is in-phase. Circuits that do not have a 100% power factor are out-of-phase. The relationship of the current sine wave to the voltage sine wave determines if a circuit is in-phase or out-of-phase. If the current and voltage sine waves start at the same time and peak at the same time, then they are in-phase. If they start at different times and do not peak at the same time, then they are out-of-phase. **See Figure 3-10.** Resistive circuits are linear, which means they are in-phase. Nonlinear loads have some amount of inductance or capacitance in them.

Chapter 3—Alternating Current Principles **31**

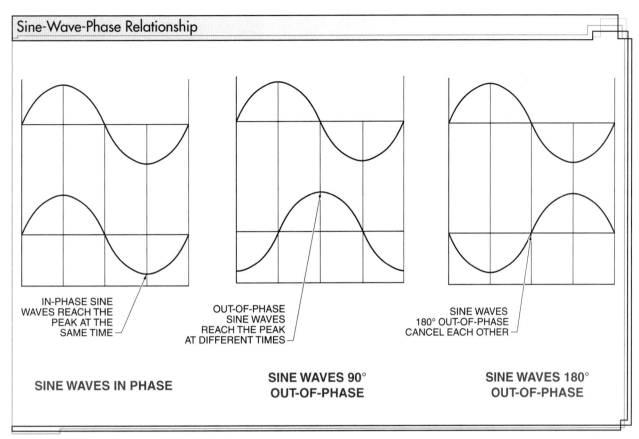

Figure 3-10. The relationship of the current sine wave to the voltage sine wave determines if a circuit is in-phase or out-of-phase.

REACTIVE POWER

Reactive power is the circulating current in a winding such as a motor or transformer. Reactive power is measured in volt-ampere reactive (VAR). A right triangle shows the relationship of reactive power, true power, and apparent power. **See Figure 3-11.**

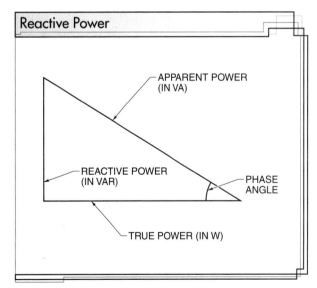

Figure 3-11. The relationship between true power, apparent power, and reactive power can be represented by a right triangle. A Pythagorean formula is used to solve for each variable.

TECH TIP

In an AC circuit in which voltage and current are in-phase, such as a circuit containing only resistance, the power in the circuit is true power. However, almost all AC circuits include capacitive reactance (from capacitors) and/or inductive reactance (from coils). Inductive reactance is more common. All motors, transformers, solenoids, and coils have inductive reactance.

The true power is always at the base of the triangle. The apparent power is always on the hypotenuse of the triangle. The reactive power is always on the side perpendicular to the base of the triangle. The true power will always be equal to or less than the apparent power of the circuit. The wattage is determined by the load being used. The apparent power will be equal to or greater than the true power and the reactive power. The reactive power will be determined by the power factor. The higher the power factor is, the lower the reactive power will be. By using trigonometry functions or the Pythagorean theorem, any one of the missing variables can be found. The Pythagorean theorem formula is as follows:

$$P_A = \sqrt{P_T^2 + P_R^2}$$

where
P_A = apparent power (in VA)
P_T = true power (in W)
P_R = reactive power (in VAR)

$$P_T = \sqrt{P_A^2 + P_R^2}$$

where
P_T = true power (in W)
P_A = apparent power (in VA)
P_R = reactive power (in VAR)

$$P_R = \sqrt{P_A^2 - P_T^2}$$

where
P_R = reactive power (in VAR)
P_A = apparent power (in VA)
P_T = true power (in W)

For example, what is the apparent power in a circuit in kilovolt-amps if the load is 10 kW and the circuit has 2 kVAR?

$$P_A = \sqrt{P_T^2 + P_R^2}$$
$$P_A = \sqrt{10,000^2 + 2000^2}$$
$$P_A = \sqrt{104,000,000}$$
$$P_A = 10,198 \text{ VA}$$
$$P_A = \mathbf{10 \text{ kVA}}$$

Practice Question

Reactive Power

1. What is the volt-amp measurement of a circuit that has 7500 W and 1750 VAR?

CHAPTER 3

Alternating Current Principles

Review Questions

Name _____ Date _____

_____ 1. The moveable coil of wire in a generator that rotates through the magnetic field is called a(n) ___.
   A. slip ring
   B. field winding
   C. brush
   D. armature

_____ 2. If the period of a sine wave is 0.02, then the frequency is ___ Hz.
   A. 0.02
   B. 4
   C. 50
   D. 60

_____ 3. An inductor that has an inductance value of 150 H and operates on a 75 Hz circuit will have an inductive reactance of ___ Ω.
   A. 11,250
   B. 35,342.9
   C. 60,215
   D. 70,650

_____ 4. A(n) ___ is an electrical device designed to store electrical energy by means of an electrostatic field.
   A. capacitor
   B. inductor
   C. battery
   D. transformer

_____ 5. A 275 µF capacitor connected to a 70 Hz supply will have a capacitive reactance of ___ Ω.
   A. 7.27
   B. 8.27
   C. 10.73
   D. 16.56

34 MASTER ELECTRICIAN'S EXAM WORKBOOK

_____ 6. If a circuit has a resistance of 10 Ω, a capacitive reactance of 25 Ω, and an inductive reactance of 75 Ω, then the impedance of the circuit is ___ Ω.
- A. 50.99
- B. 60.25
- C. 18,750
- D. 59,680

_____ 7. A 5000 W load with a rating of 7500 VA has a power factor of ___%.
- A. 37.5
- B. 60.3
- C. 66.7
- D. 77.7

_____ 8. A(n) ___ circuit is a circuit in which current leads voltage.
- A. inductive
- B. capacitive
- C. reactive
- D. parallel

_____ 9. A suitable power factor is ___% or better.
- A. 50
- B. 75
- C. 98
- D. 100

_____ 10. If a circuit is 4 kVAR with a load of 20 kW, the apparent power is ___ kVA.
- A. 20.4
- B. 22.2
- C. 50
- D. 80

Digital Resources
ATPeResources.com/QuickLinks
Access Code: 637160

CHAPTER 4

Circuits

A simple circuit consists of a voltage source, insulated conductors, and a load. Other circuit components, such as switches, fuses, or circuit breakers, may also be included. A primary concern for an electrician is the amount of current flowing in a circuit because the current flow produces heat. Heating conductors beyond their maximum allowable insulation temperature rating may cause fires and damage to the insulation. Ohm's law is used to determine the current, voltage, and resistance of most electrical circuits.

OBJECTIVES

After completing this chapter, the learner will be able to do the following:

- List the main components that make up a circuit.
- Identify the four main types of circuits used in residential, commercial, and industrial applications.
- Calculate the resistance, voltage, and current of different types of circuits.
- List several applications for series circuits.
- Describe the advantages of parallel circuits.
- Break down a combination circuit into a series circuit in order to find the resistance.
- Describe the effects of an open neutral in a multiwire circuit.

CIRCUIT REQUIREMENTS

Circuits are designed to match the needs of the load in order for the load to work properly. The four most common types of circuits are series, parallel, combination, and multiwire. Parallel circuits are by far the most common, followed by series circuits. Combination circuits are a combination of parallel and series circuits combined together to form a complete circuit. Multiwire circuits are made up of two or three ungrounded conductors that share a neutral. For any circuit to work, it must contain a voltage source, conductors, and a load.

SERIES CIRCUITS

A *series circuit* is a circuit that has two or more components connected so that there is only one current path. **See Figure 4-1.** The current leaves and then returns to the power source with the same amount of current. Circuits that use series connections include control, alarm, and signaling circuits. Many safety circuits such as emergency stop circuits are wired in series as well. Dual voltage motor windings are connected in series on high voltage.

Any break in a series circuit will stop the entire circuit from working. In series circuits, the amount of voltage at the source equals the voltage drop of the loads. If a series circuit was supplied by a 120 V power source and the circuit contained three equal-size loads, the voltage would split equally among the three loads at 40 V each. If the loads are not equal, then the voltage would divide itself proportionally to match the loads, but the total load will still equal 120 V.

A *short circuit* is any circuit in which current takes a shortcut around the normal path of current flow. A short circuit will bypass part of the circuit. This will cause the resistance to decrease, the amperage to increase, and the voltage in the remaining loads to increase. If the current is high enough, it may cause the circuit breaker to trip. If the breaker does not trip when it is supposed to, the components in the remaining part of the circuit may be damaged because the current and voltage imposed on it have exceeded their rating. The amount of current and voltage supplied to the remaining loads will depend on the location of the short circuit. There is no way of telling what the voltage will be until the circuit is shorted.

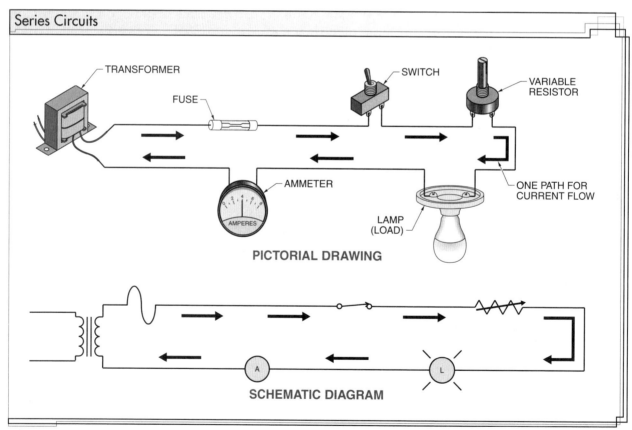

Figure 4-1. A series circuit contains only one path for current flow. Opening the circuit at any point will stop the flow of current.

Understanding Ohm's law and four basic rules will help find unknown variables of a series circuit. When working with series circuits, it is important to remember the following rules:

- Total current (I_T) stays the same throughout the circuit.
- Voltage is added together to determine total voltage (E_T).
- Resistance is added together to determine total resistance (R_T).
- Power is added together to determine total power (P_T).

Current in Series Circuits

The current in a series circuit is the same throughout the circuit. To determine total current in a series circuit, apply the following formula:

$$I_T = I_1 = I_2 = I_3$$

where
I_T = total circuit current (in A)
I_1 = current through component 1 (in A)
I_2 = current through component 2 (in A)
I_3 = current through component 3 (in A)

For example, what is the total current in a series circuit if three loads are connected in series and each load produces 5 A?

$$I_T = I_1 = I_2 = I_3$$
$$I_T = 5 = 5 = 5$$
$$I_T = \mathbf{5\ A}$$

Practice Question

Determining Total Current in a Series Circuit

1. What is the total current in a series circuit if three loads are connected in series and each load produces 0.25 A?

Voltage in Series Circuits

The sum of all the voltage in a series circuit is equal to the applied voltage because the voltage in a series circuit is divided across all the loads in the circuit. To determine total voltage in a series circuit, apply the following formula:

$$E_T = E_1 + E_2 + E_3 + \ldots$$

where
E_T = total applied voltage (in V)
E_1 = voltage drop across load 1 (in V)
E_2 = voltage drop across load 2 (in V)
E_3 = voltage drop across load 3 (in V)

For example, what is the total applied voltage of a series circuit containing 2 V, 2 V, and 8 V drops across three loads?

$$E_T = E_1 + E_2 + E_3$$
$$E_T = 2 + 2 + 8$$
$$E_T = \mathbf{12\ V}$$

Practice Question

Determining Total Voltage in a Series Circuit

1. What is the total applied voltage of a series circuit containing 10 V, 12 V, 20 V, and 25 V drops across four loads?

Resistance in Series Circuits

Resistance is the opposition to current flow. The total resistance in a series circuit equals the sum of the resistance of all loads in the circuit. To determine total resistance in a series circuit, apply the following formula:

$$R_T = R_1 + R_2 + R_3 + \ldots$$

where
R_T = total resistance (in Ω)
R_1 = resistance 1 (in Ω)
R_2 = resistance 2 (in Ω)
R_3 = resistance 3 (in Ω)

For example, what is the total resistance in a circuit containing three resistors of 25 Ω, 50 Ω, and 75 Ω connected in series?

$$R_T = R_1 + R_2 + R_3$$
$$R_T = 25 + 50 + 75$$
$$R_T = \mathbf{150\ \Omega}$$

Practice Question

Determining Total Resistance in a Series Circuit

1. What is the total resistance in a circuit containing four resistors of 10 Ω, 15 Ω, 20 Ω, and 25 Ω connected in series?

Power in Series Circuits

The total power in a series circuit is equal to the sum of the individual power produced by each individual load. To determine total power in a series circuit, apply the following formula:

$$P_T = P_1 + P_2 + P_3 + \ldots$$

where
P_T = total circuit power (in W)
P_1 = power of load 1 (in W)
P_2 = power of load 2 (in W)
P_3 = power of load 3 (in W)

For example, what is the total power in a series circuit if three loads produce 2 W each?

$P_T = P_1 + P_2 + P_3$
$P_T = 2 + 2 + 2$
$P_T = \mathbf{6\ W}$

Practice Question

Determining Total Power in a Series Circuit

1. What is the total power in a series circuit if three loads produce 4 W each?

PARALLEL CIRCUITS

A *parallel circuit* is a circuit that has two or more components connected so that there is more than one path for current to flow. **See Figure 4-2.** In a parallel circuit, energy leaves the power source and breaks up into two or more paths, and then returns to the power source with the same amount of energy. Parallel circuits are used in most electrical wiring installations such as motors, lights, and receptacles.

In a parallel circuit, the source voltage is delivered to each load. The loads are supplied with the rated voltage while they draw the required current. Even if a load is added or removed from the circuit, the voltage that each load receives will remain the same as the source voltage. The current will change as loads are added or removed from a circuit. The windings of dual voltage motors operating on low voltage are connected in parallel.

TECH TIP

Care must be taken when working with parallel circuits because current can be flowing in one part of the circuit even though another part of the circuit is turned off. Understanding and recognizing parallel-connected components and circuits enables a technician or troubleshooter to take proper measurements, make circuit modifications, and troubleshoot the circuit.

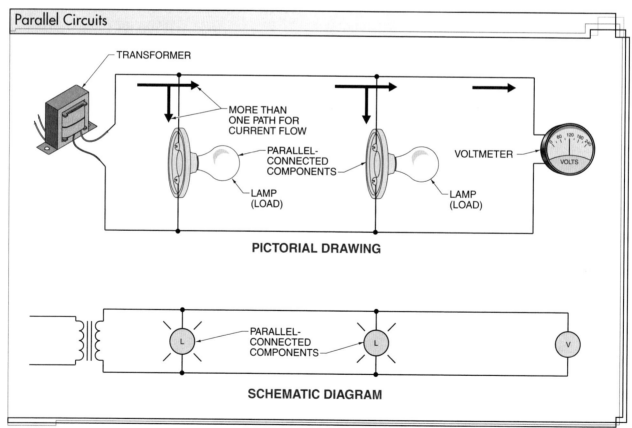

Figure 4-2. A parallel circuit has more than one path for current flow. If part of the circuit is turned off, there may still be current flowing in another part of the circuit.

If there is an opening in one branch of a parallel circuit, the remaining circuit components continue to perform as designed. This will cause the total current to drop. The only place where an opening will cause the entire circuit to stop is between the power source and the first division of the circuit or between the power source and the point at which all the branches meet. If a short or ground fault develops in the circuit, the current will increase and will open the overcurrent device.

Understanding Ohm's law and four basic rules will help find unknown variables of a parallel circuit. When working with parallel circuits, it is important to remember the following rules:
- Voltage is the same in all loads of a parallel circuit.
- Current ratings of each branch are added together.
- Total circuit resistance is always smaller than the smallest resistance load.
- Power is added together.

Current in Parallel Circuits

The total current in a parallel circuit equals the sum of the current through all the loads. To determine the total current in a parallel circuit, apply the following formula:
$$I_T = I_1 + I_2 + I_3 + \ldots$$
where
I_T = total circuit current (in A)
I_1 = current through load 1 (in A)
I_2 = current through load 2 (in A)
I_3 = current through load 3 (in A)

For example, what is the total current in a parallel circuit with four loads of 0.833 A, 6.25 A, 8.33 A, and 1.2 A?
$$I_T = I_1 + I_2 + I_3$$
$$I_T = 0.833 + 6.25 + 8.33 + 1.2$$
$$I_T = \mathbf{16.613\ A}$$

Practice Questions

Determining Total Current in a Parallel Circuit

1. What is the total current in a parallel circuit with four loads of 5 A, 2 A, 3.62 A, and 6.22 A?

TECH TIP

No material is a perfect conductor or insulator. Resistance is part of every circuit and must be considered when designing and troubleshooting circuits.

Voltage in Parallel Circuits

The voltage across each load is the same when loads are connected in parallel. To determine the total voltage in a parallel circuit, apply the following formula:
$$E_T = E_1 = E_2 = E_3 = \ldots$$
where
E_T = total applied voltage (in V)
E_1 = voltage across load 1 (in V)
E_2 = voltage across load 2 (in V)
E_3 = voltage across load 3 (in V)

For example, what is the total voltage if the voltage across each of the three loads in a parallel circuit measures 120 V?
$$E_T = E_1 = E_2 = E_3$$
$$E_T = 120 = 120 = 120$$
$$E_T = \mathbf{120\ V}$$

Practice Questions

Determining Total Voltage in a Parallel Circuit

1. What is the total voltage if the voltage across each of the three loads in a parallel circuit measures 200 V?

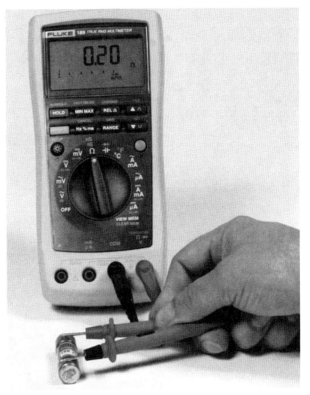

Fluke Corporation
A digital multimeter set to measure resistance can be used to test a fuse.

Resistance in Parallel Circuits

There are two different methods used to find the total resistance of a parallel circuit. The number of resistors in the circuit will determine the type of formula used. To determine the total resistance of two resistors in a parallel circuit, apply the following formula:

$$R_T = \frac{R_1 \times R_2}{R_1 + R_2}$$

where

R_T = total resistance (in Ω)
R_1 = resistance 1 (in Ω)
R_2 = resistance 2 (in Ω)

To determine the total resistance of three or more resistors in a parallel circuit, apply the following formula:

$$R_T = \frac{1}{\frac{1}{R_1} + \frac{1}{R_2} + \frac{1}{R_3} + \ldots}$$

where

R_T = total resistance (in Ω)
R_1 = resistance 1 (in Ω)
R_2 = resistance 2 (in Ω)
R_3 = resistance 3 (in Ω)

For example, what is the total resistance of a parallel circuit containing two resistors at 300 Ω and 600 Ω?

$$R_T = \frac{R_1 \times R_2}{R_1 + R_2}$$
$$R_T = \frac{300 \times 600}{300 + 600}$$
$$R_T = \frac{180{,}000}{900}$$
$$R_T = \mathbf{200\ \Omega}$$

For example, what is the total resistance in a parallel circuit containing three resistors at 5000 Ω, 2000 Ω, and 10,000 Ω?

$$R_T = \frac{1}{\frac{1}{R_1} + \frac{1}{R_2} + \frac{1}{R_3}}$$
$$R_T = \frac{1}{\frac{1}{5000} + \frac{1}{2000} + \frac{1}{10{,}000}}$$
$$R_T = \frac{1}{0.0002 + 0.0005 + 0.0001}$$
$$R_T = \frac{1}{0.0008}$$
$$R_T = \mathbf{1250\ \Omega}$$

Practice Question

Determining Total Resistance in a Parallel Circuit

1. What is the total resistance of a parallel circuit containing two resistors at 20 Ω and 25 Ω?
2. What is the total resistance in a parallel circuit containing three resistors at 100 Ω, 200 Ω, and 300 Ω?

Power in Parallel Circuits

The total power in a parallel circuit is equal to the sum of the power produced by each load. To determine the total power in a parallel circuit, apply the following formula:

$$P_T = P_1 + P_2 + P_3 + \ldots$$

where

P_T = total circuit power (in W)
P_1 = power of load 1 (in W)
P_2 = power of load 2 (in W)
P_3 = power of load 3 (in W)

For example, what is the total power of a parallel circuit with four loads that produce 100 W, 750 W, 1000 W, and 150 W?

$$P_T = P_1 + P_2 + P_3 + P_4$$
$$P_T = 100 + 750 + 1000 + 150$$
$$P_T = \mathbf{2000\ W}$$

Practice Question

Determining Total Power in a Parallel Circuit

1. What is the total power of a parallel circuit with four loads that produce 20 W, 50 W, 75 W, and 100 W?

TECH TIP

Connecting loads in parallel is the most common method used to connect loads. The voltage across each load is the same when loads are connected in parallel. The voltage across each load remains the same if parallel loads are added or removed. For example, loads such as lamps, small appliances, fans, TVs, etc. are all connected in parallel when connected to a branch circuit.

COMBINATION CIRCUITS

A *combination circuit* is a circuit that has both series and parallel connections. Most questions on the exam covering combination circuits will involve finding the total resistance. Once the resistance is determined, the current can be found as long as the voltage is provided. As with series and parallel circuits, there are rules that apply to combination circuits. When working with combination circuits, it is important to remember the following four rules:

- Start from the farthest point away from the power supply.
- Follow the series circuit rules for loads connected in series.
- Follow the parallel circuit rules for loads connected in parallel.
- Combine the loads until a single load is left.

Resistance in Combination Circuits

The total resistance in a combination circuit equals the sum of the series loads and the equivalent resistance of the parallel combinations. **See Figure 4-3.** To determine the total resistance in a combination circuit that contains two resistors connected in parallel and two resistors connected in series, apply the following formula:

$$R_T = \left(\frac{R_{P1} \times R_{P2}}{R_{P1} + R_{P2}}\right) + R_{S1} + R_{S2} + \ldots$$

where
R_T = total resistance (in Ω)
R_{P1} = parallel resistance 1 (in Ω)
R_{P2} = parallel resistance 2 (in Ω)
R_{S1} = series resistance 1 (in Ω)
R_{S2} = series resistance 2 (in Ω)

For example, what is the total resistance in a combination circuit containing a 20 Ω and 30 Ω resistor connected in parallel and a 60 Ω and 48 Ω resistor connected in series?

$$R_T = \left(\frac{R_{P1} \times R_{P2}}{R_{P1} + R_{P2}}\right) + R_{S1} + R_{S2}$$

$$R_T = \left(\frac{20 \times 30}{20 + 30}\right) + 60 + 48$$

$$R_T = \left(\frac{600}{50}\right) + 108$$

$$R_T = 12 + 108$$

$$R_T = \mathbf{120\ \Omega}$$

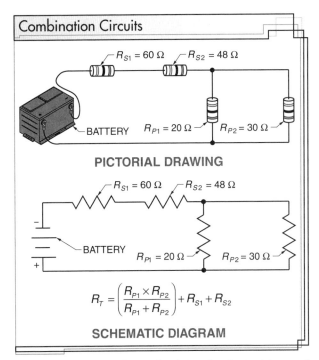

Figure 4-3. A formula is used to determine the resistance of a circuit that contains two resistors connected in parallel and two resistors connected in series. A more complex combination circuit should be broken down until it is a series circuit.

To determine the total resistance in a combination circuit that contains several series/parallel combinations, the circuit must be broken down to the basic series and parallel parts. After the circuit is broken down, the rules for series circuits apply to the series connections and the rules for parallel circuits apply to the parallel connections. To determine the total resistance of a combination circuit that contains several series/parallel combinations, apply the following procedure:

1. Combine the two or more resistance values into a single resistance value if there are two or more resistors connected in series. This is done by applying the following formula for resistance in series:

$$R_T = R_1 + R_2 + R_3 + \ldots$$

2. Combine the two or more resistors connected in parallel into a single resistance value. This is done by applying the formula for resistors in parallel. For two resistors connected in parallel, apply the following formula:

$$R_T = \frac{R_1 \times R_2}{R_1 + R_2}$$

For three or more resistors connected in parallel, apply the following formula:

$$R_T = \cfrac{1}{\cfrac{1}{R_1} + \cfrac{1}{R_2} + \cfrac{1}{R_3} + ...}$$

3. Add the sum of the series resistance values to the sum of each parallel circuit resistance value to determine the total resistance of the circuit.

For example, what is the total resistance of the following combination circuit? **See Figure 4-4.**

1. Combine the two resistors connected in series into a single resistance value.

$R_{ST1} = R_1 + R_2$
$R_{ST1} = 55 + 10$
$R_{ST1} = \mathbf{65\ \Omega}$

TECH TIP

Identifying and understanding series and parallel-circuits enables a technician to take proper measurements, make circuit modifications, and troubleshoot the circuit.

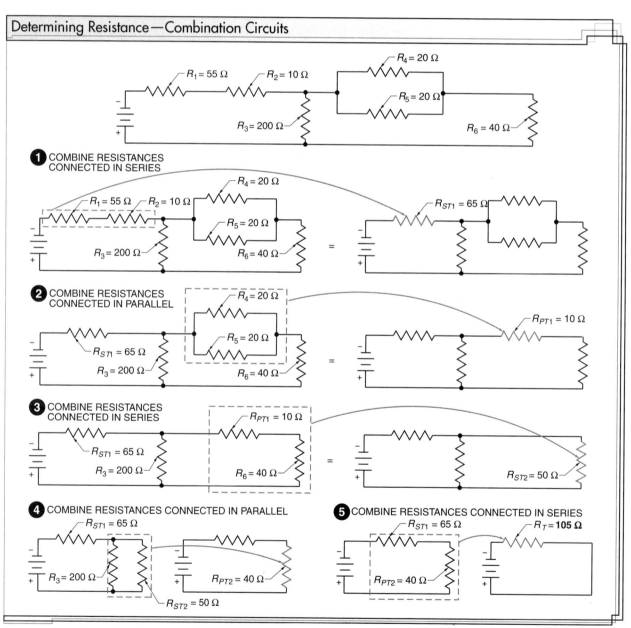

Figure 4-4. To determine the total resistance in a combination circuit that contains multiple series/parallel combinations, the circuit must be broken down to the basic series and parallel parts.

2. Combine the two or more resistors connected in parallel into a single resistance value.

$$R_{PT1} = \frac{R_4 \times R_5}{R_4 + R_5}$$

$$R_{PT1} = \frac{20 \times 20}{20 + 20}$$

$$R_{PT1} = \frac{400}{40}$$

$$R_{PT1} = \mathbf{10\ \Omega}$$

3. Combine resistors connected in series into a single resistance.

$$R_{ST2} = R_{PT1} + R_6$$
$$R_{ST2} = 10 + 40$$
$$R_{ST2} = \mathbf{50\ \Omega}$$

4. Combine the two or more resistors connected in parallel into a single resistance value.

$$R_{PT2} = \frac{R_{ST2} \times R_3}{R_{ST2} + R_3}$$

$$R_{PT2} = \frac{50 \times 200}{50 + 200}$$

$$R_{PT2} = \frac{10,000}{250}$$

$$R_{PT2} = \mathbf{40\ \Omega}$$

5. Combine resistors connected in series into a single resistance.

$$R_T = R_{ST1} + R_{PT2}$$
$$R_T = 65 + 40$$
$$R_T = \mathbf{105\ \Omega}$$

Practice Question

Determining Total Resistance in a Combination Circuit

1. What is the total resistance in a combination circuit containing a 10 Ω and 15 Ω resistor connected in parallel and a 20 Ω and 40 Ω resistor connected in series?

MULTIWIRE CIRCUITS

A *multiwire circuit* is a circuit that consists of two or more ungrounded conductors sharing a common grounded neutral conductor. On 120/240 V services, the multiwire circuit is a 3-wire circuit with two ungrounded conductors from opposite phases and one grounded conductor (neutral). On 3φ wye systems of 208/120 V or 480/277 V, a multiwire circuit can have up to three ungrounded conductors, each from a different phase, and one grounded conductor (neutral). **See Figure 4-5.** This wiring method eliminates one or two conductors depending on the voltage system. This eliminates a conductor because the neutral only has to carry the unbalanced load. The multiwire circuit must be connected to different phases in order to work properly. If the ungrounded conductors are connected to the same phase, the neutral currents will combine and may overload the current capacity of the conductor. Subsections of the NEC® that cover multiwire branch circuits include the following:

- 100 Branch Circuit, Multiwire
- 210.4 (A)
- 210.4 (B)
- 210.4 (C)
- 210.4 (D)
- 300.13 (B)

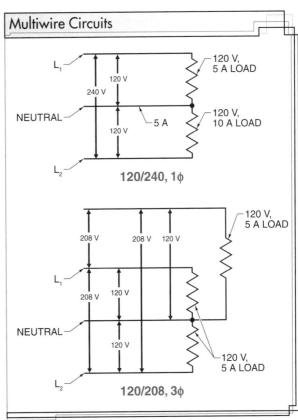

Figure 4-5. Multiwire circuits have two or more ungrounded conductors sharing a common grounded neutral conductor.

When working with multiwire circuits, it is important to remember the following rules:
- All conductors shall come from the same panel.
- The circuit shall supply only line-neutral loads.
- The grounded conductor shall not be dependent on device connections.

- The voltage between each ungrounded conductor and the grounded conductor (neutral) is the same on each phase.
- Each set of conductors shall be grouped together to identify them as working together.
- All ungrounded conductors shall be simultaneously disconnected.

Dangers of Multiwire Circuits

One of the dangers of multiwire circuits is the effect of an open neutral. If the neutral becomes open, the circuit becomes a series circuit and the voltages will divide between the two loads based on the size of the loads. If the neutral is open and the loads are balanced, the circuit will continue to work. The circuit continues to work as long as both loads are turned on. If one load is turned off, the other load will turn off as well.

An open neutral on a multiwire circuit may cause a fire or damage the loads. The load that will stop working first will always be the load with the smallest wattage. This is because the load with the smallest wattage has a higher resistance internally. Most multiwire circuit questions on the exam will include a single-phase, three-wire circuit with two ungrounded conductors and one neutral. **See Figure 4-6.** To determine the voltage on each load of a multiwire circuit with an open neutral, apply the following procedure:

1. Find the current of each load. To find the current of each load, divide the load's rated wattage by its rated voltage.

   Load 1: $\dfrac{100}{120} = 0.83$

   Load 2: $\dfrac{1000}{120} = 8.3$

2. Find the resistance of each load. To find the resistance of each load, divide the rated voltage by the current found in Step 1.

   Load 1: $\dfrac{120}{0.83} = 145$

   Load 2: $\dfrac{120}{8.3} = 14.5$

3. Find the total circuit resistance. To find the total circuit resistance, add the resistance of the two loads. Since the circuit is now a series, the circuit resistance is added together.

   $145 + 14.5 = 159.5$

4. Find the series circuit current. To find the series circuit current, divide the line-to-line voltage (usually 240 V) by the total resistance as found in Step 3.

   $\dfrac{240}{159.5} = 1.5$

5. Find the new voltage on each load. To find the new voltage on each load, multiply the current of the circuit by the resistance of the load as found in Step 2.

   Load 1: $1.5 \times 145 = 217.5$

   Load 2: $1.5 \times 14.5 = 21.75$

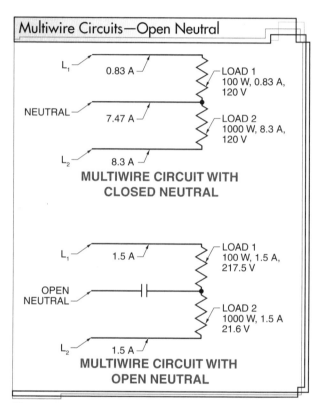

Figure 4-6. A multiwire circuit with an open neutral becomes a series circuit. The voltages will divide between the two loads based on the size of the loads.

Practice Questions

Multiwire Circuits

1. The neutral on a multiwire circuit will carry the ___ load when all loads are operating.

2. If the neutral is broken or disconnected on a single-phase multiwire branch circuit, which load will burn up first?

CHAPTER 4
Circuits
Review Questions

Name _____   Date _____

_____   **1.** The most common type of circuit is a ___ circuit.
- A. series
- B. parallel
- C. combination
- D. multiwire

_____   **2.** A series circuit with four 25 Ω resistive loads has a total resistance of ___ Ω.
- A. 6.25
- B. 25
- C. 50
- D. 100

_____   **3.** A series circuit with a 6 V and a 4 V drop across two loads has a total voltage of ___ V.
- A. 1.5
- B. 2
- C. 10
- D. 24

_____   **4.** If five loads connected in a series circuit each use 10 A, the total current is ___ A.
- A. 10
- B. 30
- C. 50
- D. 100

_____   **5.** A parallel circuit containing four resistors at 10 Ω, 15 Ω, 25 Ω, and 50 Ω has a total resistance of ___ Ω.
- A. 4.3
- B. 20
- C. 50
- D. 100

_____   **6.** A parallel circuit containing two resistors at 50 Ω and 55 Ω has a total resistance of ___ Ω.
- A. 1.1
- B. 2.2
- C. 25
- D. 26.2

7. A parallel circuit with loads of 4 A, 10 A, and 15 A has a total current of ___ A.
   A. 8.5
   B. 9.7
   C. 29
   D. 32

8. A parallel circuit containing four loads rated at 5 W, 12 W, 14 W, and 20 W has ___ W of total power.
   A. 12.6
   B. 20
   C. 49
   D. 51

9. A ___ circuit is a circuit that has both series and parallel connections.
   A. series
   B. parallel
   C. combination
   D. multiwire

10. A ___ circuit is a circuit that consists of two or more ungrounded conductors sharing a common grounded neutral conductor.
    A. series
    B. parallel
    C. combination
    D. multiwire

CHAPTER 5

Transformers

A transformer uses electromagnetism to change voltage from one level to another or to isolate one voltage from another. The changing values of a sine wave cause the magnetic field around conductors to expand and then collapse across other conductors, which induces a voltage into it. A transformer will not work on DC voltages because the expanding and collapsing magnetic field is not present in DC circuits. Voltage transformers are used to step up voltages to hundreds of thousands of volts to travel along transmission lines. The voltage is then stepped down for residential, commercial, and industrial locations. Current transformers are mainly used to step down line current to make it easier to measure.

OBJECTIVES

After completing this chapter, the learner will be able to do the following:

- Describe transformer ratios.
- Determine the volt-amps (VA) rating of a transformer.
- Calculate the efficiency of a transformer.
- Size various types of transformers.
- Describe the purpose of autotransformers and current transformers.
- Determine the proper overcurrent protective devices for transformers.
- Calculate voltage and current for delta/wye transformers.
- Describe the different types of transformer connections.

TRANSFORMER RATIOS

A transformer typically consists of two separate coils with a different number of turns of conductor wound around the same closed laminated iron core. The primary winding is the coil that is energized by the source. The secondary winding is the coil that is connected to the load. The ratio for a transformer is the ratio of the primary voltage to secondary voltage. Ratios are used to indicate different types of transformers. A ratio of 1:6 indicates a step-up transformer with 1 turn on the primary side for every 6 turns on the secondary side. A ratio of 4:1 indicates a step-down transformer with 4 turns on the primary side for every 1 turn on the secondary side.

Turns Ratios

Turns ratios are based on the number of turns on the primary side of a transformer versus the number of turns on the secondary side. The secondary side of a step-down transformer will have fewer turns than the primary side. On a step-up transformer, the secondary side will have more turns than the primary side. **See Figure 5-1.** To determine the turns ratio of a transformer, apply the following formula:

$$R = \frac{primary\ V}{secondary\ V}$$

where
R = ratio
primary V = primary voltage
secondary V = secondary voltage

For example, what is the turns ratio of a transformer with a primary voltage of 480 V and a secondary voltage of 240 V?

$$R = \frac{primary\ V}{secondary\ V}$$

$$R = \frac{480\ V}{240\ V}$$

$$R = 2:1$$

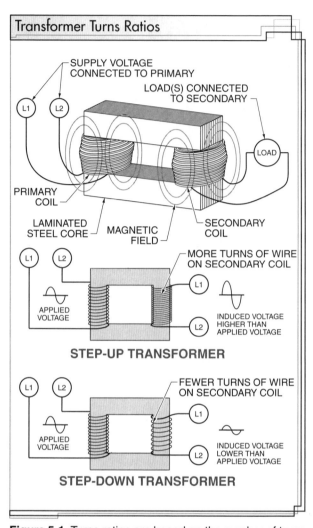

Figure 5-1. Turns ratios are based on the number of turns on the primary side of a transformer versus the number of turns on the secondary side.

Voltage Ratios

For voltage transformers, the voltage ratio will equal the turns ratio. If the turns ratio is 4:1, then the voltage ratio will be 4:1. Efficiency does not affect the voltage output of a transformer. To determine the relationship between the number of turns and the voltage, apply the following formula:

$$\frac{N_P}{N_S} = \frac{E_P}{E_S}$$

where
N_P = number of turns in primary coil
N_S = number of turns in secondary coil
E_P = voltage applied to primary coil (in V)
E_S = voltage induced in secondary coil (in V)

Practice Questions

Determining Transformer Turns Ratio

1. What is the ratio of a transformer with a primary voltage of 12,500 V and a secondary voltage of 4160 V?

2. What is the turns ratio of a transformer with a primary voltage of 13,800 V and a secondary voltage of 69,000 V?

If one of the four parts of the ratio is missing, cross multiplication is used to determine the missing value. This is done by multiplying the two given values that are diagonal from each other and dividing by the remaining value. **See Figure 5-2.**

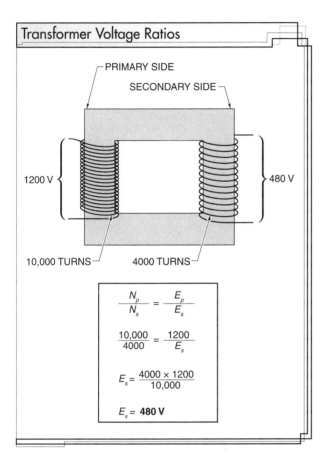

Figure 5-2. Cross multiplication is used to find the missing value when one of the four parts of the ratio is missing.

For example, what is the voltage on the secondary side of a transformer if the primary side has 10,000 turns and 1200 V and the secondary side has 4000 turns?

$$\frac{N_P}{N_S} = \frac{E_P}{E_S}$$

$$E_S = \frac{N_S \times E_P}{N_P}$$

$$E_S = \frac{4000 \times 1200}{10,000}$$

$$E_S = \mathbf{480\ V}$$

For example, what is the required number of turns on the secondary side if the primary side has 1200 V and 10,000 turns and the secondary is 480 V?

$$\frac{N_P}{N_S} = \frac{E_P}{E_S}$$

$$N_S = \frac{N_P \times E_S}{E_P}$$

$$N_S = \frac{10,000 \times 480}{1200}$$

$$N_S = \mathbf{4000\ turns}$$

Practice Questions

Determining Transformer Voltage Ratio

1. What is the number of turns on the primary side of a transformer that is connected to a 120 V source? The secondary side has 60 turns and delivers 24 V.

2. A power transformer has 85 turns on the primary side and 255 turns on the secondary side. What voltage will the secondary side deliver if the primary voltage is 120 V?

Current Ratios

Most transformers are voltage transformers. Electricians are more concerned with the voltage from the secondary side of a transformer compared to the primary input voltage. As the voltage in a transformer changes, the current changes. The current ratio will be the opposite of the voltage or turns ratio. This means the current will do the exact opposite of what the voltage does. If the voltage is stepped up, then the current is stepped down. If the voltage is stepped down, then the current is increased. This makes the current inversely proportional to the voltage.

While efficiency does not affect the voltage of a transformer, it does affect the current output of a transformer. This is because the volt-amp rating of the transformer is affected. There may be questions on the exam that will give the amperage of one side of a transformer and not the other. If this is the case, the amperage of the side given is used. To determine the current or voltage of a given side of a transformer, apply the following formula:

$$\frac{I_P}{I_S} = \frac{E_S}{E_P}$$

where
I_P = current in primary coil (in A)
I_S = current in secondary coil (in A)
E_S = voltage induced in secondary coil (in V)
E_P = voltage applied to primary coil (in V)

For example, what is the amperage on the primary side if a transformer has 800 V on the primary side and 200 A and 200 V on the secondary side (4:1 ratio)?

$$\frac{I_P}{I_S} = \frac{E_S}{E_P}$$

$$I_P = \frac{I_S \times E_S}{E_P}$$

$$I_P = \frac{200 \times 200}{800}$$

$$I_P = \frac{40{,}000}{800}$$

$$I_P = \mathbf{50\ A}$$

Practice Questions

Determining Current Ratios

1. If a transformer with 240 turns on the primary side and 30 turns on the secondary side draws 0.25 A from a 120 V line, what is the secondary current?
2. If a 9:2 step-down transformer draws 1.8 A on the primary side, what is the current on the secondary side?
3. If a doorbell transformer draws 20 VA from a 120 V source, what is the current on the secondary side if it delivers 10 V?

RATING TRANSFORMERS

It is standard practice to rate transformers for their output capabilities because it is the output of the transformer to which the loads are connected. Transformers are usually rated in volt-amps (VA) or kilovolt-amps (kVA).

The volt-amp rating of a a 1ϕ transformer is found by multiplying the voltage of the primary side by the current of the primary side or by multiplying the voltage of the secondary side by the current of the secondary side. **See Figure 5-3.** To determine the volt-amp rating of a 1ϕ transformer, apply the following formula:

$$VA_P = V_P \times I_P$$
where
VA_P = volt-amps of primary side
V_P = primary voltage (in V)
I_P = primary current (in A)

or

$$VA_S = V_S \times I_S$$
where
VA_S = volt-amps of secondary side
V_S = secondary voltage (in V)
I_S = secondary current (in A)

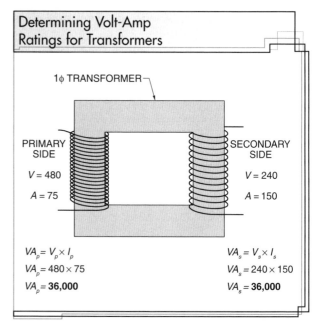

Figure 5-3. Transformers are rated in volt-amps or kilovolt-amps depending on the size of the transformer.

To determine the volt-amp rating of a 3ϕ transformer, apply the following formula:

$$VA_P = V_P \times I_P \times 1.732$$
where
VA_P = volt-amps of primary side
V_P = primary voltage (in V)
I_P = primary current (in A)
1.732 = square root of 3

or

$$VA_S = V_S \times I_S \times 1.732$$
where
VA_S = volt-amps of secondary side
V_S = secondary voltage (in V)
I_S = secondary current (in A)
1.732 = square root of 3

For example, what is the volt-amp input for a 1ϕ transformer that delivers 240 V at 150 A?

$$VA_P = V_P \times I_P$$
$$VA_P = 240 \times 150$$
$$VA_P = \mathbf{36{,}000\ VA}$$

What is the volt-amp input for a 3ϕ transformer that delivers 240 V at 150 A?

$$VA_P = V_P \times I_P \times 1.732$$
$$VA_P = 240 \times 150 \times 1.732$$
$$VA_P = \mathbf{62{,}352\ VA}$$

Practice Questions

Determining Transformer Volt-Amps

1. What is the volt-amp output of a 1φ transformer that delivers 480 V and 150 A?
2. What is the volt-amp output of a 3φ transformer that delivers 480 V and 150 A?

To convert the volt-amp rating of a transformer into kilovolt-amps, the volt-amp rating is divided by 1000.

For example, what is the kilovolt-amp rating of a 150,000 VA transformer?

$$kVA = \frac{VA}{1000}$$

$$kVA = \frac{150,000}{1000}$$

$$kVA = \mathbf{150}$$

Practice Questions

Determining Transformer Kilovolt-Amp Rating

1. What is the kilovolt-amp output of a transformer that operates at 480 V, 3φ with 400 A on the secondary side of the transformer?
2. What is the kilovolt-amp input of a transformer that operates at 2400 V and 35 A on the primary side of the transformer?

TRANSFORMER EFFICIENCY

No transformer is 100% efficient. All transformers lose some power to heat. This loss of power should be considered when performing calculations for questions that provide the efficiency. When performing transformer efficiency calculations, the output volt-amp rating of a transformer should always be less than the input volt-amp rating. **See Figure 5-4.** The efficiency of the transformer depends on the core of the transformer. A transformer with an air core should have a lower efficiency than a transformer with an iron core. An iron core transformer is made of thin sheets called laminations to help reduce eddy currents. To determine the efficiency of a transformer, apply the following formula:

$$Eff = \frac{P_S}{P_P} \times 100$$

where
Eff = efficiency (in %)
P_S = power of secondary circuit (in W)
P_P = power of primary circuit (in W)

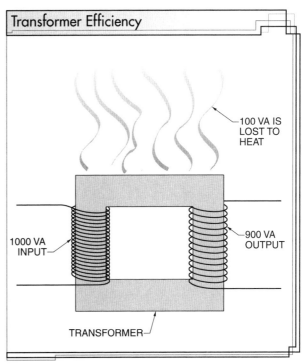

Figure 5-4. All transformers lose some power to heat. The amount of power lost depends on the core of the transformer.

When efficiency is provided, the input or output volt-amp rating of a transformer can be found. To determine the input or output volt-amp rating of a transformer, apply one of the following formulas:

$$P_S = P_P \times Eff$$

where
P_S = power of secondary circuit (in W)
P_P = power of primary circuit (in W)
Eff = efficiency (in %)

or

$$P_P = \frac{P_S}{Eff}$$

where
P_P = power of primary circuit (in W)
P_S = power of secondary circuit (in W)
Eff = efficiency (in %)

For example, what is the output volt-amp rating of an 80% efficient transformer that has a 1000 VA input?

$$P_S = P_P \times Eff$$
$$P_S = 1000 \times 0.80$$
$$P_S = \mathbf{800\ VA}$$

> **Practice Questions**
>
> *Determining Transformer Efficiency*
> 1. What is the volt-amp output rating of a transformer that is 95% efficient with 500 VA on the primary side?
> 2. What is the volt-amp input rating of a transformer that has an output of 2500 VA and is 78% efficient?
> 3. What is the volt-amp input rating of a transformer that is 95% efficient if the secondary side is delivering 1 kVA?
> 4. What is the efficiency of a transformer that has a 2500 kVA rating on the secondary side and requires 2750 kVA on the primary side?

SIZING TRANSFORMERS

Transformers are required to carry the load being served. This is called the calculated load. Electric utility companies often size transformers for the actual load of the building. The utilities are not required to follow the NEC®, but they do have other standards to follow. According to the NEC®, electricians are required to size transformers for the calculated load.

Sizing Single-Phase (1ϕ) Transformers

Single-phase (1ϕ) transformers are sized to handle the load that is applied to them. Most 1ϕ transformers are used for dwellings and are 120/240 V. These loads are sized at 100%. After all loads are sized at 100%, they are added to determine the total load in kilovolt-amps. To size 1ϕ transformers of 120 V or 240 V, apply the following formula:

$L = kVA \times 100\%$

where

L = load

kVA = kilovolt-amps

100% = constant

For example, what size transformer is required for a 1ϕ system if the loads are 120 V at 10 kVA and 240 V at 14 kVA?

Load one:
$L = kVA \times 100\%$
$L = 10 \times 1.00$
$L = 10\ kVA$

Load two:
$L = kVA \times 100\%$
$L = 14 \times 1.00$
$L = 14\ kVA$

$14 + 10 = \mathbf{24\ kVA}$

> **Practice Question**
>
> *Determining Transformer Size—Single-Phase (1ϕ)*
> 1. What size transformer is required for a 1ϕ system if the loads are 120 V at 20 kVA and 240 V at 20 kVA?

Sizing Three-Phase (3ϕ) Transformers

Three-phase (3ϕ) transformers are sized differently than 1ϕ transformers. With a 3ϕ transformer, a set of three 1ϕ transformers are connected together, which divides the load up three ways. There are several different types of 3ϕ transformers. A special 3ϕ system called an open delta uses only two sets of coils but still supplies 3ϕ power.

Sizing Wye Transformers. Three-phase wye transformers are sized based on a balanced load. Each phase has the same voltage-to-grounded conductor and the same voltage between phases. Each of the three transformers receives ⅓ of the load. **See Figure 5-5.** To size a single 3ϕ wye transformer, the three loads are added together. To determine the load to size a single wye-connected transformer, apply the following formula:

$L = L_1 + L_2 + \dots$

where

L = load

L_1 = first load

L_2 = second load

For example, what size transformer is required for a 3ϕ wye system if the 1ϕ loads are 35 kVA and the 3ϕ loads are 56 kVA?

$L = L_1 + L_2$
$L = 35 + 56$
$L = \mathbf{91\ kVA}$

If three 1ϕ transformers were connected in a wye configuration, the loads would be added together and then divided by three. This is done to balance the load so that each transformer gets ⅓ of the total load. To determine the load to size three 1ϕ transformers connected in a wye configuration, apply the following formula:

$$L = \frac{L_1 + L_2 + \dots}{3}$$

where

L = load per transformer

L_1 = first load

L_2 = second load

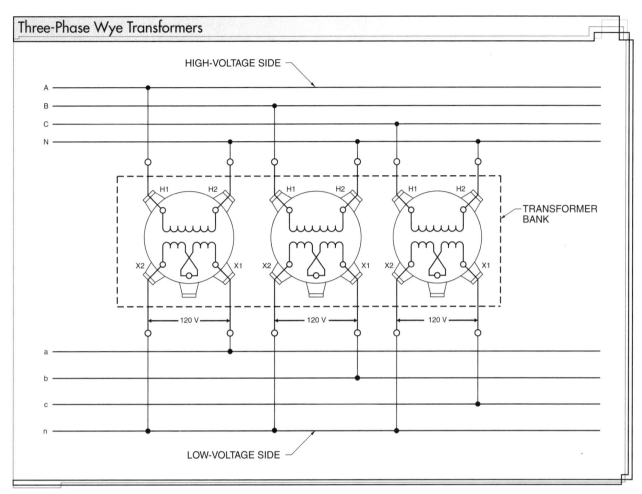

Figure 5-5. Three-phase wye transformers are sized based on a balanced load. Each of the three transformers in a transformer bank receives ⅓ of the load.

For example, what size 1ϕ transformer is required if three 1ϕ transformers are connected in a wye configuration to a 1ϕ load of 35 kVA and a 3ϕ load of 56 kVA?

$$L = \frac{L_1 + L_2 + \ldots}{3}$$

$$L = \frac{35 + 56}{3}$$

$$L = \frac{91}{3}$$

$$L = \mathbf{30.33 \text{ kVA}}$$

TECH TIP

For every 10°C rise in temperature above the transformer's rated limit, the life of the transformer is reduced by approximately 50%.

Practice Questions

Determining Transformer Size—Wye

1. What size transformer is required for a 3ϕ wye system if the 1ϕ loads are 20 kVA and the 3ϕ loads are 23 kVA?

2. What size 1ϕ transformer is required if three 1ϕ transformers are connected to a 1ϕ load of 20 kVA and a 3ϕ load of 23 kVA?

Sizing Closed Delta Transformers. A 3ϕ closed delta transformer does not output the same voltage to the grounded conductor from all three phases. This type of transformer cannot be balanced like a wye transformer. Within a 3ϕ closed delta system, one transformer will be larger than the other two transformers. The larger transformer is called the lighting transformer. The two smaller transformers are called power transformers.

The lighting transformer is sized at 67% for 1φ loads and 33% for 3φ loads. The power transformers are sized at 33% for both 1φ and 3φ loads. When delta transformers are built into a single package, the lighting transformer is sized larger than the other windings, but the total kilovolt-amp rating is still the kilovolt-amp rating of all three transformers added together. **See Figure 5-6.**

For example, what size lighting and power transformer is required for a 3φ closed delta system if the 1φ load is 30 kVA and the 3φ load is 58 kVA?

Lighting Transformer:
1φ
$L = kVA \times 67\%$
$L = 30 \times 0.67$
$L = 20.1$ kVA

3φ
$L = kVA \times 33\%$
$L = 58 \times 0.33$
$L = 19.14$ kVA

$20.1 + 19.14 = $ **39.24 kVA**

Power Transformers:
1φ
$L = kVA \times 33\%$
$L = 30 \times 0.33$
$L = 10$ kVA

3φ
$L = kVA \times 33\%$
$L = 58 \times 0.33$
$L = 19.14$ kVA

$10 + 19.14 = $ **29.14 kVA**

Distribution substations contain large utility transformers.

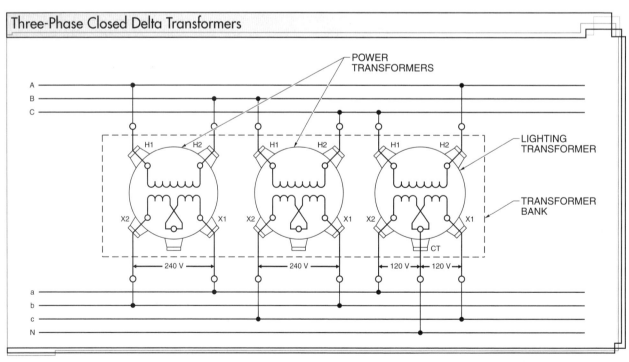

Figure 5-6. When sizing three-phase closed delta transformers, the kilovolt-amp rating of all the windings are added to determine the total kilovolt-amp rating.

Practice Question

Determining Transformer Size—Three-Phase (3ϕ) Closed Delta

1. What size of lighting and power transformers is required for a closed delta system if the 1ϕ load is 18 kVA and the 3ϕ load is 30 kVA?

Sizing Open Delta Transformers. Open delta connections should have the primary and secondary connected in the same open connection. It is not possible to use an open delta connection with a closed delta connection or with a wye connection. One advantage of the open delta system is the ability to keep the minimum loads running if a transformer of a closed delta system stops working. The nonfunctioning transformer would be removed and the remaining two would be reconnected into the open delta system to maintain some power.

Open delta systems provide the same voltages as the closed delta system, but they can only handle about 58% of the 3ϕ loads that a closed delta can handle. This system requires two transformers to produce the three phases. The two transformers are the power transformer and the lighting transformer. **See Figure 5-7.** The open delta system is commonly used where there are more 1ϕ loads than 3ϕ loads. The lighting transformer is sized at 100% for 1ϕ loads and 58% for 3ϕ loads. The power transformer is sized for 3ϕ loads only. The 3ϕ loads for the power transformer are sized at 58%.

For example, what size of lighting and power transformers is required for an open delta system if the 1ϕ loads are 20 kVA and the 3ϕ loads are 15 kVA?

Lighting Transformer:

1ϕ
$$L = kVA \times 100\%$$
$$L = 20 \times 1.00$$
$$L = 20 \text{ kVA}$$

3ϕ
$$L = kVA \times 58\%$$
$$L = 15 \times 0.58$$
$$L = 8.7 \text{ kVA}$$

$$20 + 8.7 = \mathbf{28.7 \text{ kVA}}$$

Power Transformers:

3ϕ
$$L = kVA \times 58\%$$
$$L = 15 \times 0.58$$
$$L = \mathbf{8.7 \text{ kVA}}$$

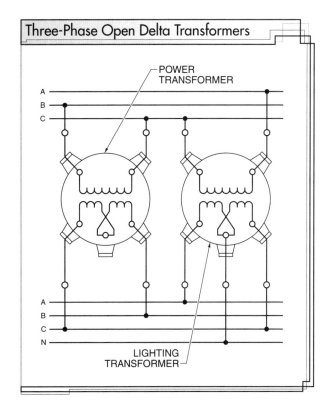

Figure 5-7. A power transformer and a lighting transformer make up a three-phase open delta system.

Practice Question

Determining Transformer Size—Three-Phase (3ϕ) Open Delta

1. What size transformer is required in an open delta system if the 1ϕ load is 15 kVA and the 3ϕ load is 20 kVA?

AUTOTRANSFORMERS

Autotransformers are different from most transformers because they have only one winding. The primary side is not isolated from the secondary side. Autotransformers can be used for stepping up voltage but are typically used for stepping down voltage. Autotransformers are typically used where the change in voltage is relatively small, such as 208 V up to 230 V, or 240 V down to 208 V. Because the voltage change is small, the efficiency of an autotransformer is greater than the efficiency of a two-winding transformer.

To step up the voltage in an autotransformer, the primary is connected to one end of the winding, and the other conductor is connected somewhere along the winding. The secondary is connected to the same end that the first primary wire is connected to, and the other conductor is connected to the other end of the winding. To step down the voltage, the primary is connected to both ends of the winding, and the secondary is connected to one common end and somewhere between the two primary conductors. **See Figure 5-8.** *Note:* The NEC® (210.9) does not allow an autotransformer to be used for lighting and appliance branch circuits unless there is a common grounded conductor to both the primary and secondary sides.

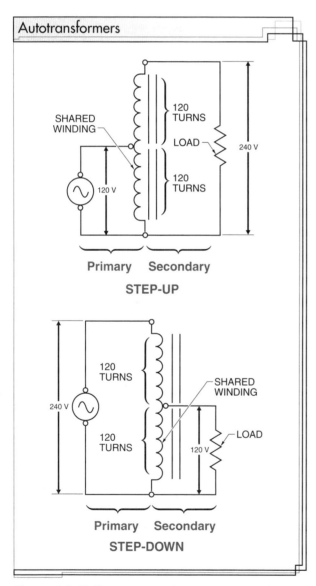

Figure 5-8. Unlike most transformers, autotransformers have only one winding. The primary and secondary sides share the same winding.

CURRENT TRANSFORMERS

A *current transformer (CT)* is a transformer that is used to isolate an ammeter to prevent the hazards caused by connecting the meter to current. A CT is used to step current down. The standard output of a CT is 5 A. The CT uses the power conductor as the primary and is installed through the middle of the coil, which is used for the secondary. **See Figure 5-9.** Some of the standard sizes of CTs are 500:5, 300:5, and 1000:5. The output of the CT is proportional to the input. If the current on the primary were 500 A, then the output would be 5 A. If 350 A were on the primary, then the secondary would be 3.5 A.

When the metering equipment is disconnected from the CT, current is still flowing through the primary wire. Because of this, the secondary terminals will have a high voltage across them. When CT metering equipment is disconnected, the terminals shall be shorted together for safety.

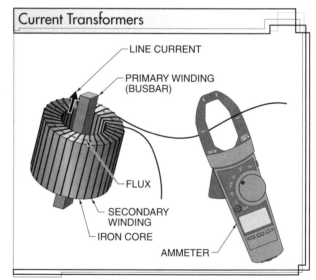

Figure 5-9. Current transformers are used to step current down for metering purposes.

POTENTIAL TRANSFORMERS

A *potential transformer*, also known as an instrument transformer, is a precision two-winding transformer that is used to step down high voltage to allow safe voltage measurement. It is used to reduce high voltages to voltages that are safe for metering systems. A potential transformer can reduce voltages as high as 2400/4160 V or 7200/13,800 V down to about 120 V, which allows standard voltage meters to be used. Potential transformers can be used on single-phase or three-phase systems.

TRANSFORMER OVERCURRENT PROTECTIVE DEVICES

Transformers should be protected from overcurrent, short circuits, and ground faults. Article 450 in the NEC® covers the protection requirements for transformers below and above 1000 V. Table 450.3(A) covers transformers that are above 1000 V on the primary side, and above and below 1000 V on the secondary side.

Transformer Protection—Over 1000 Volts

Table 450.3(A) is divided into two sections that are used to size overcurrent protective devices for transformers with over 1000 V on the primary side, and over 1000 V or 1000 V or less on the secondary side. The percentages selected to size overcurrent protective devices are based on the location of the transformer in an unsupervised or supervised area. An unsupervised area is a location where a maintenance electrician or an electrical engineer is not present or employed. After the primary and secondary sides of the transformer have been rated, refer to Table 240.6 to find a standard protective device size. **See Figure 5-10.**

For example, what size circuit breaker is required to protect a transformer with 12,500 V and 10 A on the primary side and 4,160 V and 30 A on the secondary side with an impedance of 3%?

Any location with an impedance of no more than 6% equals 600% on the primary side and 300% on the secondary side. See Table 450.3(A).

Primary side = 10 A × 600% = 60 A

Secondary side = 30 A × 300% = 90 A

Refer to 240.6 for a standard size. The standard sizes are 60 A for the primary side and 90 A for the secondary side.

> **Practice Question**
>
> *Determining Transformer Protective Device Size—Over 1000 Volts*
>
> 1. What size fuse is required to protect a transformer with 4160 V and 40 A on the primary side and 2400 V and 69 A on the secondary side with an impedance of 7%?

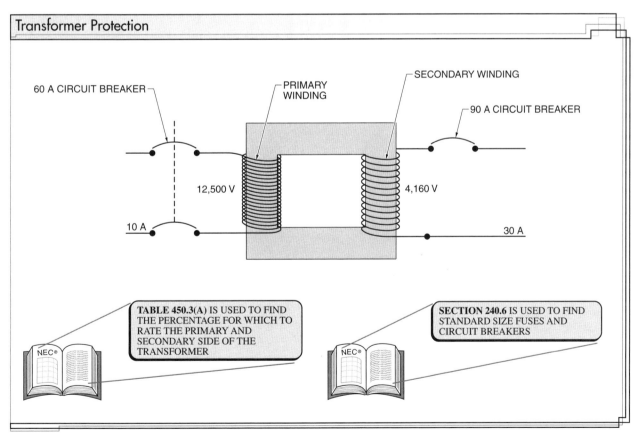

Figure 5-10. Circuit breakers or fuses are used to protect transformers on both the primary and secondary sides.

Transformers are used to reduce the voltage to a level required by the load.

Transformer Protection— 1000 Volts or Less

Table 450.3(B) covers transformers that have both the primary and secondary voltage 1000 V or less. Table 450.3(B) provides the percentages for sizing the fuses or circuit breakers that protect the transformer on the primary and secondary sides.

For example, what size of circuit breakers is required to protect the primary and secondary side of a transformer with 480 V and 8 A on the primary side and 120 V and 36 A on the secondary side?

The primary and secondary side protection method shows 250% for the primary side and 125% for the secondary side. See Table 450.3(B).

    Primary side = 8 A × 250% = 20 A

According to Note 3, if the circuit breaker is not a standard size, go down to the next standard size. See Table 240.6.

    Primary side = 20 A
    Secondary side = 36 A × 125% = 45 A

According to Note 1, if the circuit breaker is not a standard size, go up to the next standard size. See Table 240.6.

    Secondary side = 45 A

The standard size on the primary side is 20 A, and the standard size on the secondary side is 45 A.

> **Practice Question**
>
> *Determining Transformer Protective Device Size—1000 Volts or Less*
>
> 1. What size circuit breaker is required to protect the primary and secondary sides of a transformer with 480 V and 50 A on the primary side and 240 V and 100 A on the secondary side?

DELTA-WYE CALCULATIONS

Three-phase transformers transform phase voltages and phase currents. They do not transform line voltages and currents. The turns ratio, voltage ratio, and current ratios that were used for 1ϕ transformers will also work for individual windings or phases.

Delta and wye connections can be used on either the primary or secondary side of the transformer. The voltage is determined by the turns ratio and the connection type. The most common example is the delta primary and wye secondary.

There is a difference in both voltage and current between the wye and delta connections. Delta and wye transformers transform voltages and currents similar to a 1ϕ transformer, but the currents transform differently due to the fact that there are three phases. Understanding the values of the phase line (phase to phase) and phase coil (phase to phase) is essential when making conversions. The phase line is the reading from one phase to another phase. The phase coil is the reading across one complete winding of the transformer. **See Figure 5-11.** To determine the voltage and current of a delta system, apply the following formulas:

    Voltage:
        Line volts = Phase volts

    Current:
        Line amps = Phase amps × 1.732

$$Phase\ amps = \frac{Line\ amps}{1.732}$$

To determine the voltage and current of a wye system, apply the following formulas:

    Voltage:
        Line volts = Phase volts × 1.732

$$Phase\ volts = \frac{Line\ volts}{1.732}$$

    Current:
        Line amps = Phase amps

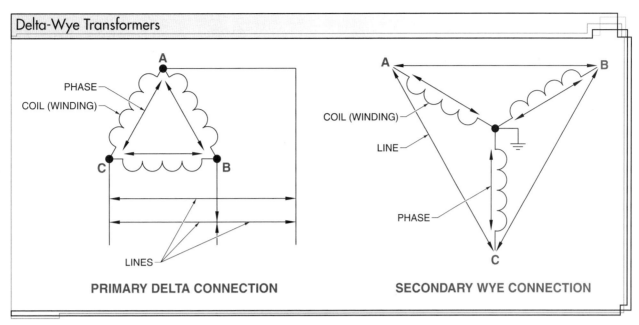

Figure 5-11. Understanding phase line and phase coil readings is essential when performing delta-wye conversions.

For example, what is the secondary line current and voltage of a 4:1 delta-wye transformer with 480 V and 43.3 A on the primary side? The turns ratio is 4:1, which means the voltage will go down 4 times and the current will go up 4 times.

Primary Side (Delta) Voltage:
Line volts = Phase Volts
Line volts = 480 V

Current:
$$Phase\ amps = \frac{Line\ amps}{1.732}$$
$$Phase\ amps = \frac{43.3}{1.732}$$
Phase amps = 25 A

Secondary Side (Wye) Voltage:
$$Voltage\ ratio = \frac{480}{4}$$
Voltage ratio = 120

Line volts = Phase Volts × 1.732
Line volts = 120 × 1.732
Line volts = **208 V**

Current:
Line amps = Phase amps
Line amps = 25 × 4
Line amps = **100 A**

Practice Question

Delta-Wye Conversions

1. What is the secondary line current of a delta-wye transformer set? The transformer's primary voltage is 480 V with 50 A on the line. The secondary voltage is 120/208 V.

TRANSFORMER CONNECTIONS

Transformer connections are determined based on the voltage supplied and voltage required for the loads being served. Some transformers use a common connection point on one side while the other side is connected to different tap leads. Transformers also have tap connections that will increase or decrease the turns ratio to adjust the output voltage if it is too high or to low.

ABB, Inc.
Three-phase transformers can be wired in delta or wye configurations.

Individual transformers can be used to transform the high-voltage power from transmission lines.

TECH TIP

The low-voltage winding of a standard distribution transformer connected to a single-phase circuit is normally made with two equal coils. These coils are arranged so that they may be connected in series or parallel. This arrangement permits current to be delivered at two different voltages.

Single-Phase (1ϕ) Transformer Connections

Single-phase (1ϕ) transformer connections usually use 240 V or 480 V on the primary side and 120 V or 240 V on the secondary side. Transformers are connected in series for higher voltages and parallel for lower voltages. **See Figure 5-12.** It is the electrician's responsibility to check the nameplate and wiring diagram of each transformer to ensure the proper connection.

Three-Phase (3ϕ) Transformer Connections

A 3ϕ transformer is made up of 1ϕ transformers connected in different configurations. The common types of 3ϕ transformers are wye, closed delta, and open delta. **See Figure 5-13.** The wye and closed delta transformers can be used for the primary or secondary sides with both sides being the same or opposite of each other. The open delta transformers must be used on the primary and secondary sides of the transformer. The most common is a delta-wye transformer, which is usually a step-down transformer.

Wye Transformer Connections. For a 3ϕ wye transformer connection, one side of each winding is connected together. This common connection is the grounding point of the transformer. The common voltages from a wye configuration are 120/208 V for low voltage and 277/480 for high voltage. Electric companies usually use wye transformer connections because it is easier to balance the loads on all three phases.

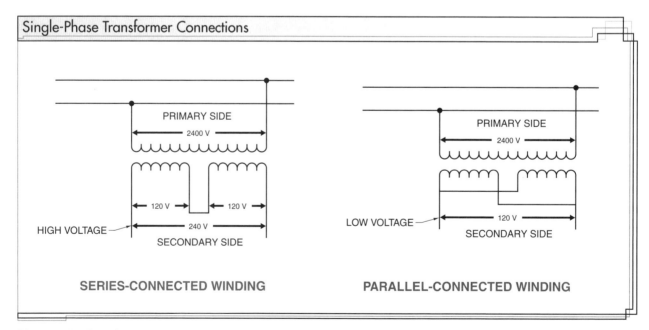

Figure 5-12. Transformers are connected in series for high voltage and in parallel for low voltage.

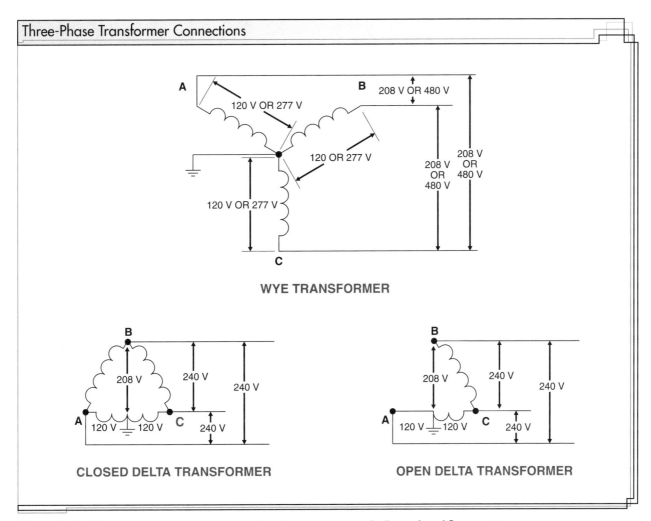

Figure 5-13. A three-phase system originates from three separate windings of an AC generator.

Closed Delta Transformer Connections. For a 3ϕ delta-connected transformer, the windings are connected in series to make a complete loop with the three windings. One of the windings is connected in the center to produce the neutral connection. The phases on each side will produce 120 V to the neutral conductor. The other phase will produce a voltage of 208 V to the neutral. This is called the wild leg, or high leg. The voltage on the high leg can be found by multiplying the line to neutral voltage of phases A or C times 1.732. The NEC® requires this conductor to be marked orange and connected to the far right in a voltage watt-hour meter. The high leg conductor should be connected to the B phase in the panel.

Open Delta Transformer Connections. An open delta transformer uses two 1ϕ transformers. One transformer is called the lighting transformer and the other is called the power transformer. These transformers produce the same voltage as closed delta transformers, but they will only handle about 58% of the 3ϕ power compared to closed delta transformers.

Buck-Boost Transformer Connections. A *buck-boost transformer* is a type of transformer designed to buck (lower) or boost (raise) line voltage. These transformers are usually used to raise or lower the voltage no more than 20% of the original circuit voltage. If equipment is rated at 208 V and the supply is 230 V, a buck connection is used. If the equipment is rated at 230 V and the supply is 208 V, a boost connection is used. There are multiple connections used to boost or buck a transformer. **See Figure 5-14.**

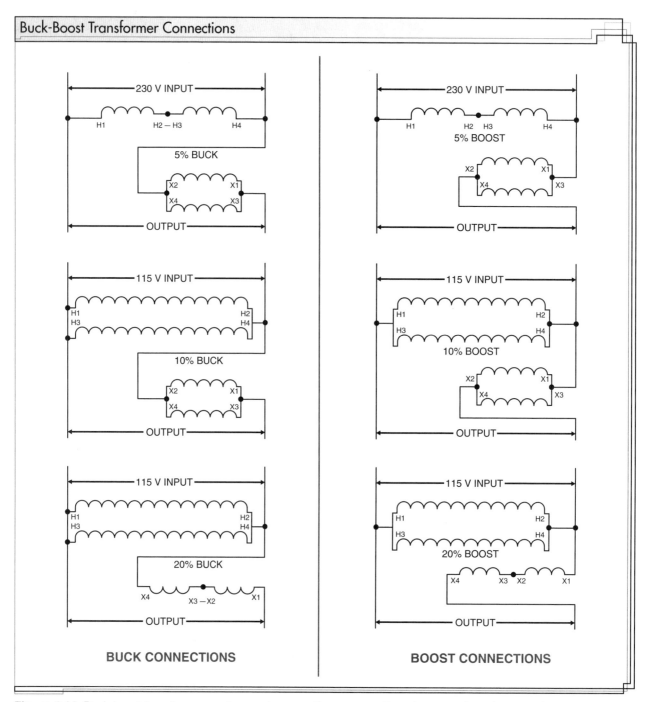

Figure 5-14. Buck-boost transformers reduce or increase the source voltage because the primary and secondary sides are wired in series with each other.

CHAPTER 5

Transformers

Review Questions

Name _____   Date _____

_____ 1. If a transformer has a 5:1 turn ratio and the voltage on the secondary is 120 V, then the voltage on the primary is ___ V.
   A. 24
   B. 120
   C. 240
   D. 600

_____ 2. If the secondary side of a transformer is 500 A and 24 V and the primary is 120 V, then the primary current is ___ A.
   A. 4.17
   B. 20.8
   C. 50
   D. 100

_____ 3. When a transformer steps up the voltage 4 times, the current steps down ___ times.
   A. 2
   B. 4
   C. 6
   D. 8

_____ 4. A 1ϕ, 240 V, 50 A transformer will be ___ VA.
   A. 4.8
   B. 1200
   C. 12,000
   D. 15,000

_____ 5. A 3ϕ transformer that has 240 V and 200 A on the secondary will be ___ VA.
   A. 24,000
   B. 27,714
   C. 48,000
   D. 83,136

_____ 6. A transformer with 480 V and 100 A will be ___ kVA.
   A. 4.8
   B. 48
   C. 83.136
   D. 48,000

63

_____ 7. If the primary side of a transformer is 1000 VA and the transformer is 90% efficient, then the secondary is ___ VA.
　　A. 900
　　B. 1000
　　C. 1100
　　D. 1111

_____ 8. An autotransformer has ___ winding(s).
　　A. one
　　B. two
　　C. three
　　D. four

_____ 9. A current transformer has a standard output of ___ A.
　　A. 2
　　B. 4
　　C. 5
　　D. 10

_____ 10. Transformer protection is found in Article ___ of the NEC®.
　　A. 240
　　B. 430
　　C. 450
　　D. 460

Digital Resources
ATPeResources.com/QuickLinks
Access Code: 637160

CHAPTER 6

Grounding, Bonding, and Neutrals

Electrical systems are grounded to provide a direct connection to the earth. One conductor in the systems shall be grounded and an intentional connection to the earth shall be made. Bonding is required to interconnect all non-current-carrying metal parts to the electrical source to ensure an effective ground-fault current path. The neutral conductor is used to carry unbalanced current in a circuit and provides an effective fault current path back to equipment. It is critical to correctly bond and size neutral conductors in order to have a safe working circuit and to prevent damage to equipment.

OBJECTIVES

After completing this chapter, the learner will be able to do the following:

- Define grounding and describe its importance in an electrical system.
- Describe how to size grounded conductors.
- Describe the function of the main bonding jumper.
- Identify the exceptions for using a grounding electrode conductor that is smaller than 3/0.
- Explain the purpose of bonding jumpers.
- Describe the process for sizing system grounding electrode conductors.
- Explain the function of the neutral conductor for 1ϕ and 3ϕ systems.
- Identify the two restrictions that prevent the neutral wire from being reduced in size.

GROUNDING

Grounding is the connection of all exposed non-current-carrying metal parts to the earth. Equipment, circuits, or systems that have been grounded are connected to the earth directly or to some type of electrode that is connected to the earth. Per 250.80, service raceways and enclosures shall be grounded. These raceways and enclosures shall be connected to the earth or a conducting body connected to the earth. **See Figure 6-1.** The purpose of grounding is to limit the extra voltage that could be imposed by lightning, line surges, or contact with a higher voltage system. It can also help stabilize the voltage to earth during normal operations.

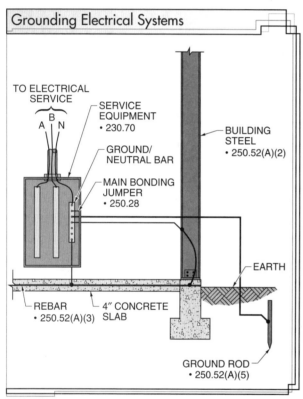

Figure 6-1. Raceways and enclosures should be connected to the earth or a conducting body that is connected to the earth.

BONDING

Bonding is the process of establishing continuity and conductivity through a connection. Bonding is required to interconnect all non-current-carrying metal parts to the electrical source to ensure an effective ground-fault current path. An effective ground-fault current path will help in the operation of OCPDs, such as fuses and circuit breakers, if a ground fault occurs.

Electrical services have special bonding requirements as stated in 250.92. The primary reason for these special requirements for services is the absence of adequate overcurrent protection on the line side of service equipment. Electrical utility power systems typically do not provide secondary overcurrent protection. The only protection provided by the electrical utility is for the primary of the transformer. Care should be exercised when bonding electrical service equipment because the available fault current is typically greater at this location than anywhere else in the electrical distribution system.

GROUNDED CONDUCTOR

A *grounded conductor* is a conductor that has been intentionally grounded to earth. The grounded conductor can be a system conductor or a circuit conductor. **See Figure 6-2.** The grounded conductor is commonly a neutral conductor. Not all electrical distribution systems use the grounded conductor as a neutral. For example, corner-grounded delta systems contain a grounded conductor that is not a neutral conductor. While not all grounded conductors are neutral conductors, they are in the majority of electrical distribution systems.

The grounded conductor becomes part of the effective ground-fault current path and has to be sized accordingly for that purpose. It should also be sized to carry the maximum unbalanced current of the electrical system. Subsections 250.24(C), 250.24(C)(1), and 250.24(C)(2) require supplying a grounded conductor to a service. Any service that is less than 1000 V and grounded at any point is required to have the grounded conductors connected to each service disconnecting means grounded conductor terminal or bus.

Sizing the Grounded Conductor

Per 250.24(C)(1), the grounded conductor shall not be smaller than the required grounding electrode conductor from Table 250.102(C)(1). If the phase conductors are larger than 1100 kcmil for copper or 1750 kcmil for aluminum, the conductor shall not be smaller than 12.5% of the largest phase conductor.

Per 250.24(C)(2), if the service-entrance conductors are in parallel and the conductors are all in one conduit, the circular mils of the conductors that make up each phase are added together. The phase with the largest area in circular mils is used to size the grounded conductor per 250.24(C)(1) and Table 250.102(C)(1). If the phase conductors are larger than 1100 kcmil copper or 1750 kcmil aluminum, the grounded conductor shall not be smaller than 12.5%. **See Figure 6-3.**

If the conductors are installed in parallel and in separate conduits, per 250.24(C)(2) the grounded conductor shall be sized based on the size of the service-entrance ungrounded conductors in each raceway per Table 250.102(C)(1). The grounded conductor shall not be smaller than 1/0 AWG according to 310.10(H). **See Figure 6-4.** If the phase conductors are larger than 1100 kcmil copper or 1750 kcmil aluminum, the grounded conductor in each conduit shall be at least 12.5% of the phase conductors.

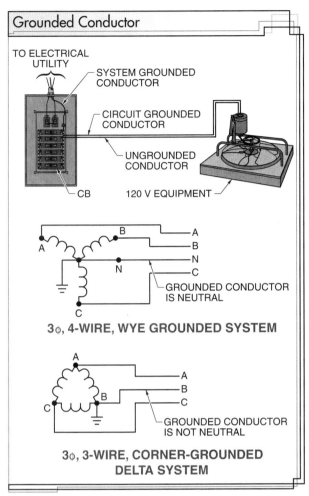

Figure 6-2. A grounded conductor is a conductor that has been intentionally grounded to earth.

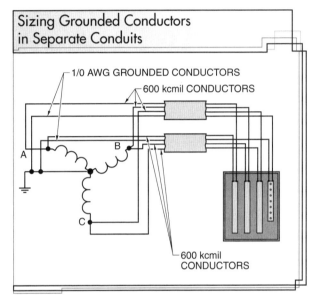

Figure 6-4. When conductors are installed in separate conduits, the grounded conductor shall be sized based on the service-entrance conductors in each raceway.

For example, what is the minimum size of the grounded conductor for a service that has 250 kcmil copper conductors connected to a 225 A circuit breaker?

See 250.24(C)(1) and Table 250.102(C)(1).

250 kcmil = No. 2 AWG Cu

Grounded conductor size = **No. 2 AWG copper conductor**

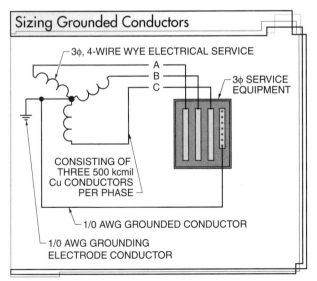

Figure 6-3. The grounded conductor shall never be smaller than 12.5% of the largest service-entrance conductor.

Practice Question

Determining Grounded Conductor Size

1. What size of grounded conductor is required for a service that has 500 kcmil conductors on the ungrounded phases protected by a 400 A overcurrent protective device?

MAIN BONDING JUMPER

A *main bonding jumper (MBJ)* is the connection at the service equipment that bonds together the equipment grounding conductor, grounded conductor, and grounding electrode conductor. The main bonding jumper provides a fault current path from the equipment grounding conductor to the grounded conductor. The MBJ can be a simple machine screw, as in the case of a 100 A panelboard, or it can be a No. 4/0 conductor for a 1000 A service. **See Figure 6-5.**

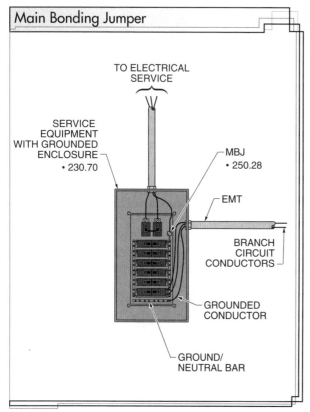

Figure 6-5. The main bonding jumper provides the fault current path from the equipment grounding conductor to the grounded conductor.

The MBJ is attached by a green screw, exothermic welds, listed pressure connectors, listed clamps, or other listed means. See 250.8 for applicable provisions. When the MBJ is a screw only, it shall be green. This distinguishes it from the other screws on the grounded conductor terminal block. The MBJ shall be constructed of copper or other corrosion-resistant materials per 250.28(A). The main bonding jumper is installed in the first main disconnect that is served from the utility power. It should not be installed downstream or in subpanels located in the same building. This could result in multiple or parallel grounding paths, posing the threat of electrical shock and causing objectionable current on the grounding system per 250.6.

Sizing the Main Bonding Jumper

Per 250.28(D), if the main bonding jumper is a wire, it must not be smaller than the sizes shown on Table 250.102(C)(1). If the service-phase conductors are above 1100 kcmil copper or 1750 kcmil aluminum, then the main bonding jumper must be at least 12.5% of the phase conductors. If the phases are parallel, the areas (in circular mils) of the conductors of each phase are added together. The total area measurement in circular mils is used to size the MBJ.

For example, what is the required size of the main bonding jumper that is made of an aluminum conductor for a service that has 750 kcmil service-entrance conductors?

See 250.28(D) and Table 250.102(C)(1).
750 kcmil Al = No. 3/0 AWG Al
Main bonding jumper size = **No. 3/0 AWG aluminum conductor**

Practice Question

Determining Main Bonding Jumper Size

1. What size of main bonding jumper is required for a service that has three 500 kcmil copper conductors per phase supplying a panel?

TECH TIP

One of the most used and misunderstood articles in the NEC® is Article 250, Grounding & Bonding. Unfortunately, when a learner or electrician misunderstands an NEC® rule or requirement, it can result in an incorrect and sometimes unsafe electrical installation. The key to understanding Article 250 and correctly applying its provisions lies in obtaining a fundamental understanding of the terminology used. The best first step any apprentice or student of the NEC® can take is to review the Article 100 and Article 250 definitions of grounding and bonding.

GROUNDING ELECTRODE CONDUCTOR

A *grounding electrode conductor (GEC)* is a conductor that connects the grounding electrode(s) to the system-grounded conductor and/or the equipment grounding conductor. This connection can occur at the service equipment or at the source of a separately derived system. The connection is also permitted to be made to a point on the grounding electrode system. The GEC may be constructed of aluminum, copper, or copper-clad aluminum; may be solid or stranded; and may be insulated, covered, or bare according to 250.62.

Sizing the Grounding Electrode Conductor

Section 250.66 is used to size grounding electrode conductors. According to Table 250.66, the maximum size for a grounding electrode conductor is a No. 3/0 AWG copper conductor. Many construction projects use No. 4/0 bare copper for all of their grounding electrode conductors, which is a common size carried by most supply houses. The following are exceptions to Table 250.66 that would allow a smaller conductor to be used:

- Connections to ground rod, pipe, or plate electrodes. Per 250.66(A), conductors are not required to be larger than a No. 6 copper conductor.
- Connections to concrete-encased electrodes. Per 250.66(B), when this is the sole connection, the grounding electrode shall not be required to be larger than a No. 4 AWG copper wire.
- Connections to ground rings. Per 250.66(C), the conductor does not have to be larger than the ground ring, but it must be at least a No. 2 AWG.

For example, what size of grounding electrode conductor is required for a service that is using No. 4/0 AWG copper conductors as the entrance conductors?
See 250.66 and Table 250.66.
No. 4/0 AWG = No. 2 AWG
Grounding electrode conductor size = **No. 2 AWG copper conductor**

Practice Question

Determining Grounding Electrode Conductor Size

1. What size of grounding electrode conductor is required for a service of 750 kcmil copper conductors and an overcurrent protective device of 600 A?

Per 250.53(G), grounding rod and pipe electrodes must be installed to a minimum depth of 8' (2.44 m) unless a rock bottom is encountered. The angle of rod and pipe electrodes must not exceed 45°.

GROUNDING ELECTRODES

Subsection 250.52(A) lists the types of electrodes that can be used along with the specifications. Section 250.50 states that all grounding electrodes that are present in a building or structure shall be used to form the electrode system. The electrodes listed from 250.52(A)(1) through 250.52(A)(7) should be used to create a grounding electrode system if available. If any of these electrodes listed are not available, then one or more of the electrodes specified in 250.52(A)(4) through 250.52(A)(8) must be installed and used. **See Figure 6-6.** Aluminum electrodes or underground metal gas lines are not permitted to be used as grounding electrodes.

EQUIPMENT GROUNDING CONDUCTORS

An *equipment grounding conductor (EGC)* is an electrical conductor that provides a low-impedance path between electrical equipment and enclosures and the system-grounded conductor and grounding electrode conductor. This connection usually occurs at the service equipment or at the source of power and ensures the operation of the overcurrent protective device (OCPD) when fault conditions occur. **See Figure 6-7.**

EGCs may be in the form of a separate wire, which can be bare, covered, or insulated with green insulation. They may also be metal conduits. Article 250.118 lists 14 equipment ground conductors that may be used. Per 250.8, connections cannot depend on solder only.

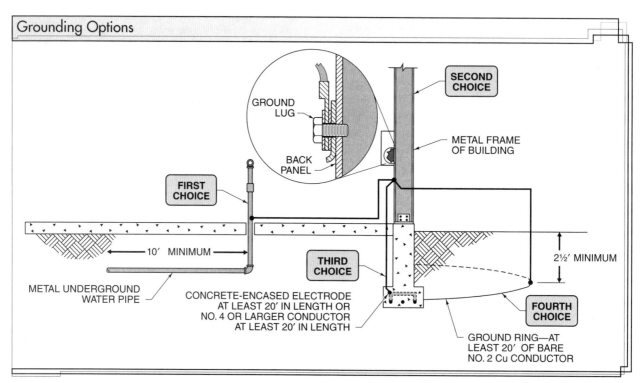

Figure 6-6. Grounding is accomplished by connecting a circuit to a metal underground pipe, metal frame of a building, concrete-encased electrode, or ground ring.

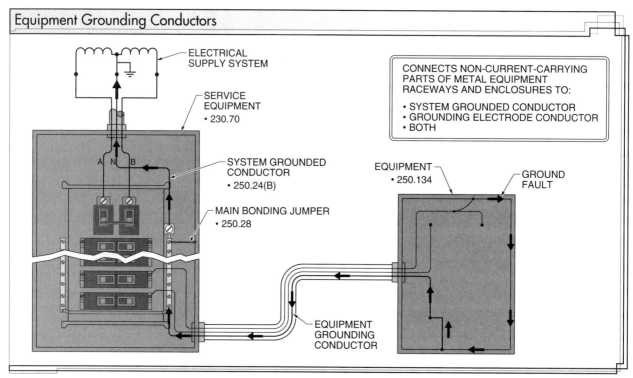

Figure 6-7. An equipment grounding conductor is an electrical conductor that provides a low-impedance path between electrical equipment and enclosures and the system-grounded conductor and grounding electrode conductor.

Sizing Equipment Grounding Conductors

The size of the EGC shall be based upon the size or rating of the OCPD ahead of the equipment, which is supplied. Table 250.122 lists the OCPD ratings and the appropriate-size EGC for copper, aluminum, or copper-clad aluminum. The equipment grounding conductor does not have to be larger than the phase conductors. For instance, the EGC sized for a 30 A branch-circuit in a raceway is based upon the 30 A OCPD rating and requires either a No. 10 copper or No. 8 aluminum EGC.

Per 250.122(B), if the circuit conductors are increased in size due to a voltage drop, a similar adjustment shall be made proportionally, based on the conductor's area in circular mils. The new larger conductor's circular mils is divided by the circular mils of the usually sized conductor to determine the multiplier for the circuit. This multiplier is then used to increase the EGC circular mils as well.

If multiple circuits are run in the same raceway, cable, or cable trays, one EGC can be installed. The size of the EGC shall be based on the largest OCPD protecting the circuit conductors installed in the raceway per 250.122(C). **See Figure 6-8.**

For motor circuits that use a wire-type EGC, two methods of sizing the EGC can be used per 250.122(D)(1) and (D2). If the OCPD is an inverse time breaker, non-time-delay fuse, or dual-element fuse, the EGC shall be sized from Table 250.122 based on the OCPD size. If the OCPD used is an instantaneous trip breaker or motor short-circuit protector, the EGC is permitted to be sized from the maximum time-delay fuse used for short-circuit and ground-fault protection.

Where circuit conductors are installed in parallel, as allowed in 310.10(H), the EGC shall be installed according to 250.122(F)(1) or (2). For conductors installed in single raceways, auxiliary gutters, or cable trays, the EGC of the wire type shall be sized according to Table 250.122 based on the OCPD size. If in multiple raceways, the EGC shall be installed in each raceway and sized according to Table 250.122 based on the size of the OCPD used to protect the circuit.

For multiconductor cables, the EGC shall be in each cable and shall be connected in parallel. If the multiconductor cables are installed in parallel in the same raceway, cable tray, or auxiliary gutter, a single EGC shall be permitted and sized according to Table 250.122. This EGC shall be used in combination with the EGC provided in the cables and connected together to form the EGC of the circuit. If a single full size EGC is not used in the raceway, cable tray, or auxiliary gutter, then a full-size EGC shall be installed in each multiconductor cable based on Table 250.122.

For example, what size of equipment grounding conductor is required for a parallel set of feeder conductors in separate raceways that uses 350 kcmil copper conductors and is supplied by a 600 A overcurrent device?

See 250.122(F) and Table 250.122.

600 A OCPD = No. 1 AWG

Equipment grounding conductor size = **No. 1 AWG in each raceway**

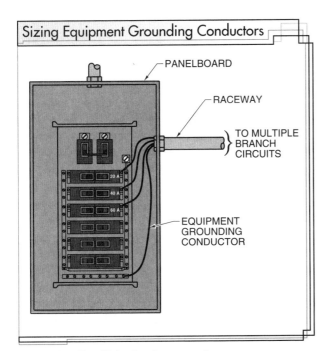

Figure 6-8. If multiple circuits are run in a common raceway, only the equipment grounding conductor, based on the largest-sized overcurrent protective device protecting the conductors, is required.

Practice Question

Determining Equipment Grounding Conductor Size

1. What size of equipment grounding conductor is required for a circuit that is protected by a 150 A inverse-time circuit breaker and is using No. 1/0 copper THW conductors?

BONDING JUMPERS (CONDUCTORS)

A *bonding jumper* is a conductor that electrically connects two or more metal parts. **See Figure 6-9.** Bonding jumpers are required to be constructed of copper or other corrosion-resistant materials. Section 250.102 contains the provisions for bonding jumpers.

Bonding jumpers are used to create a ground-fault current path around loose-fitting raceways on the supply or load side of the service. They can also be used to bond the raceways of the conduit system. The method of sizing the bonding jumper is based on its location in the electrical system.

Supply-Side Bonding Jumpers

Subsection 250.102(C) sizes bonding jumpers on the supply side of the service. These bonding jumpers are sized per Table 250.102(C)(1). If the phase conductors are over 1100 kcmil copper or 1750 kcmil aluminum, the bonding jumper must be at least 12.5% of the phase conductors. If the conductors are in parallel and in one conduit, then the conductors' areas in circular mils are added together and treated as one conductor that is sized based on the total area in circular mils as found in Table 8 of chapter 9. If the parallel service conductors are installed in two separate conduits, then the bonding jumper is sized for the service conductors in each raceway. **See Figure 6-10.**

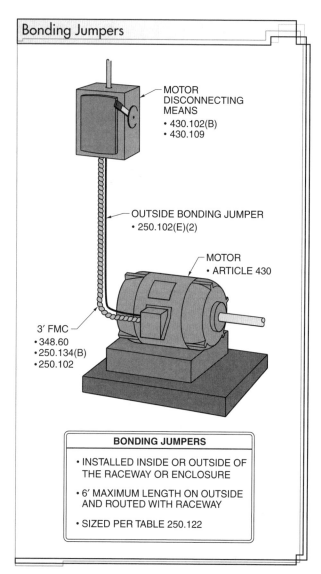

Figure 6-9. A bonding jumper is a conductor that electrically connects two or more metal parts.

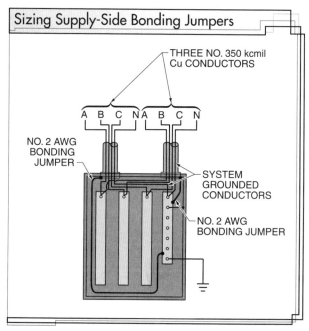

Figure 6-10. When service conductors are installed in two separate conduits, the supply-side bonding jumper is sized for the service conductors in each raceway.

TECH TIP

Bonding jumpers are allowed to be installed either on the inside or outside of the raceway or enclosure. When installed on the outside of the raceway the total length shall not exceed 6' and they shall be installed so that they closely follow the raceway.

For example, what size bonding jumper is required for an 800 A service that has three 350 kcmil copper parallel conductors per phase installed in separate conduits?

See 250.102(C) and Table 250.102(C)(1).

350 kcmil = No. 2 AWG Cu

Bonding jumper size = **No. 2 AWG copper conductor for each conduit**

Practice Question

Determining Supply-Side Bonding Jumper Size

1. What size bonding jumper is required on the supply side of a main disconnect that has 600 kcmil THHN conductors that terminate into a 400 A circuit breaker?

Equipment Bonding Jumpers—Load Side (Feeders or Branch Circuits)

An *equipment bonding jumper (EBJ)* is a conductor that connects two or more parts of the equipment grounding conductor. Subsection 250.102(D) sizes the equipment bonding jumpers on the load side of the service. This includes feeders and branch circuits. These bonding jumpers are used with flexible metal conduit, or expansion joints. The size of the equipment bonding jumpers on the load side is determined by the OCPD size or rating from Table 250.122. **See Figure 6-11.**

For example, what size of equipment bonding jumper is required for a 400 A feeder that is using 600 kcmil copper conductors?

See 250.102(D) and Table 250.122.

400 = No. 3 AWG jumpers

Equipment bonding jumper size = **No. 3 AWG copper conductor**

Practice Question

Determining Equipment Bonding Jumper Size—Load Side

1. What size of equipment bonding jumper is required for a 250 A feeder that has two 300 kcmil XHHW conductors installed in a PVC raceway?

SYSTEM BONDING JUMPERS

Section 250.30 shows how a separately derived system must be grounded and bonded into an electrical system. A *system* is any separately derived power source such as a transformer, standby generator, fuel cell, or other power production system. As with the main utility service, the grounding connections shall not be made on the load side of the separately derived system. This means there shall not be any system bonding jumpers in subpanels fed from the main disconnect.

The system bonding jumper provides the same function as the main bonding jumper but is installed in a separately derived system. The location of this jumper is usually in the transformer but can also be installed in the first main disconnect of the system. The system grounding electrode conductor must be connected at the same place where the system bonding jumper is installed. The connection can be made at both locations if it does not create a parallel path for the grounded conductor. It is usually required to be installed at only one location.

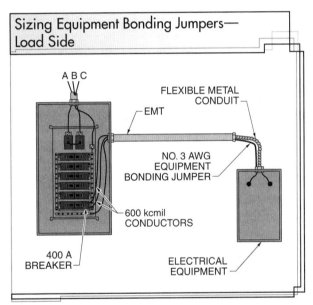

Figure 6-11. The size of the equipment bonding jumpers on the load side is determined by the OCPD size or rating from Table 250.122.

Sizing System Bonding Jumpers

System bonding jumpers shall comply with 250.28(A) through (D). Refer to 250.30(A)(1) for sizing system bonding jumpers. If the system bonding jumper is a conductor, it shall be sized in accordance with 250.28(D). System bonding jumpers should not be smaller than the sizes shown in Table 250.102(C)(1) for conductors up to 1100 kcmil copper or 1750 kcmil aluminum. Use 12.5% of the largest phase conductor for conductors sized over 1100 kcmil copper or 1750 kcmil aluminum. **See Figure 6-12.**

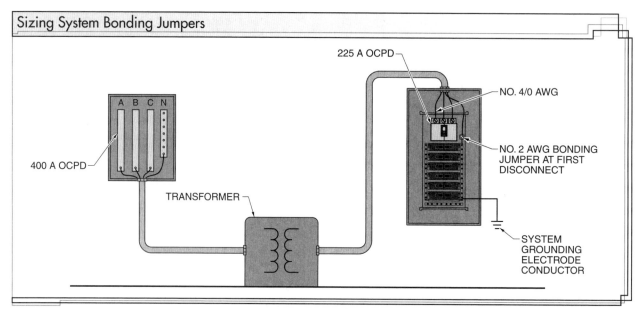

Figure 6-12. System bonding jumpers should not be smaller than the sizes shown in Table 250.102(C)(1) for conductors up to 1100 kcmil copper or 1750 kcmil aluminum.

For example, what size of system bonding jumper is required for a separately derived transformer that has 4/0 copper conductors feeding a 225 A panel?
See 250.30(A)(1), 250.28(D), and Table 250.102(C)(1).
No. 4/0 AWG = No. 2 AWG
System bonding jumper = **No. 2 AWG copper conductor**

Practice Question

Determining System Bonding Jumper Size

1. What size of system bonding jumper is required for a separately derived system that has 500 kcmil conductors feeding a 300 A panelboard?

SYSTEM GROUNDING ELECTRODE CONDUCTORS

Per 250.30(A)(5), for single separately derived systems, a GEC is required to connect the newly established grounded conductor to the grounding electrode. This connection must be made at the same point on the separately derived system where the system bonding jumper is connected. In some installations, such as a panelboard with main lugs only, the connection is made at the source of the separately derived system. The GEC is sized per 250.66. The GEC size is based on the size of the largest derived-system phase conductors.

Per 250.30(A)(4), the GEC for separately derived systems can consist of one or two components. The component is selected based upon its proximity to the GEC connection to the system. Section 250.30(A)(4) requires that the grounding electrode be as near as practicable to and preferably in the same area as the grounding electrode connection to the system. For example, if the separately derived system is created by a transformer on the 17th floor of a building, the grounding electrode selected should be as close as possible to that system on the 17th floor. The actual grounding electrode should be the nearest metal water pipe or the structural metal grounding electrode. The metal water pipe must be as specified in 250.52(A)(1) and the structural steel must conform with 250.52(A)(2). If neither of these two methods is available, then any of the other methods allowed in 250.52(A) should be used.

Sizing System Grounding Electrode Conductors

The system grounding electrode conductor shall be sized using Table 250.66 and is based on the size of the phase conductors feeding the disconnecting means from the separately derived system. The size of the system grounding electrode conductor does not have to be larger than a No. 3/0 AWG conductor. Where a location might have more than one separately derived system, a common grounding electrode conductor shall be permitted to be used. Sections 250.30(A)(6)(a) through (c) describe the conditions for using common grounding electrodes. **See Figure 6-13.**

For example, what size of system grounding electrode conductor is required for a transformer feeding a panelboard using 500 kcmil copper conductors on the secondary side?

See 250.30(A)(5), 250.66, and Table 250.66.

500 kcmil = No. 1/0 AWG

System grounding electrode conductor size = **No. 1/0 AWG copper conductor**

Practice Question

Determining System Grounding Electrode Conductor Size

1. What size of system grounding electrode conductor is required for a service that has three 500 kcmil conductors per phase?

SYSTEM-GROUNDED CONDUCTOR

The system-grounded conductor shall be brought to the first main disconnect as required in 250.24(C). The grounded conductor shall be sized to help carry the fault current of the separately derived systems and shall be able to carry the maximum unbalanced amperage of the system. If the grounded conductor is connected to a 3ϕ, 3-wire system, the grounded conductor shall have an ampacity of not less than the ungrounded conductors.

Sizing System-Grounded Conductors

Section 250.30(A)(3) provides the regulations for sizing the system-grounded conductor. The system-grounded conductor shall be sized the same way that the grounded conductor was sized for the main service. Table 250.102(C)(1) shall be used based on the largest phase conductor and shall not be smaller than the system grounding electrode conductor. Where the phase conductors are over 1100 kcmil copper or 1750 kcmil aluminum, the jumper is sized at 12.5% of the phase conductor's circular mil area.

The grounded conductor is connected to ground at the main electrical panel.

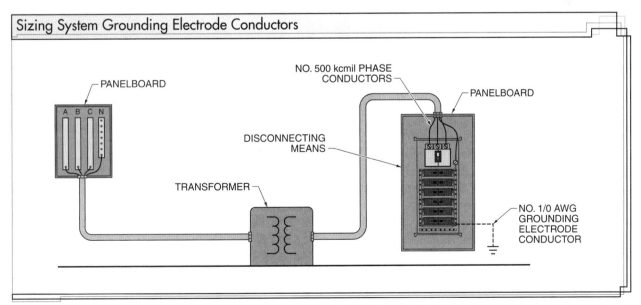

Figure 6-13. The system grounding electrode conductor is sized according to Table 250.66 and is based on the size of the phase conductors feeding the disconnecting means.

If the phase conductors are in parallel and in separate conduits, the conductors in each conduit shall be used to size the grounded conductor. The grounded conductor shall not be smaller than No. 1/0 AWG per 310.10(H). **See Figure 6-14.**

For example, what is the minimum size of grounded conductor required for a transformer with two 250 kcmil copper conductors per phase in separate conduits supplying a 400 A panel?

See 250.30(A)(3), 250.66, Table 250.102(C)(1), and 310.10(H).

250 kcmil = No. 2 AWG

Per 310.10(H), the smallest conductor to be parallel = No. 1/0 AWG

Minimum size grounded conductor = **No. 1/0 AWG copper conductor**

Practice Question

Determining Grounded Conductor Size

1. What is the minimum size of system-grounded conductor required for a service that has 350 kcmil conductors for a 250 A service?

TECH TIP

The grounded conductor is commonly a neutral conductor. However, not all electrical systems use the grounded conductor as a neutral.

A continuous grounding conductor may be fastened to the frame of an electrical device.

NEUTRAL CONDUCTORS FOR SINGLE-PHASE (1ϕ) AND THREE-PHASE (3ϕ) SYSTEMS

Under normal conditions, a *neutral conductor* is a current-carrying conductor that is intentionally grounded and connected to the neutral point of a system. The neutral conductor must be sized correctly to carry the current that is imposed on it. Per 220.61, the neutral must be able to carry the maximum unbalanced current of the circuit and it must also help provide an effective fault-current path back to the transformer as sized per 250.24(C)(1) and (2). The neutral conductor shall be the larger of the maximum unbalanced current per 220.61 or the grounded conductor brought to the service as determined in 250.24(C)(1) and (2). The maximum unbalance load is the phase carrying the highest line to neutral load. The unbalanced current is the difference in current between the ungrounded phases line to neutral loads.

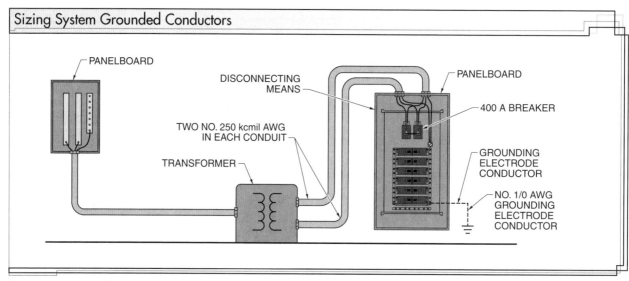

Figure 6-14. Per 310.10(H), the grounded conductor shall not be smaller than No. 1/0 AWG.

The neutral conductor is connected to the neutral point of the supplying transformer. The neutral point is the midpoint tap of a 1φ system, the midpoint tap of one transformer of a 3φ delta system, or the common point of the wye transformer connection. **See Figure 6-15.**

Single-Phase (1φ) Neutral Conductors

For 1φ systems, the neutral carries the difference in current between the two phases. This is called the unbalanced current. If one ungrounded phase is completely shut off, then the neutral must carry the total line-to-neutral current of the other phase. This would be the maximum unbalanced current. The maximum unbalanced current will be the phase carrying the highest current. **See Figure 6-16.** To determine the unbalanced current for a 1φ system, apply the following formula:

$$UC = A - B$$

where

UC = unbalanced current (in A)

A = current of phase A (in A)

B = current of phase B (in A)

For example, what is the unbalanced current on the neutral if phase A has a 50 A and 120 V load and phase B has a 30 A and 120 V load?

$$UC = A - B$$
$$UC = 50 - 30$$
$$UC = \mathbf{20\ A}$$

Practice Question

Determining Unbalanced and Maximum Unbalanced Current for 1φ Neutral Conductors

1. If phase A has a 30 A and 120 V load and phase B has a 15 A and 120 V load, what is the unbalanced current on the neutral?
2. Phase A has a 30 A and 120 V load and phase B has a 15 A and 120 V load. If phase B is completely shut off, what will be the current on the neutral?
3. If phase A has a 125 A and 277 V load and phase B has a 65 A and 277 V load, what is the unbalanced current on the neutral?
4. If phase A has a 125 A and 277 V load and phase B has a 65 A and 277 V load, what is the maximum unbalanced current?

TECH TIP

High neutral current is dangerous because it causes overheating in the neutral conductor. Because there is no circuit breaker in the neutral conductor to limit current as in the phase conductors (A, B, and C), overheating of the neutral conductor can become a fire hazard. Excessive current in the neutral conductor can also cause higher than normal voltage drops between the neutral conductor and ground conductor at 120 V outlets.

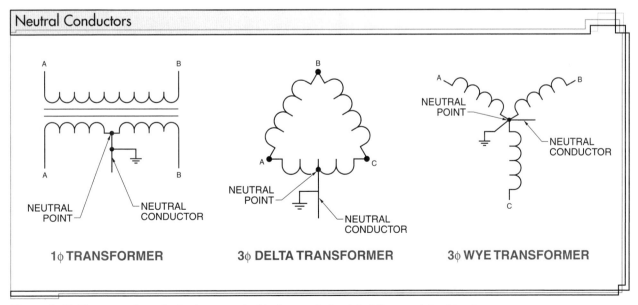

Figure 6-15. The neutral conductor is connected to the neutral point of the supplying transformer.

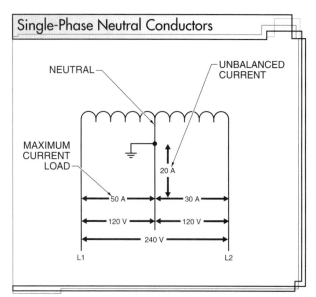

Figure 6-16. With 1ϕ systems, the neutral carries the difference between phase A and phase B.

Three-Phase (3ϕ) Delta Neutral Conductors

Three-phase (3ϕ) delta systems are treated like 1ϕ systems. This is because the neutral carries the unbalanced and maximum unbalanced current from phase A and phase C. The B phase is the high leg with a voltage of 208 V. **See Figure 6-17.**

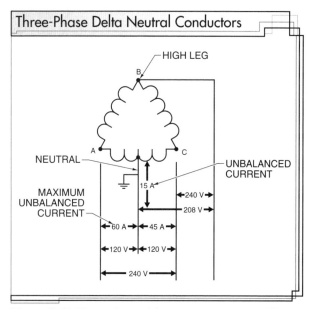

Figure 6-17. Three-phase delta systems are treated like 1ϕ systems because the neutral carries the unbalanced and maximum unbalanced current from phase A and phase C.

Three-Phase (3ϕ) Wye Neutral Conductors

The 3ϕ wye system can have all three phases used with the neutral. The common service voltages are 120/208 V where each phase to the neutral is 120 V or 277/480 V where each phase has 277 V to the neutral. **See Figure 6-18.** As with 1ϕ systems, the neutral shall be able to carry the unbalanced current as well as the maximum unbalanced current.

To determine the maximum unbalanced current for a 3ϕ wye system, use the largest amount of line to neutral current. If just two of the phase conductors and the neutral are being used, then the neutral will carry close to the same current as the phase conductors. See 310.15(B)(4)(b). To determine the unbalanced current in a 3ϕ wye system, apply the following formula:

$$UC = \sqrt{(A^2 + B^2 + C^2) - (A \times B + B \times C + A \times C)}$$

where

UC = unbalanced current (in A)
A = current of phase A (in A)
B = current of phase B (in A)
C = current of phase C (in A)

For example, what is the unbalanced load of a 120/208 V, 3ϕ wye system that has 50 A on phase A, 30 A on phase B, and 65 A on phase C?

$$UC = \sqrt{(A^2 + B^2 + C^2) - (A \times B + B \times C + A \times C)}$$
$$UC = \sqrt{(50^2 \times 30^2 + 65^2) - (50 \times 30 + 30 \times 65 + 50 \times 65)}$$
$$UC = \sqrt{(2500 + 900 + 4225) - (1500 + 1950 + 3250)}$$
$$UC = \sqrt{7625 - 6700}$$
$$UC = \sqrt{925}$$
$$UC = \mathbf{30\ A}$$

Practice Question

Determining Unbalanced and Maximum Unbalanced Load for 3ϕ Wye Neutral Conductors

1. What is the unbalanced current if phase A is 40 A, phase B is 60 A, and phase C is 80 A?
2. What is the maximum unbalanced current if phase A is 40 A, phase B is 60 A, and phase C is 80 A?
3. What is the unbalanced current if phase A is 20 A, phase B is 60 A, and phase C is 100 A?
4. What is the maximum unbalanced current if phase A is 20 A, phase B is 60 A, and phase C is 100 A?
5. What is the unbalanced current if phase A is 40 A, phase B is 60 A, and phase C is 0 A?

Chapter 6—Grounding, Bonding, and Neutrals

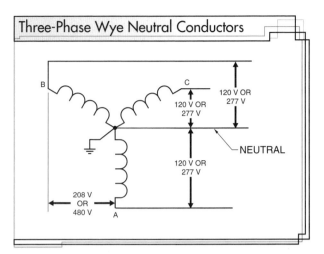

Figure 6-18. Common service voltages for 3ϕ wye systems are 120/208 V where each phase to the neutral is 120 V or 277/480 V where each phase has 277 V to the neutral.

Derating Neutral Conductors

Per 220.61(B)(2), the neutral conductor may have a demand factor of 70% applied when the unbalanced load is in excess of 200 A. Per 220.61(C), there are two restrictions that do not allow for any reduction in the size of the neutral conductor. These restrictions are as follows:

- no reduction for nonlinear loads, such as computers, printers, copy machines, etc., operating from a 4-wire wye system
- no reduction if the load is supplied from a 3-wire circuit using two phases and a neutral conductor from a 4-wire, 3ϕ wye system

Electrical equipment located in wet or damp locations must be grounded.

For example, what is the total demand for a system that has 400 A of unbalanced current?

First, 200 A is rated at 100%.
$200 \times 1.00 = 200$ A
Remaining 200 A is rated at 70%.
$200 \times 0.70 = 140$ A
Total demand = 200 + 140
Total demand = **340 A**

For example, what is the total demand for a neutral that has 690 A of unbalanced current that includes 250 A of nonlinear current?

First, 200 A is rated at 100%.
$200 \times 1.00 = 200$ A
Nonlinear current is rated at 100%.
$250 \times 1.00 = 250$ A
Remainder of current = 690 – 200 – 250
Remainder of current = 240 A
Remaining 240 A is rated at 70%.
$240 \times 0.70 = 168$ A
Total demand = 200 + 250 + 168
Total demand = **618 A**

Practice Question

Determining Neutral Demand

1. What is the neutral demand if the unbalanced current on the neutral is 566 A?
2. If the unbalanced current on the neutral is 566 A with 200 A of nonlinear loads supplied from a wye system, what is the neutral demand?

UNGROUNDED SYSTEMS

An *ungrounded system* is a system that does not have a winding grounded at the supply transformer. This allows the system to continue operating when a ground fault occurs on one of the phases. This system is commonly used in factories where an orderly shutdown is required. Per 250.21(B), ungrounded systems have ground-fault detection systems installed on them. These detectors provide a warning when a fault occurs on one of the phases. The fault is then cleared. If the fault is not cleared, then a second fault on a different phase will shut down the system in an uncontrolled manner.

High-Resistance Ground Systems

High-resistance ground systems (HRGSs) are being used to replace ungrounded systems that were used in industrial locations, which required an orderly shutdown. With a solidly grounded system, if a ground fault were to occur in any one of the ungrounded conductors, the circuit would shut down, which could cause issues such as shutting down parts of equipment during a process, which could damage the equipment or product and make the product unusable.

An HRGS provides a method in which the ground-fault detectors that are installed on the system will indicate when a ground fault has occurred. A high-impedance device is installed in the location where the main bonding jumper would be installed, which will limit the fault current to between 5 A and 7 A. This will not trip an overcurrent device, and the detectors will be used to indicate whether a fault has occurred.

Section 250.36 states the requirements for this type of system. It is permitted on three-phase systems of 480 V to 1000 V in which detectors are installed and servicing is performed by qualified personnel. Since the fault current is limited, the maximum size grounded conductor required is a #8 AWG copper or #6 aluminum conductor. **See Figure 6-19.**

TECH TIP

High-resistance ground systems use ground fault detectors to signal when a ground fault occurs and prevents disorderly shutdowns of equipment which may result in production delays, hazards due to immediate loss of power and financial losses.

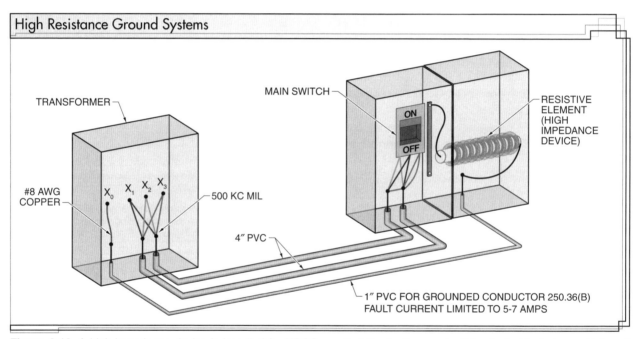

Figure 6-19. A high-impedance device is installed for HRGSs so that the fault current will be limited to between 5 A and 7 A and the overcurrent device will not trip.

Grounding, Bonding and Neutrals

CHAPTER 6

Review Questions

Name _____   Date _____

_____ 1. When used outside, aluminum or copper-clad aluminum grounding conductors shall not be terminated within ___″ of the earth.
   A. 6
   B. 12
   C. 18
   D. 24

_____ 2. For grounding systems of over 1 kV, the minimum insulation level for neutral conductors of solidly grounded systems shall be ___ V.
   A. 600
   B. 1000
   C. 5000
   D. 15,000

_____ 3. A metal underground water pipe in direct contact with the earth for ___″ or more shall be used as a grounding electrode.
   A. 10
   B. 12
   C. 15
   D. 20

_____ 4. A service entrance conductor of 250 kcmil requires a No. ___ grounding electrode conductor.
   A. 2
   B. 2/0
   C. 3/0
   D. 4

_____ 5. An overcurrent protection device rated at 1000 A requires a No. ___ equipment grounding conductor.
   A. 2
   B. 2/0
   C. 3/0
   D. 4/0

6. When a main bonding jumper or a system bonding jumper is a screw only, the screw shall be identified with a ___ finish that shall be visible with the screw installed.
   A. white
   B. black
   C. green
   D. silver

7. Alternating-current systems of less than 50 V shall be grounded when ___.
   A. supplied by transformers, if the transformer supply system exceeds 150 V to ground
   B. supplied by transformers, if the transformer supply system is ungrounded
   C. installed as overhead conductors outside of buildings
   D. all of the above

8. A bonding jumper on the supply side of a service shall not be smaller than the sizes shown in Table ___ for grounding electrode conductors.
   A. 250.122
   B. 250.102(C)(1)
   C. 310.15
   D. 310.16

9. A ground ring encircling a building or structure, in direct contact with the earth, must consist of at least 20″ (6.0 m) of bare copper conductor not smaller than No. ___ AWG.
   A. 2
   B. 4
   C. 4/0
   D. 6

10. When the service consists of more than one enclosure as permitted in 230.71(A), the main bonding jumper for each enclosure shall be sized based on the ___ ungrounded conductor(s) in each enclosure.
    A. largest
    B. average size of the
    C. smallest
    D. total circular mils of the

11. A premises wiring system supplied by a grounded AC service shall have a grounding electrode conductor, or be connected to the grounded service conductor, at each service and at the ___.
    A. load end of the service drop or lateral
    B. subpanel
    C. first main disconnect
    D. A and C

12. Grounding electrode conductors shall be permitted to be ___ and suitable for the area in which it is installed.
    A. aluminum
    B. copper
    C. copper-clad aluminum
    D. all of the above

_____ 13. A concrete-encased electrode can be a conductor consisting of at least ___' of bare copper conductor not smaller than No. 4 AWG.
- A. 10
- B. 20
- C. 25
- D. 50

_____ 14. Rod and pipe electrodes shall be installed such that at least ___' of length is in contact with the soil.
- A. 6
- B. 8
- C. 10
- D. 12

_____ 15. A ground ring shall be buried at a depth below the earth's surface of not less than ___".
- A. 24
- B. 30
- C. 36
- D. 42

Digital Resources
ATPeResources.com/QuickLinks
Access Code: 637160

CHAPTER 7

Conductor Ampacity and Protective Devices

All circuits require conductors to complete the circuit. Most conductors use insulation to prevent the circuit from shorting out or producing a ground-fault condition. The lifespan of a conductor will depend on the insulation, the operating conditions, and the amount of load. The conductors of a circuit must be protected against overloading, short circuits, and ground faults with overcurrent protection devices.

OBJECTIVES

After completing this chapter, the learner will be able to do the following:

- Describe what is considered a current-carrying conductor.
- Explain the most common method used to determine ampacity.
- Describe how ambient temperature affects conductor size.
- Identify when a neutral conductor is considered a current-carrying conductor and when it is not.
- Determine how many current-carrying conductors are allowed in auxiliary gutters before derating is required.
- Explain how to determine the ampacity of copper and aluminum busbars.
- Identify the three main types of overcurrent protective devices.

CONDUCTORS FOR GENERAL WIRING

Article 310 covers the basic requirements for conductors installed in dwellings, offices, and industrial sites. Subsection 310.106(D) requires conductors to be insulated. Conductors may be covered or bare when stated elsewhere in the NEC®. Conductor materials that are recognized by the NEC® include copper, aluminum, and copper-clad aluminum. **See Figure 7-1.** Per 110.5, copper is used unless stated otherwise. Per 310.106(C), all conductors that are No. 8 AWG and larger shall be stranded if installed in a raceway. Conductors smaller than a No. 8 AWG can be solid or stranded when installed in a raceway.

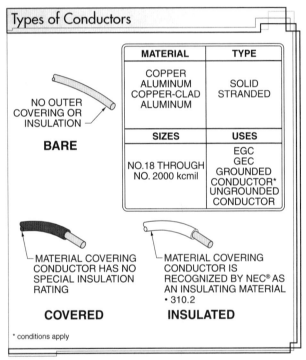

Figure 7-1. Electrical conductors may be bare, covered, or insulated.

Current-Carrying Conductors

All ungrounded conductors that carry current are considered current-carrying conductors. Conductors that are used only for signaling or controls and used for starting duty are not considered current-carrying conductors. There are a few exceptions for signal and control conductors. Per 725.51, if the signal and control conductors have to carry more than 10% of their ampacity continuously when used for fire alarms or for class 1, 2, or 3 circuits, they are considered current-carrying conductors. Information for determining when the neutral or grounded conductor counts as a current-carrying conductor is provided in 310.15(B)(5).

Parallel Conductors

Parallel conductors are two or more conductors that are electrically connected at both ends to form a single conductor. Section 310.10(H) covers the requirements for parallel conductors. Conductors cannot be in parallel if they are No. 1 AWG or smaller. Conductors can be connected in parallel by phase, polarity, neutral, or grounded conductor. Each individual set of parallel conductors shall meet the following guidelines:
- have the same length
- have the same conductor materials
- have the same size
- have the same insulation
- terminate in the same manner

Each cable or conduit shall have the same number of conductors installed to maintain balance in the electrical system. **See Figure 7-2.** The impedance of each conductor of the parallel set shall carry the same amount of current. Because each conductor will carry current, each separate conductor shall be counted as a current-carrying conductor. Equipment grounding conductors shall be installed in parallel if the ungrounded conductors are in more than one raceway. Equipment grounding conductors in parallel can be smaller than No. 1/0 AWG, but are sized from Table 250.122 based on the OCPD.

Minimum Size of Conductors

Table 8 of chapter 9 in the NEC® lists copper, aluminum, and copper-clad aluminum conductors in sizes from 18 AWG through 2000 kcmil. In general, the NEC® does not permit conductors smaller than No. 14 copper, No. 12 aluminum, or copper-clad aluminum to be used for conductors for general wiring in applications up to 2000 V. Table 310.106(A) lists the minimum conductor size for both copper and aluminum conductors. This size changes depending upon the voltage rating of the conductor. These are minimum sizes, and there are many instances in which a particular NEC® provision requires a specific-size conductor.

Insulated Conductor Applications

Section 310.104 provides the information for insulated conductors and their applications. The insulation on conductors may be limited based on conditions such as temperature and location (wet, dry, or damp). Table 310.104(A) lists the standard type of conductors used today and their limitations. Standard types of flexible cords or fixture wires are provided in Table 400.4.

It is important to know the type of location when determining conductor ampacity. Locate the THHW conductors from Table 310.104(A). Notice that the maximum operating temperature depends on whether the location is wet or dry. When the conductor is located in a wet location, the maximum operating temperature will be 75°C. When the conductor is located in a dry location, the maximum operating temperature will be 90°C.

This is because the wet location reduces the conductor's ability to dissipate heat. The NEC® has recently recognized new insulations that are capable of maintaining their higher ampacities in both wet and dry locations. Table 310.104(A) identifies these insulations with the suffix "-2." For example, RHW-2 and XHHW-2 are capable of maintaining their higher ampacities in both wet and dry locations.

Conductors installed in dry and damp locations may be subjected to only a moderate degree of moisture; thus, no adjustment of the conductor's ampacity is necessary. Conductors installed in wet locations, however, may be unprotected from the weather and subject to continuous wet conditions. Wet locations include the following:
- buried in the earth
- inside a concrete slab that is in contact with the earth
- on a roof or outside a building
- any wash-down areas

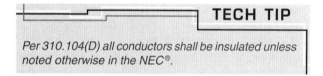

TECH TIP

Per 310.104(D) all conductors shall be insulated unless noted otherwise in the NEC®.

Conductor Identification

The methods for identifying conductors in a circuit are provided in 310.110. Per 310.110(A), insulated or covered grounded conductors shall be identified according to 200.6. Per 200.6 insulated or covered grounded conductors No. 6 AWG or smaller must be identified by a continuous white or gray color that covers the entire length of the conductor. If the conductor is larger than No. 6 AWG, it can be marked with tape or other methods. Per 310.110(B) the equipment grounding conductor shall be identified according to 250.119. Per 250.119, equipment grounding conductors can be bare, green, or green with one or more yellow strips. If the grounding conductor is insulated or covered and larger than No. 6 AWG it can be marked with tape or by other methods.

Ungrounded conductors shall be clearly distinguished from the grounded or equipment grounding conductors. Per 210.5(C), if more than one voltage system is located in a building, each system and its phases shall be identified separately from the other systems. The same requirements for feeders are provided in 215.12(C). The method of identifying the system and phases shall be posted at the panels that supply the conductors.

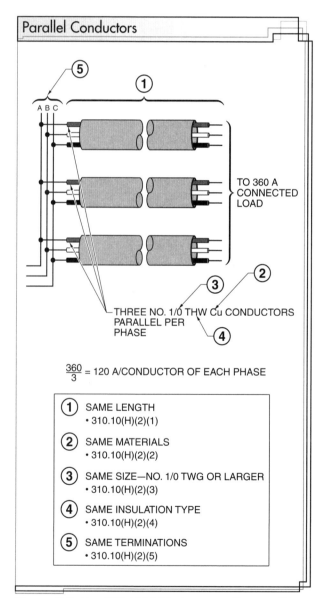

Figure 7-2. Per 310.10(H)(2) there are five conditions parallel conductors shall meet.

AMPACITY

Ampacity is the maximum amount of current a conductor can carry continuously without exceeding its temperature rating. Conditions that may affect a conductor's allowable ampacity, and require derating, include the type of insulation, ambient temperature surrounding the conductor, number of current-carrying conductors in a raceway or cable, and temperature rating.

The conductor insulation type has a direct effect on the allowable ampacity of the conductor. Generally, the higher the temperature rating of the conductor insulation, the higher the allowable ampacity. Section 310.15 provides the means for finding the allowable ampacity of conductors. The two methods that can be used to find the ampacity for conductors are as follows:

- The use of the ampacity tables found in 310.15(B)(16), 310.15(B)(17), 310.15(B)(18), 310.15(B)(19), and 310.15(B)(20). Tables 310.15(B)(16) and (18) are based on three current-carrying conductors in a raceway, cable, or in the earth and in an ambient temperature of 30°C (86°F). Tables 310.15(B)(17) and (19) are for single conductors in free air, and Table 310.15(B)(20) is for not more than three insulated conductors supported by a messenger wire in the air. All of these tables are for conductors of 0 V to 2000 V and for a limited number of insulation types.
- The use of engineering supervision as stated in 310.15(C). This is not a common method of calculating the ampacity as it requires the use of a computer program to determine the ampacity of each conductor.

Allowable Conductor Ampacity

All insulated conductors have a maximum operating temperature at which the insulation of the conductor is not adversely affected. Conductor ampacities are directly related to these operating temperature limitations. Table 310.15(B)(16) lists conductor ampacities for insulated conductors under 2000 V, installed in a raceway, cable, or earth. These values assume an ambient temperature of 30°C (86°F), and not more than three current-carrying conductors in a raceway or cable. While there are many different types of insulations listed, all of them are assigned to one of three basic classifications: 60°C (140°F), 75°C (167°F), and 90°C (194°F). **See Figure 7-3.**

When none of the conditions on which the table is based are exceeded, the ampacity listed in the table under the temperature and insulation type is the allowable ampacity that the conductor is able to carry without damaging the insulation. It is the insulation on the conductors that is being protected. Heat is generated by current flow because of the resistance in the wire. If the temperature exceeds its rating, the insulation will become hard and brittle over time and will crack, allowing voltage and current to take an alternative current path.

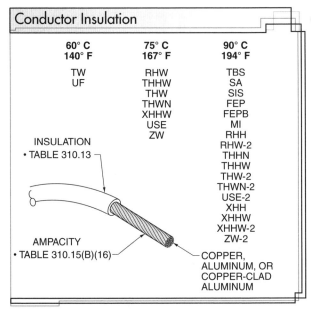

Figure 7-3. Electrical insulation is classified according to temperature rating.

For example, what is the allowable ampacity of a No. 3 THW copper conductor?
See Table 310.15(B)(16).
75°C column = 100 A
Allowable ampacity = **100 A**

For example, what is the allowable ampacity of a No. 6 THHN aluminum conductor?
See Table 310.15(B)(16).
90°C column = 60 A
Allowable ampacity = **60 A**

Practice Questions

Determining Allowable Ampacity

1. What is the allowable ampacity for a No. 1 THW copper conductor?
2. What is the allowable ampacity of a No. 2/0 THW aluminum conductor?
3. What is the allowable ampacity of a No. 4 XHHW used in a conduit that is installed in the ground?

AMBIENT TEMPERATURE

Table 310.15(B)(16) ampacities are based on an assumed ambient temperature of 30°C (86°F). The ambient temperature is the maximum temperature that can be found anywhere along the conductor length. It directly affects the conductor's ability to dissipate heat. If the ambient temperature is too high, the heat dissipation rate decreases, resulting in an increased conductor and insulation temperature.

Generally, the higher the ambient temperature, the more difficult it is for a conductor to dissipate heat. When the ambient temperature is above 30°C (86°F), the ampacity of the conductor shall be adjusted to compensate for the decreased heat dissipation. The ambient temperature correction factors are provided in Table 310.15(B)(2)(a) for temperatures of 30°C (86°F). Table 310.15(B)(2)(b) provides temperature correction for locations based on 40°C (104°F). The correction factors depend upon the type of insulation on the conductor, the type of conductor material (copper, aluminum, or copper-clad aluminum), and the ambient temperature. Correction factors are cross-listed for both Celsius and Fahrenheit to make the application easier.

There may be instances in which the ambient temperature remains below the table value of 30°C (86°F). In these cases, an increased ampacity is permitted by the correction factors. The correction table allows the ampacity to be increased for temperatures as low as 10°C (50°F). Tables 310.15(18), (19), and (20) are based on an ambient temperature of 40°C. Table 310.15(B)(2)(b) provides the temperature correction factors for conductors located in areas of 40°C.

For example, what is the correction factor for THHN conductors in a 50°C location?

See the Correction Factors section of Table 310.15(B)(2)(a).

THHN conductors in a 50°C location = 0.82

Correction factor = **0.82 A**

For example, what is the allowable ampacity for a No. 3/0 THW conductor in a 50°C location?

See Table 310.15(B)(16).

No. 3/0 THW conductor = 200

See the Correction Factors section of Table 310.15(B)(2)(a).

THW conductors in a 50°C location = 0.75

200 × 0.75 = 150

Ampacity = **150 A**

Practice Questions

Determining Allowable Ampacity Using Correction Factors

1. What is the ampacity for a No. 1 THW copper conductor in a 40°C location?
2. What is the ampacity of a No. 2/0 THW aluminum conductor in a 105°F location?
3. What is the allowable ampacity of a No. 4 XHHW conductor used in a conduit that is installed in the raceway that is located in a 40°C location?

Raceways Exposed to Sunlight on or above Rooftops

When raceways and conductors are installed on a roof in direct sunlight, the temperature inside the raceway can increase due to the heat of the sun and reflection off the rooftop. If the raceway or cable installation is less than ⅞″ above the surface of the roof, the average high temperature for the location shall be increased 33°C (60°F). **See Figure 7-4.** If installed more than the required ⅞″, temperature is not required to be added to the average high temperature of the installation location. Rooftop temperature additions are not required for XHHW-2 insulated conductors even if installed at ⅞″ or lower.

TECH TIP

When conductors within raceways exposed to sunlight on or above rooftops are derated, the average high temperature for the location can be determined from the weather service or other sources. When taking the master electrician's exam, the average high temperature for the locations will be provided in the question.

Conduit Fill

When conductors are installed in a conduit system, the heat is generated by the current flowing through the conductors. This heat is held in by the conduit and will raise the air temperature inside the conduit. The size of the conduit does not make a difference.

Per 310.15(B)(3)(a), when more than three current-carrying conductors are placed in a conduit, they need to be derated. Table 310.15(B)(3)(a) gives the derated percentages based on the number of conductors or cables, including spare and current-carrying neutral conductors, but not including grounding conductors, in the raceway. Per 310.15(B)(3)(a)(2), the derating based on conduit fill does not apply to raceways that are 24″ or less in length.

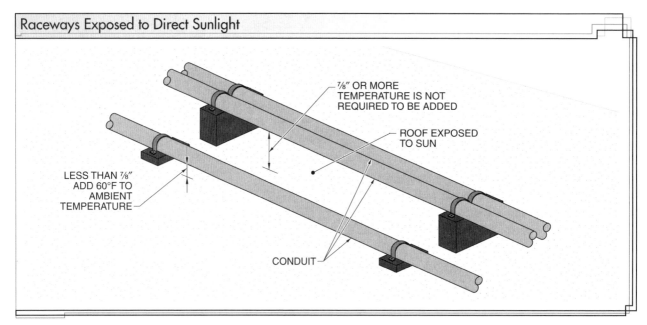

Figure 7-4. Raceways that are installed on rooftops and exposed to direct sunlight may need to be derated. The amount of temperature added depends on the distance the conduit is from the roof.

Conductor information is printed continuously along the conductor's insulation.

Neutral as a Current-Carrying Conductor

Section 310.15(B)(5) covers the conditions under which the neutral conductor shall be counted as a current-carrying conductor. Per 310.15(B)(5)(a), if the neutral only carries the unbalanced current, it should not be considered a current-carrying conductor. Per 310.15(B)(5)(b), in a 3-wire circuit with two ungrounded conductors and the neutral of a 3ϕ, 4-wire, wye system, the neutral is considered a current-carrying conductor. Per 310.15(B)(5)(c), if the neutral is part of a 3ϕ, 4-wire, wye circuit from a wye system carrying non-linear loads, the neutral is considered a current-carrying conductor because of the harmonics that the grounded conductor is carrying. **See Figure 7-5.**

For example, what is the allowable ampacity of a No. 6 THHW conductor installed above a ceiling with a total of nine current-carrying conductors in the conduit?

See Table 310.15(B)(16).
No. 6 THHW conductors in a dry location = 75 A
See Table 310.15(B)(3)(a).
Nine current-carrying conductors = 70%
$75 \times 0.7 = 52.5$ or 53 A
Ampacity = **53 A**

Practice Questions

Determining Ampacity Using Current-Carrying Conductors

1. What is the ampacity for No. 1 THW copper conductors that have 6 current-carrying conductors in the conduit?

2. What is the ampacity of a No. 2/0 THW aluminum conductor in a conduit that is 20′ long?

3. What is the allowable ampacity of No. 4 XHHW conductors used in a conduit that has 13 current-carrying conductors in it?

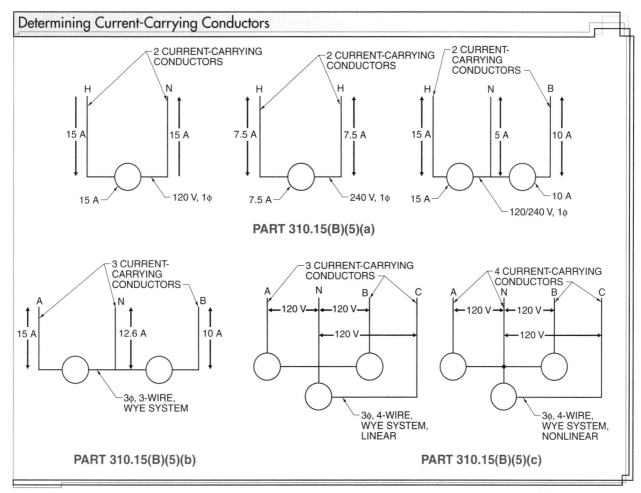

Figure 7-5. Section 310.15(B)(5) covers the conditions under which the neutral conductor shall be counted as a current-carrying conductor.

1φ Residential Service and Feeder Conductors

Reference 310.15(B)(7) to determine the rating and size for 1φ, service entrance conductors or feeder conductors that carry the entire load of the dwelling unit for 100 to 400 A services. The service entrance or feeder conductors are sized based on 83% of the standard service rating of the service equipment after the loads are calculated. Take into consideration any temperature correction or derating of the conductors due to raceway fill.

Table 310.15(B)(16) is used to size the ungrounded conductor for services over 400 A. The grounded conductor shall be permitted to be smaller than the ungrounded conductors, but shall be sized to carry the maximum unbalanced load as calculated and shall meet the requirements of 220.61. Derating and temperature correction shall be applied if the installation requires it.

For example, what size of service entrance conductors is required for a service with a calculated load of 180 A and an actual service rating of 200 A?

$200 \times 0.83 = 166$ A

See Table 310.15(B)(16), 75°C column

175 A = **No. 2/0 conductors**

Practice Questions

Determining Conductor Size for 1φ Residential Services

1. The service calculated for a one-family dwelling is 156 A. Using the next standard size overcurrent protective device, what size of THWN service entrance conductors is required?

2. The service calculated for a multifamily dwelling is 280 A. What size of THHN, aluminum service entrance conductors is required?

Temperature and Conduit-Fill Derating Factors

The allowable ampacity question on the exam may include both temperature and conduit-fill derating factors. If this is the case, then the conductors are derated by using the correction factors in Table 310.15(B)(2)(a) or (b) for temperature correction, and Table 310.15(B)(3)(a) for conduit fill. The order in which the calculations are performed does not matter. Keep in mind that temperature derating always applies, and conduit-fill derating applies only to conduit systems longer than 24″.

Sometimes the conductors in a circuit may go through more than one temperature condition, which could provide two different allowable ampacities for the circuit. The lowest ampacity calculated based on conduit fill and ambient temperature is used. There is an exception according to 310.15(A)(2). This exception is based on the factor of 10% of the circuit length or 10′, whichever is less.

For instance, a circuit of No. 6 THW conductors 150′ long passes through a 30°C location and a 40°C location. Since 10% of the circuit length would be 15′, the maximum length that is allowed for the higher temperature is 10′. The ampacity from Table 310.15(B)(16) for No. 6 THW conductors in a 30°C location is 65 A. In a 40°C location, the allowable ampacity would be 65 multiplied by the correction factor of 0.88, which equals 57.2 A. If the part of the circuit in the 40°C location is 10′ or less, then the ampacity of the 30°C location can be used.

For example, what is the allowable ampacity of No. 2 THW aluminum conductors that are installed in a 36°C location with a total of 12 current-carrying conductors in the conduit?

See Table 310.15(B)(16).
No. 2 THW Al conductors at 75°C = 90 A
See Table 310.15(B)(2)(a), Correction Factors.
Temperature Correction = 0.88
Conduit Fill 310.15(B)(3)(a) = 50%
$90 \times 0.88 \times 0.5 = 44$ A
Allowable ampacity = **39.6 A**

Practice Question

Determining Ampacity Using Temperature and Conduit-Fill Derating

1. What is the ampacity for No. 1 THW copper conductors that are installed in a 38°C location with a total of 6 current-carrying conductors in the conduit?

Conductors from 2001 V to 35,000 V

NEC® 310.60 provides ampacity information for conductors rated from 2001 V to 35,000 V. The ampacity ratings can be determined from Tables of 310.60(67) through (86). These conductors are usually insulated. Underground conductors must be installed in electrical ducts.

Auxiliary Gutters and Wireways

Per 366.23 and 376.22, up to 30 current-carrying conductors are allowed in auxiliary gutters or metal wireways before derating is required. If there are more than 30 current-carrying conductors in any cross-sectional area of the wireway or gutter, the derating is based on Table 310.15(B)(3)(a). Control and signaling conductors do not count as current-carrying conductors.

Conductors in nonmetallic wireways are derated differently from metal wireways. This is because nonmetallic (plastic) wireways do not dissipate heat as fast as metal wireways. Per 378.22, the sum of all the conductors in a cross-sectional area in a nonmetallic wireway shall not exceed 20%. Conductors in nonmetallic wireways are derated when there are more than 3 current-carrying conductors installed. The derating percentage is based on the number of current-carrying conductors. See Table 310.15(B)(3)(a).

Terminal Temperature Rating

The temperature rating of the equipment terminations also affects a conductor's allowable ampacity. The ampacity of the conductors shall be coordinated so as not to exceed the lowest temperature rating of any component in the electrical circuit, including the electrical equipment per 110.14(C). Although conductors may be selected that are suitable for 90°C conditions, the termination will probably have a lower temperature rating. In these cases, the ampacity of the conductor shall be selected so that it does not exceed the lowest rating of the equipment in the circuit.

Most electrical equipment terminals are rated at 60°C, 75°C, or 90°C. Some equipment contains dual ratings such as 60°C/75°C. Such equipment is permitted to use the 75°C rating provided all other components of the circuit are suitable for 75°C. When the conductor insulation rating exceeds the equipment termination temperature rating, the ampacity of the conductor shall be based on the equipment rating. At present, there is very little equipment available that is suitable for use with 90°C conductors. Conductors rated at 90°C are permitted to be installed with the lower-rated equipment, but their ampacities shall be based on the temperature rating of the equipment. Conductors rated for 90°C are used for their higher ampacity so that they can be derated for the ambient temperature of the terminals. The size of the conductor cannot be less than the sizes listed in the 60°C and 75°C columns of Table 310.15(B)(16).

If a terminal is rated for a conductor smaller than a No. 1 AWG conductor or less than 100 A, then the terminal is a 60°C terminal and the conductor cannot be smaller than the conductors shown in the 60°C column of 310.15(B)(16). If a terminal is rated for a conductor larger than a No. 1 AWG conductor or more than 100 A, then the terminal is a 75°C terminal and the wire cannot be smaller than what is shown in the 75°C column of Table 310.15(B)(16). **See Figure 7-6.**

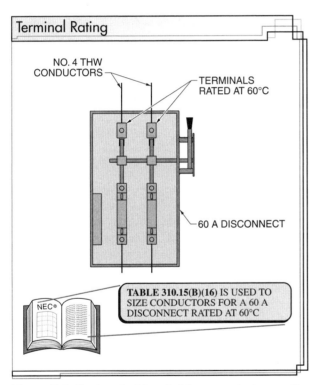

Figure 7-6. If a terminal is rated for a conductor smaller than a No. 1 AWG or less than 100 A, then the terminal is a 60°C terminal and the conductor cannot be smaller than the conductors shown in the 60°C column of 310.15(B)(16).

For example, what size of THW conductors is required for a 60 A device with terminations rated at 60°C?
See 110.14(C)(1)(a) and Table 310.15(B)(16).
Terminations rated at 60°C at 60 A = No. 4 THW conductors
Minimum conductor size = **No. 4 THW conductors**

For example, what size of THHN conductors is required for a 200 A panel with terminations rated at 75°C?
See 110.14(C)(1)(b) and Table 310.15(B)(16).
Terminations rated at 75°C at 200 A = No. 3/0 THHN conductors
Minimum conductor size = **No. 3/0 THHN conductors**

> *Practice Questions*
>
> *Determining Conductor Size Using Terminal Temperature Rating*
> 1. What size of conductor is required for a 50 A circuit if the device is marked for 75°C?
> 2. What size of conductor is required for a 150 A circuit using THHN wire?

SIZING CONDUCTORS FOR BRANCH CIRCUITS

Conductors are sized to carry the load to which they are connected. This must be done without overloading the conductor or damaging the insulation around the conductor. Per 210.19(A)(1), conductors used for branch circuits must be sized at 125% of the continuous loads plus 100% for the noncontinuous loads. The grounded conductor must follow this same rule if connected to an overcurrent device. If it is not connected to the overcurrent device, then it is sized at 100% for both the continuous and noncontinuous loads.

Sizing Branch Circuit Conductors for Continuous Loads

A continuous load is one that will run for 3 hr or more at a time. Per 210.19(A), ungrounded conductors shall be sized at 125% of the rated current for continuous loads or 80% of the conductor ampacity.

For example, what is the maximum current for a No. 2 THHN copper conductor supplying a continuous load?
See Table 310.15(B)(16).
No. 2 THHN copper conductor = 130 A
See 210.19.
130 A = 125% of current or 80% of conductor ampacity
130 A × 80% = 104 A
Maximum current = **104 A**

For example, what size of THW conductors is required to carry a continuous load rated at 45 A with terminals rated at 75°C?

See 210.19.

45 × 125% = 56.25 A

See Table 310.15(B)(16).

Terminals rated at 75°C = 56.25 A minimum

Minimum size conductors = **No. 6 THW conductors**

Practice Questions

Determining Current for Continuous-Load Conductors

1. What is the maximum current a No. 1 THW aluminum conductor can carry on a continuous load?
2. What amount of current is required for sizing a conductor for a 34 A continuous load?

Busbar Ampacity

A *busbar* is a copper or aluminum bar, usually rectangular in shape, that conducts electricity within an electrical device. Busbars are commonly used in motor control centers but are also used on large current services and feeders. Per 366.23, a copper busbar can carry 1000 A per square inch. An aluminum busbar can carry 700 A per square inch. To determine the size of a busbar in square inches, multiply the width of the bar times the thickness. To determine the ampacity of the busbar, multiply the square inches of the busbar by 1000 for copper or by 700 for aluminum. **See Figure 7-7.** The numbers must be in decimal form before multiplying.

For example, what is the ampacity of a copper busbar that measures 4″ × ½″?

Sq in. of busbar = width × thickness
Sq in. of busbar = 4″ × 0.5″
Sq in. of busbar = 2

Ampacity = sq in. of busbar × 1000
Ampacity = 2 × 1000
Ampacity = **2000 A**

For example, what is the ampacity of an aluminum busbar that measures 4″ × ½″?

Sq in. of busbar = width × thickness
Sq in. of busbar = 4″ × 0.5″
Sq in. of busbar = 2

Ampacity = sq in. of busbar × 700
Ampacity = 2 × 700
Ampacity = **1400 A**

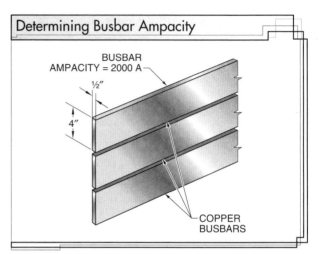

Figure 7-7. Copper busbars can carry 1000 A per square inch and aluminum busbars can carry 700 A per square inch.

Practice Questions

Determining Busbar Ampacity

1. What is the ampacity of a copper busbar that measures ⅜″ × 2″?
2. What is the ampacity of an aluminum busbar that measures 1″ × 4″?

OVERCURRENT PROTECTION

The two most common overcurrent protective devices used in electrical distribution systems are fuses and circuit breakers. **See Figure 7-8.** Overcurrent protection of conductors ensures that conductors are protected in accordance with their allowable ampacity. All conductors shall be protected against overcurrent in accordance with their allowable ampacities unless otherwise required or permitted by 240.4 (A through G). This applies to standard building conductors as well as fixture wires, flexible cords, or cables. The size of the overcurrent protective device is determined after derating for temperature correction and conduit fill. Overcurrent protection for flexible cords and fixture wires should be in compliance with 240.5.

Section 310.15 requires that conductor ampacities be obtained from Tables 310.15(B)(16) through 310.15(B)(19) or under engineering supervision by calculation. The process of selecting overcurrent protection requires that first the conductor's allowable ampacity be calculated and then the overcurrent protection be selected. See 240.4(A through G) for alternate methods to the general rule of protecting conductors in accordance to their allowable ampacity.

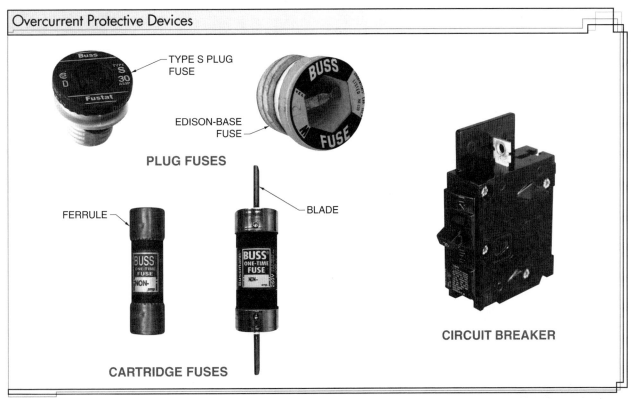

Figure 7-8. Overcurrent protective devices protect conductors and equipment by opening the circuit if the current reaches a specific value.

All switches that contain fuses or circuit breakers shall be readily accessible. Article 240.24(A) requires that the center of the grip of the operating handle of the switch or circuit breaker be no more than 6′ 7″ in the highest position above the floor or working platform. There are a few allowed exceptions as listed in 240.24 (A)(1), (2), (3), and (4).

Overcurrent Protection for Devices Rated 800 A or Less

Often, the calculated ampacity does not correspond to a standard-size fuse or circuit breaker. The next higher standard device rating may be used in these cases per 240.4(B), provided the conductors are not part of a multioutlet branch circuit for cord-and-plug-connected portable loads and the next higher standard device rating does not exceed 800 A. In cases where the cord-and-plug-connected portable loads are supplied by multioutlet receptacle circuits, and the ampacity of the conductor does not correspond to a standard rating for a fuse or circuit breaker, the next lower standard device rating shall be selected. **See Figure 7-9.**

For example, what is the maximum-size fuse for branch circuit conductors that have an allowable ampacity of 315 A?

See 240.4(B) and Table 240.6(A).
Per Table 240.6(A), the next standard size up from 315 = 350
Maximum size fuse = **350 A**

Practice Question

Determining Overcurrent Protection for Devices Rated 800 A or Less

1. What is the maximum-size circuit breaker allowed for a branch circuit conductor carrying 380 A?

Overcurrent Protection for Devices Rated over 800 A

In installations where the calculated conductor ampacity does not correspond to a standard size rating for fuses or circuit breakers, and the rating of the overcurrent device required is over 800 A, the conductor must be greater or equal in ampacity to the overcurrent protective device rating per 240.4(C). **See Figure 7-10.**

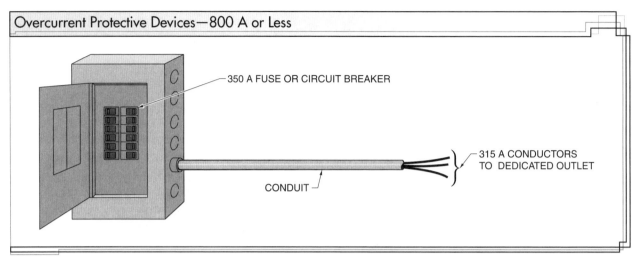

Figure 7-9. Overcurrent protective devices of 800 A or less that do not correspond to a standard-size device are rounded up.

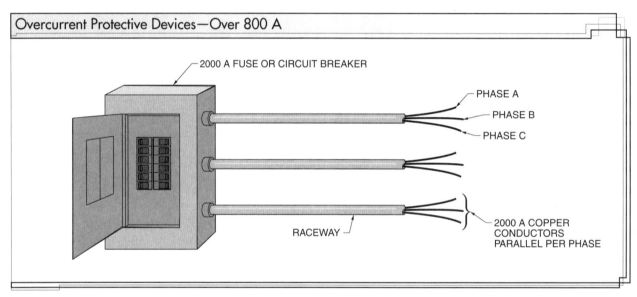

Figure 7-10. Per 240.4(C), conductors used to distribute power from an overcurrent protection device rated over 800 A must be greater or equal in ampacity.

For example, what is the required ampacity for conductors connected to a circuit breaker rated at 2000 A?
  Per 240.4(C), the ampacity of conductors shall equal or exceed the overcurrent protective device.
  Overcurrent protective device = 2000 A
  Ampacity = **2000 A**

Practice Question

Determining Overcurrent Protection for Devices Rated over 800 A

1. What is the required ampacity for conductors that are connected to a 1500 A circuit breaker?

Overcurrent Protection for Small Conductors

See Table 310.15(B)(16) and notice the double asterisk next to the AWG sizes of No. 18, 16, 14, 12 and 10 conductors. Refer to 240.4(D) for the ampacity rating of these conductors. Per 240.4(D), a No. 14 AWG = 15 A, a No. 12 AWG = 20 A, and a No. 10 AWG = 30 A for copper conductors. A No. 12 AWG = 15 A and a No. 10 AWG = 25 A for aluminum conductors. The sizes of No. 16 AWG and No. 18 AWG are used for control conductors. The No. 16 AWG can handle 10 A under certain conditions and the No. 18 AWG can handle 7 A under certain conditions. **See Figure 7-11.**

Ampacity Ratings for Small Conductors

CONDUCTOR SIZE	COPPER CONDUCTORS*	ALUMINUM CONDUCTORS*
18	7†	—
16	10‡	—
14	15	—
12	20	15
10	30	25

* in A
† See 240.4(D)(1)
‡ See 240.4(D)(2)

Figure 7-11. The ampacity rating for small conductors is found in 240.4(D).

TECH TIP

For small conductors, the general overcurrent provisions limit the overcurrent protection to 15 A for No. 14 copper, 20 A for No. 12 copper, and 30 A for No. 10 copper.

The No. 16 AWG and No. 18 AWG are not allowed for use as standard building conductors but can be used for control circuits. The sizing of the overcurrent protective device is applied to the conductor after any derating. This rule does not apply to fixture wires for flexible cords or cables nor does it apply to Table 240.4(G).

For example, what is the maximum rating for an overcurrent protective device for 4 No. 10 THHN copper conductors in a conduit feeding a lighting circuit (HID lighting)?

See Table 310.15(B)(16)
No. 10 THHN = 40 A
Per 310.15(B)(5), the neutral counts as current-carrying conductor.
See Table 310.15(B)(3)(a).
Adjustment factor = 80%
$40 \times 0.80 = 32$ A
Per 240.4(D), go to 30 A as required.
Maximum rating for an overcurrent protective device = **30 A**

Practice Question

Determining Overcurrent Protection Size for Small Conductors

1. What is the maximum overcurrent protective device allowed on a No. 12 THW conductor with 10 current-carrying conductors in the conduit?

STANDARD SIZES FOR OVERCURRENT PROTECTIVE DEVICES

The standard ratings for overcurrent protective devices are given in Table 240.6. This list includes both circuit breakers and fuses. **See Figure 7-12.** Part A covers the standard ratings for fuses and inverse-time circuit breakers (ITCBs). Part B covers adjustable-trip circuit breakers (ATCBs). Section 240.6(C) contains three provisions which, if met, permit an adjustable-trip circuit breaker to have an ampere rating equal to the adjusted current setting rather than the maximum setting of the circuit breaker.

Fuses

The standard ratings for 15 A to 50 A fuses increase in increments of 5 (15, 20, 25, 30, 35, 40, 45, and 50). For 50 A to 110 A fuses, the ratings increase in increments of 10 (50, 60, 70, 80, 90, 100, and 110). For 125 A to 250 A fuses, the ratings increase in increments of 25 (125, 150, 175, 200, 225, and 250). For 250 A to 500 A fuses, the ratings increase in increments of 50 (250, 300, 350, 400, 450, and 500). Beyond 500 A, the ratings are not as uniform. The maximum standard rating for a fuse is 6000 A. Additional standard ratings for fuses are 1 A, 3 A, 6 A, 10 A, and 601 A.

Fuses and Inverse-Time Circuit Breakers (ITCBs)

INCREMENT	STANDARD AMP RATINGS
5	15, 20, 25, 30, 35, 40, 45, 50
10	50, 60, 70, 80, 90, 100, 110
25	125, 150, 175, 200, 225, 250
50	250, 300, 350, 400, 450, 500
100	500, 600, 700, 800
200	1000, 1200
400	1600, 2000
500	2500, 3000
1000	3000, 4000, 5000, 6000

Figure 7-12. Additional standard-rated fuses include 1 A, 3 A, 6 A, 10 A, and 601 A.

Inverse-Time Circuit Breakers

An *inverse-time circuit breaker (ITCB)* is a circuit breaker with an intentional delay between the time when the fault or overload is sensed and the time when the circuit breaker operates. The standard ratings for ITCBs are the same as those for fuses with the exception of 1 A, 3 A, 6 A, 10 A, and 601 A, which apply only to fuses. ITCBs are the most commonly used circuit breakers in the electrical industry. **See Figure 7-13.**

Adjustable-Trip Circuit Breakers

An *adjustable-trip circuit breaker (ATCB)* is a circuit breaker with a trip setting that can be changed by adjusting the amp setpoint, trip-time characteristics, or both within a particular range. In general, the rating of these devices shall be determined from the maximum setting available in the ATCB per 240.6(B). For instance, if the setting is adjustable over a range of 600 A to 800 A, the rating is based on 800 A. However, 240.6(C) permits the actual setting, not the maximum setting, to be utilized if the circuit breaker has restricted access to the adjusting means.

Restricted access is defined as being behind removable and sealable covers, being behind bolted equipment enclosure doors, or being located in locked rooms accessible only to qualified personnel. If these conditions are met, a circuit breaker with an 800 A frame size (maximum rating) that has a long-time pickup setting of 600 A can be considered to have a 600 A rating, and the wire ampacity could be selected based on the 600 A setting instead of the 800 A rating.

For example, which of the following is not a standard-size fuse: 601 A, 50 A, 325 A, or 1000 A?

Per 240.6, the 325 A fuse is not on the list.

Non-standard-size fuse = **325 A**

> **Practice Question**
>
> *Determining Standard Overcurrent Protection Size*
>
> 1. Is a 120 A breaker a standard size for an overcurrent protective device?

TYPES OF OVERCURRENT PROTECTIVE DEVICES

There are three main types of overcurrent protective devices: plug fuses, cartridge fuses, and circuit breakers. Overcurrent protective devices are rated for current. Fuses and circuit breakers, unless listed for continuous use, can only handle 80% of their current rating for continuous loads.

Plug Fuses

A *plug fuse* is a fuse that uses a metallic strip that melts when a predetermined amount of current flows through it. Plug fuses incorporate a screw configuration that is interchangeable with fuses of other amp ratings. Part V of 240 covers plug fuse devices. Plug fuses, also called Edison-base fuses, are not used in new installations and can only be used for replacement purposes. Also available are the Type S fuses. The Type S fuse uses an adapter that will not allow for oversizing of the fuses on a circuit once it is installed. It has a smaller base than the Edison-base fuse and it is not interchangeable. **See Figure 7-14.**

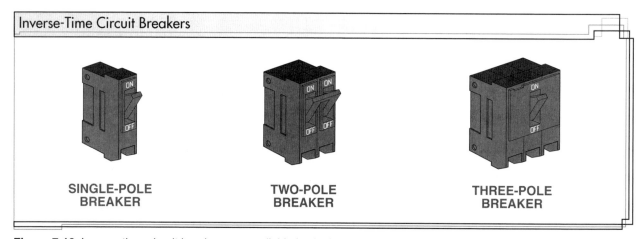

Figure 7-13. Inverse-time circuit breakers are available in single-, two-, and three-pole configurations.

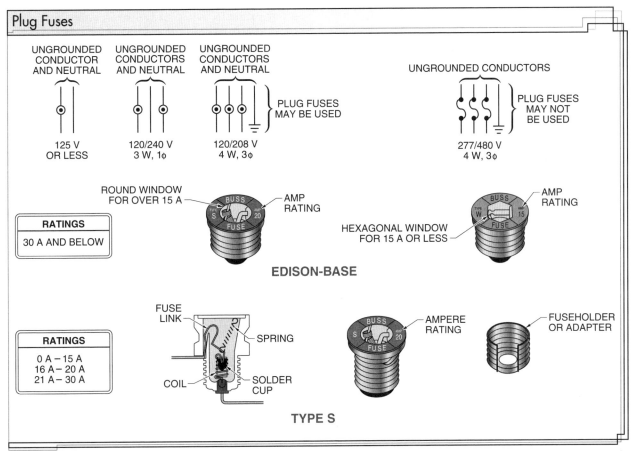

Figure 7-14. Plug fuses shall not be installed in circuits operating at over 150 V to ground.

Cartridge Fuses

A *cartridge fuse* is a fuse constructed of a metallic link or links that are designed to open at predetermined current levels to protect circuit conductors and equipment. **See Figure 7-15.** Part VI of 240 covers cartridge fuses that are found in many disconnects. There are different types of cartridge fuses such as nontime delay or time delay, standard, and current limiting. They may also be rated for 250 V and 600 V circuits. These fuses have an interrupting rating of 10,000 A or more.

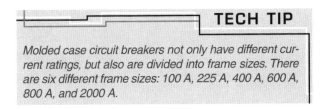

TECH TIP

Molded case circuit breakers not only have different current ratings, but also are divided into frame sizes. There are six different frame sizes: 100 A, 225 A, 400 A, 600 A, 800 A, and 2000 A.

Circuit Breakers

A *circuit breaker (CB)* is a reusable overcurrent protective device that can be reset after it trips due to a short or a ground fault in the circuit. Part VII of 240 covers circuit breakers. Circuit breakers have the same 250 V and 600 V ratings. The interrupting rating is 5000 A unless otherwise marked. A circuit breaker may also be used as a switch if marked as SWD or HID.

OVERCURRENT PROTECTION FOR BRANCH CIRCUITS

Overcurrent protection size for conductors depends on the loads being continuous or noncontinuous. Overcurrent protective devices rated for 100% are sized based on the ampacity rating regardless of whether the load is rated for continuous duty or noncontinuous duty.

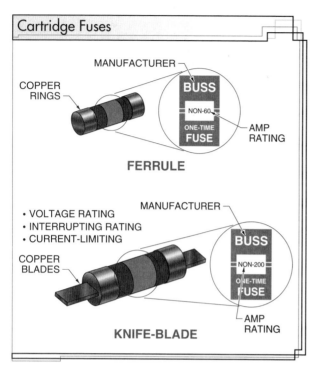

Figure 7-15. Cartridge fuses are designed with either a ferrule or knife-blade configuration.

The lighting circuits in an office building are considered continuous loads.

Overcurrent Protection for Noncontinuous Loads

Where circuits are connected to loads that are considered to be noncontinuous, the overcurrent protective device is sized based on the ampacity rating or 100% per 210.20(A). The size of the overcurrent protective device shall be determined after derating for conduit fill or temperature correction.

For example, what size of overcurrent protective device is allowed for an 18 A circuit?

Per 210.20(A), the overcurrent protective device must be equal to the load or the next standard size up.
Per 240.6, the next standard size up from 18 = 20
Overcurrent protective device = **20 A**

Practice Question

Determining Overcurrent Protection Size for Noncontinuous Loads

1. What is the required fuse size for a load of 53 A?

Overcurrent Protection for Continuous Loads

If a load is a continuous load running for 3 hr or more, the overcurrent protective devices must be sized at 125% per 210.20(A). The size of the overcurrent protective devices is determined after derating for conduit fill or temperature correction.

For example, what size of circuit breaker is required for a load that draws 23 A for 8 hr in a row?

Per 210.20(A), the overcurrent protective device must be sized at 125%.
$23 \times 1.25 = 28.75$
The next standard size up from 28.75 = 30
Overcurrent protective device = **30 A**

Practice Questions

Determining Overcurrent Protection Size for Continuous Loads

1. What is the maximum continuous load that may be placed on a 200 A overcurrent protective device?
2. If a continuous load is rated at 40 A, what size of breaker is required?
3. What size of fuse is required for a circuit that draws 30 A for 6 hr at a time?

TECH TIP

The time required for a fusible element to open varies inversely with the magnitude of the current that flows through the fuse or circuit breaker.

Overcurrent Protection for Noncontinuous and Continuous Loads

If continuous loads are combined with noncontinuous loads, each load is rated individually and then added together to determine the size of the overcurrent protective device. Noncontinuous loads are rated at 100%, and continuous loads are rated at 125%. The size of the overcurrent protective device is determined after derating for conduit fill or temperature correction.

For example, what is the required circuit breaker size for a noncontinuous load of 17 A and a continuous load of 25 A?

Noncontinuous load rated at 100%
$17 \times 1.00 = 17$
Continuous load rated at 125%
$25 \times 1.25 = 31.25$ A
Circuit breaker size = Noncontinuous load + Continuous load
$17 + 31.25 = 48.25$ A
Per 240.6, the next standard size circuit breaker from 48.25 = 50 A
Circuit breaker size = **50 A**

Practice Questions

Determining Overcurrent Protection Size for Noncontinuous and Continuous Loads

1. What is the fuse rating for a continuous load of 32 A and a noncontinuous load of 20 A?
2. What size of circuit breaker is required for a circuit that has 15 A of noncontinuous loads and 20 A of continuous loads?

INTERRUPTING CAPACITY

Per 110.9, an overcurrent protective device must have an interrupting rating at nominal circuit voltage at least equal to the current that is available at the line terminals of the equipment. If the circuit can deliver a fault current that is above the rating of the overcurrent protective device, then it may cause the fuse or circuit breaker to explode when under a faulted condition. A standard circuit breaker, unless otherwise marked, is rated at 5000 AIC. A standard fuse is rated at 10,000 A, unless otherwise marked. This information on fuses and circuit breakers can be found in 240.60(C) and 240.83(C).

CHAPTER 7
Conductor Ampacity and Protective Devices
Review Questions

Name _____ Date _____

_____ 1. If seven No. 12 THHN current-carrying conductors are in a raceway and the ambient temperature is 86°F, the ampacity of the conductors is ___ A.
   A. 15
   B. 21
   C. 25
   D. 30

_____ 2. The maximum current allowed on a No. 10 THW conductor in a conduit with five other No. 10 THW current-carrying conductors and two bare No. 10 grounding conductors is ___ A.
   A. 20
   B. 21
   C. 28
   D. 30

_____ 3. A No. 250 kcmil THWN aluminum conductor installed in a metal wireway with 15 other current-carrying conductors with an ambient temperature of 40°C will have an allowable ampacity of ___ A.
   A. 90.2
   B. 180
   C. 205
   D. 224

_____ 4. A No. ___ THHN conductor is required for a 50 A circuit with terminals marked 75°C.
   A. 4
   B. 6
   C. 8
   D. 10

_____ 5. The minimum size for a THWN conductor in a conduit with a total of 15 current-carrying conductors supplying a 48 A load is No. ___ THWN.
   A. 2
   B. 3
   C. 4
   D. 8

6. No. 250 KCMIL THWN aluminum conductors installed in a nonmetal wireway with 15 other current-carrying conductors with an ambient temperature of 40°C will have an allowable ampacity of ___ A.
   A. 90.2
   B. 180
   C. 224
   D. 225

7. The ampacity of a No 12 THHN conductor in a raceway with an ambient temperature of 28°C is ___ A.
   A. 15
   B. 20
   C. 30
   D. 31.2

8. A 4/0 THHN copper conductor installed in free air on a 480 V circuit at a temperature of 30°C will have an allowable ampacity of ___ A.
   A. 205
   B. 260
   C. 300
   D. 405

9. A TC cable with No. 6 THHN wire in a cable tray and with a spacing of one cable width between the multiconductor cables will have an allowable ampacity of ___ A.
   A. 75
   B. 85
   C. 105
   D. 110

10. A No.12 THHN copper conductor in a conduit with a total of six current-carrying conductors with an ambient temperature of 122°F will have a maximum current of ___ A.
    A. 19.68
    B. 20
    C. 24.6
    D. 32.8

CHAPTER 8

Voltage Drop

Digital Resources
ATPeResources.com/QuickLinks
Access Code: 637160

Voltage drop should be taken into consideration when feeder or branch circuit conductors span a large distance from the service equipment to the load. This is to ensure that the proper voltage is maintained throughout the circuit. The longer the run between equipment and its protective devices, the larger the voltage drop will be. If the voltage drop in a run is too great, then larger conductors may have to be used. There is no code requirement in the NEC®, only a recommendation that limits the amount of voltage drop in a circuit. Most electrical projects require a voltage drop of no more than 3% in their specifications.

OBJECTIVES

After completing this chapter, the learner will be able to do the following:

- Identify the maximum voltage drop recommended by the NEC®.
- List the factors that affect voltage drop.
- Explain ways to reduce voltage drop.
- Determine the size of conductors in circular mils using Table 8 of chapter 9 in the NEC®.
- Calculate voltage drop using Ohm's law.
- Determine how to use the voltage drop formula to find voltage drop, conductor size, maximum load, and maximum distance for 1ϕ circuits.
- Determine how to use the voltage drop formula to find voltage drop, conductor size, maximum load, and maximum distance for 3ϕ circuits.
- Calculate power loss in a circuit using voltage drop.
- Determine the proper voltage drop for combined feeder and branch circuits.

VOLTAGE DROP RECOMMENDATIONS FROM THE NEC®

Voltage drop is the voltage that is lost due to the resistance of conductors. Even though there are no requirements for maximum voltage drop, the NEC® gives recommendations to help ensure that circuits operate efficiently. Per 210.19 Informational Note No. 4, it is recommended that branch circuits have a maximum voltage drop of 3%. When combined with a feeder, the total maximum voltage drop should not exceed 5%. Per 215.2(A)(1)(b) Informational Note No. 2, it is recommended that feeders have a maximum voltage drop of 3%. When combined with a branch circuit, the total maximum voltage drop should not exceed 5%.

FACTORS THAT AFFECT VOLTAGE DROP

The four factors that affect the amount of voltage drop are conductor size, conductor material, the length of the conductors, and the load. The size of a conductor is determined by its diameter, which is measured in circular mils. Smaller diameter conductors have more resistance per 1′ than larger diameter conductors. The best conductor materials are silver, copper, gold, and aluminum. Silver and gold are not feasible for use with conductors because of their cost, but they are sometimes used for plating of contacts. Copper and aluminum are commonly used for conductors in electrical wiring. Copper conductors have less resistance than aluminum conductors. **See Figure 8-1.**

Conductor Resistance	
CONDUCTOR MATERIAL	RESISTANCE AT 75°C FOR 1 kcmil PER 1000′*
Silver	12.3
Copper	12.9
Gold	17.8
Aluminum	21.2

* in Ω

Figure 8-1. Materials that make the best conductors are silver, copper, gold, and aluminum. Because of cost, copper and aluminum are most commonly used for conductors in construction.

The amount of resistance in a conductor also varies with the length of the conductor. The longer the conductor, the more resistance there is in the total circuit. The load is not a part of a conductor, but it does affect the voltage drop of the circuit. The larger the load in amps, the greater the voltage drop will be.

REDUCING VOLTAGE DROP

Voltage drop occurs in both 1ϕ and 3ϕ circuits. An increase in current or resistance will increase the amount of voltage drop in the circuit. This is represented using Ohm's law ($E = I \times R$). If the resistance or the amperage in a circuit increases, the voltage drop increases. There are several methods used to decrease voltage drop. The most common and practical method is to increase the conductor size. The larger the diameter of the conductor, the lower the resistance. Other methods used to decrease voltage drop include the following:

- Reduce the load. The fewer amps in the circuit, the lower the voltage drop will be. This method is generally not practical because all equipment shall operate at the voltage and amperage for which it is rated. Per 110.3(B), equipment shall be installed by its labeling and listing.
- Shorten distances of the circuit. Shorter conductors decrease the amount of resistance in the circuit.
- Change conductor materials from aluminum to copper. Aluminum has a higher resistance than copper. Using copper conductors instead of aluminum will reduce the voltage drop of the circuit. Changing a copper conductor to aluminum is not an option.

TABLE 8—CONDUCTOR PROPERTIES

Table 8 of chapter 9 contains conductor properties for both copper and aluminum conductors. This table can be used to find the area of a conductor in circular mils. Knowing the size of a conductor in circular mils is important when calculating voltage drop. The conductor sizes are on the left-hand side of the table with the area in circular mils listed in the third column. The areas in circular mils are given for the conductors from No. 18 AWG to No. 4/0 AWG. Conductors 250 to 2000 are named to represent the area of the conductor in circular mils. For example, 250 kcmil is actually 250,000 circular mils, and 1000 kcmil is actually 1,000,000 circular mils.

In Table 8, under the quantity column, the number of strands per conductor is given as either 1 or 7 or more. The 1 represents a solid conductor, and the 7 or more represents the number of strands in a stranded conductor. While there are two different types of conductors for each size, the areas in circular mils are the same.

The uncoated and coated columns in Table 8 list the resistance of both copper and aluminum conductors per 1000′. All aluminum conductors are uncoated. Older copper conductors contained a thin coating that was used to bind the insulation to the conductor. This coating increased the resistance of the conductor. Copper conductors used today no longer have this coating. It is important to differentiate coated conductors from uncoated conductors, especially in older installations, in order to determine proper resistance calculations.

SINGLE-PHASE (1ϕ) BRANCH CIRCUIT CALCULATIONS

Circuits that are 1ϕ will have either two ungrounded conductors per circuit or a single ungrounded conductor and a grounded conductor. Either way, there are at least two conductors required to make the circuit work.

The two methods used to determine voltage drop for 1ϕ circuits are using Ohm's law or the voltage drop formula.

Using Ohm's Law to Calculate Voltage Drop

To determine voltage drop using Ohm's law, the resistance and current must be known. The voltage drop is determined by multiplying the resistance times the current.

Most questions on the exam will either give the resistance of a conductor per 1000′, or the size of the conductor will be given, which requires the resistance to be found by using Table 8 in chapter 9 of the NEC®. In Table 8, the resistance for copper conductors is listed in two columns. One column is for coated conductors and the other column is for uncoated conductors. The uncoated conductors will be used on the exam.

Once the resistance per 1000′ is found, the resistance is divided by 1000 to get the resistance for 1′ of conductor. The resistance for 1′ of conductor is multiplied by the total length of the two conductors in the circuit. To determine the voltage drop of the circuit, the total resistance for the length of both conductors is multiplied by the current of the load. **See Figure 8-2.** To determine the total resistance of a circuit using Table 8 of chapter 9, apply the following formula:

$$TR = \left(\frac{R}{1000}\right) \times (D \times 2)$$

where
TR = total resistance (in Ω)
R = resistance (in Ω) per 1000′
1000 = constant
D = distance (in ft) of one of the conductors in the circuit
2 = constant

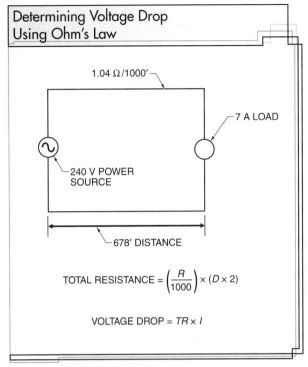

Figure 8-2. In order to determine the voltage drop using Ohm's law, the resistance and current must be known.

For example, what is the voltage drop of a circuit that has 1.04 Ω per 1000′ and is 678′ from the power source with a load of 7 A?

$$TR = \left(\frac{R}{1000}\right) \times (D \times 2)$$

$$TR = \left(\frac{1.04}{1000}\right) \times (678 \times 2)$$

$TR = 0.00104 \times 1356$

$TR = 1.41$

Use Ohm's law to determine voltage drop (VD).
$VD = TR \times I$
$VD = 1.41 \times 7$
$VD = \mathbf{9.87}$

Practice Question

Determining Voltage Drop Using Ohm's Law

1. A 120 V, 1ϕ circuit is 300′ away from the power source and has a load current of 20 A. The conductors are copper with a resistance of 0.87 Ω/1000′. What is the voltage drop of the circuit?

Using the Voltage Drop Formula to Determine Voltage Drop

The second method used to determine voltage drop for 1φ systems is to use the voltage drop formula. The voltage drop formula uses the specific resistance of the conductor, the distance the load is placed from the source, the conductor size, and the circuit load to determine the voltage drop. **See Figure 8-3.** To determine voltage drop for a 1φ system, apply the following formula:

$$VD = \frac{2 \times K \times I \times D}{CM}$$

where
VD = voltage drop (in V)
2 = constant
K = direct current (in Ω), 12.9 for copper conductors or 21.2 for aluminum conductors
I = current (in A) of the load
D = one way distance (in ft)
CM = conductor size (in cmil) from Table 8 of chapter 9

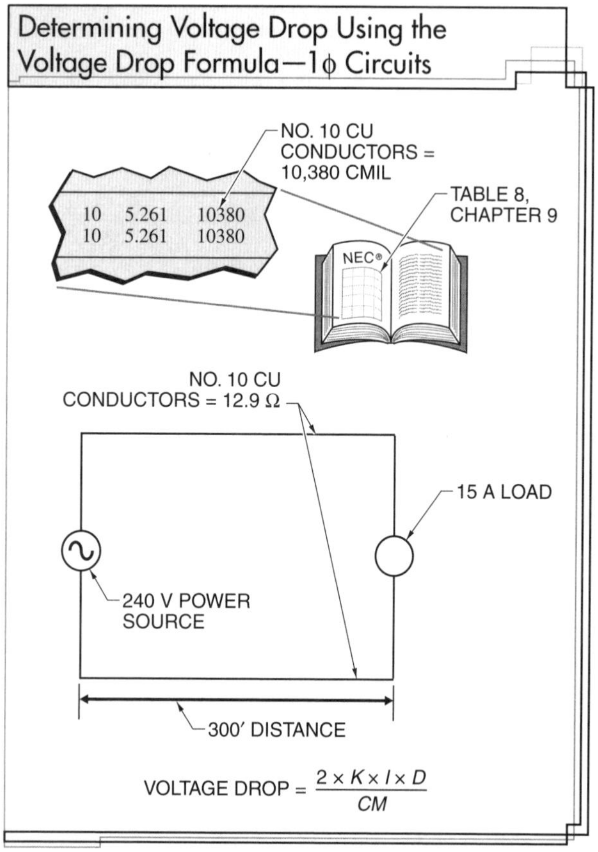

For example, what is the voltage drop for a circuit located 300′ from the source, that is carrying 15 A and using No. 10 copper conductors?

$$VD = \frac{2 \times K \times I \times D}{CM}$$

$$VD = \frac{2 \times 12.9 \times 15 \times 300}{10,380}$$

$$VD = \frac{116,100}{10,380}$$

$$VD = \mathbf{11.2\ V}$$

Practice Questions

Determining Voltage Drop for 1φ Circuits

1. What is the voltage drop for a 120 V load that is 200′ from the voltage source and draws 12 A on a No. 8 THHN copper conductor?

2. What is the voltage drop for a 480 V circuit that has No. 3 AWG aluminum conductors carrying 40 A for a distance of 300′?

Sizing Conductors for 1φ Circuits

One of the methods that may be used to reduce voltage drop is to increase the size of the conductors. The larger the diameter of the conductor, the lower the resistance on the circuit. The voltage drop formula variables can be rearranged to solve for the conductor size in circular mils. This formula will be used to find the correct conductor size for a circuit while keeping the recommended voltage drop of 3% or less.

In order to find the correct conductor size, the voltage drop must be known. Some questions on the exam will give a set percentage for the voltage drop. Other questions may give the source voltage and the load voltage. If this is the case, the difference of these two values is calculated to determine the voltage drop of the circuit. If there is no amount given in the problem, then 3% of the voltage source is used. **See Figure 8-4.**

Figure 8-3. Using the voltage drop formula is one way to determine the voltage drop of a 1φ circuit.

TECH TIP

The voltage drop of a circuit can be decreased by increasing conductor size, reducing the load, shortening the distance of the circuit, or changing the material of the conductor. The most common and practical method used to decrease voltage drop is to increase the conductor size. The larger the diameter of the conductor, the lower the resistance.

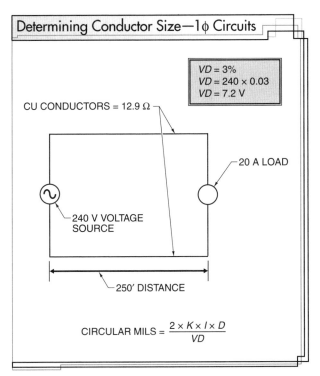

Figure 8-4. Variables from the voltage drop formula can be rearranged to determine the proper conductor size of 1φ circuits.

After the cmil of the conductor is known, the conductor size can be determined by using Table 8 in chapter 9. Check Table 310.15(B)(16) to see if the conductor is allowed to carry the current determined in the problem. With short distances, the conductor determined by the formula may be too small when compared to Table 310.15(B)(16). To determine the proper conductor size for a 1φ circuit, apply the following formula:

$$CM = \frac{2 \times K \times I \times D}{VD}$$

where
- CM = conductor size (in cmil) from Table 8 of chapter 9
- 2 = constant
- K = direct current (in Ω), 12.9 for copper conductors or 21.2 for aluminum conductors
- I = current of the load (in A)
- D = one way distance (in ft)
- VD = voltage drop (in V)

For example, what is the required THW conductor size for a circuit that is supplied from a 240 V source, has a load of 20 A, and is 250′ away from the power supply?

$$CM = \frac{2 \times K \times I \times D}{VD}$$

$$CM = \frac{2 \times 12.9 \times 20 \times 250}{240 \times 0.03}$$

$$CM = \frac{2 \times 12.9 \times 20 \times 250}{7.2}$$

$$CM = \frac{129,000}{7.2}$$

$$CM = 17,917 \text{ cmil}$$

See Table 8 from chapter 9.
17,917 cmil = **No. 6 conductors at 26,240 cmil**
Per Table 310.15(B)(16), No. 6 THW conductors can carry up to 65 A.

Practice Questions

Determining Conductor Size for 1φ Circuits

1. What size of copper conductors is needed to keep the voltage drop to 3% for a 120 V, 1φ circuit that is drawing a load of 17 A and is 150′ from the source?

2. What size of conductor is required to carry a 50 A, 480 V load that is 500′ from the panelboard? The circuit only allows a 2% voltage drop in order to operate properly.

Determining Maximum Load for 1φ Circuits

The maximum load that can be put on a circuit without passing the recommended 3% voltage drop can be determined using the voltage drop formula. The voltage drop formula variables can be rearranged to solve for the maximum load of a 1φ circuit. This formula will be used to determine the maximum load that can be put on a circuit with a predetermined conductor size and distance while keeping the recommended voltage drop of 3% or less. **See Figure 8-5.** To determine the maximum load a 1φ circuit can carry, apply the following formula:

$$I = \frac{CM \times VD}{2 \times K \times D}$$

where
- I = current of the load (in A)
- CM = conductor size (in cmil) from Table 8 of chapter 9
- VD = voltage drop (in V)
- 2 = constant
- K = direct current (in Ω), 12.9 for copper conductors or 21.2 for aluminum conductors
- D = one way distance (in ft)

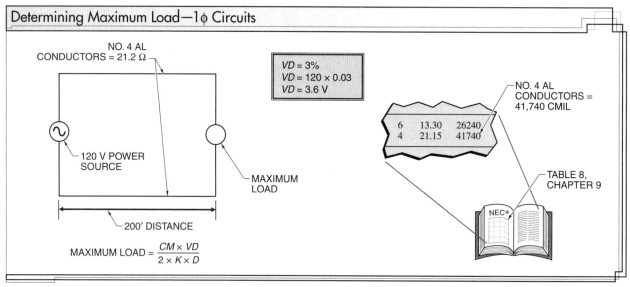

Figure 8-5. The maximum load that can be placed on a circuit without passing the recommended 3% voltage drop can be determined using the voltage drop formula. The variables from the voltage drop formula can be rearranged to determine the maximum load of 1φ circuits.

For example, what is the maximum load that can be placed on a No. 4 aluminum conductor that is located 200′ from a 120 V power source?

$$I = \frac{CM \times VD}{2 \times K \times D}$$

$$I = \frac{41{,}740 \times (120 \times 0.03)}{2 \times 21.2 \times 200}$$

$$I = \frac{41{,}740 \times 3.6}{2 \times 21.2 \times 200}$$

$$I = \frac{150{,}264}{8480}$$

$$I = \mathbf{17.7\ A}$$

Practice Question

Determining Maximum Load on a 1φ Circuit

1. What is the maximum load that can be placed on a circuit that is 300′ away from the source using No. 6 aluminum conductors on a 240 V circuit?

Determining Maximum Distance from the Power Source to Load for 1φ Circuits

The maximum distance that is allowed for a circuit without passing the recommended 3% voltage drop can be determined using the voltage drop formula. The voltage drop formula variables can be rearranged to determine the maximum distance a 1φ load can be from a power source. This formula will be used to determine the maximum distance a 1φ load can be from the power source with a predetermined conductor size, resistance, and current, while keeping the recommended 3% or less voltage drop. **See Figure 8-6.** To determine the maximum distance for a 1φ circuit, apply the following formula:

$$D = \frac{CM \times VD}{2 \times K \times I}$$

where
D = one way distance (in ft)
CM = conductor size (in cmil) from Table 8 of chapter 9
VD = voltage drop (in V)
2 = constant
K = direct current (in Ω), 12.9 for copper conductors or 21.2 for aluminum conductors
I = current of the load (in A)

TECH TIP

If the voltage drop variable is not given in an exam question, assume a maximum voltage drop of 3%. To determine the voltage drop value, multiply the voltage source by 3%.

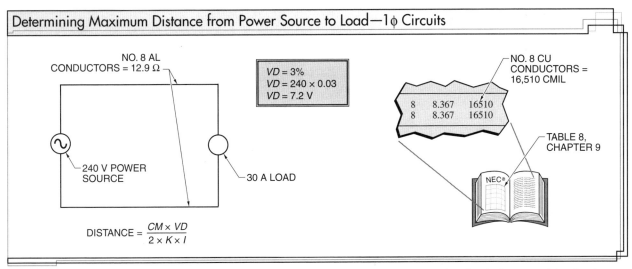

Figure 8-6. Variables from the voltage drop formula can be rearranged to determine the circuit distance of 1φ circuits.

For example, what is the maximum distance a load of 30 A can be placed from a 240 V power source when No. 8 conductors are used?

$$D = \frac{CM \times VD}{2 \times K \times I}$$

$$D = \frac{16{,}510 \times (240 \times 0.03)}{2 \times 12.9 \times 30}$$

$$D = \frac{16{,}510 \times 7.2}{2 \times 12.9 \times 30}$$

$$D = \frac{118{,}872}{774}$$

$$D = \mathbf{154'}$$

Practice Question

Determining Maximum Distance between a 1φ Power Source and Load

1. What is the maximum distance between a power supply and a load when the load is 240 V and draws 15 A on No. 12 AWG copper conductors?

THREE-PHASE (3φ) BRANCH CIRCUIT CALCULATIONS

The voltage drop on a 3φ circuit is calculated differently from that of a 1φ circuit because the 3φ circuit has three current-carrying conductors instead of two. The Ohm's law method cannot be used for 3φ circuits, so the voltage drop formula is used to perform calculations. The formula is the same except the constant number 2 is replaced with the square root of 3 (1.732). **See Figure 8-7.**

To determine the voltage drop for a 3φ circuit, apply the following formula:

$$VD = \frac{1.732 \times K \times I \times D}{CM}$$

where

VD = voltage drop (in V)
1.732 = constant
K = direct current (in Ω), 12.9 for copper conductors or 21.2 for aluminum conductors
I = current of the load (in A)
D = one way distance (in ft)
CM = conductor size (in cmil) from Table 8 of chapter 9

For example, what is the voltage drop for a 15 A circuit that is 300' from the power source and uses No. 10 copper conductors?

$$VD = \frac{1.732 \times K \times I \times D}{CM}$$

$$VD = \frac{1.732 \times 12.9 \times 15 \times 300}{10{,}380}$$

$$VD = \frac{100{,}542.6}{10{,}380}$$

$$VD = \mathbf{9.7\ V}$$

Practice Question

Determining Voltage Drop for 3φ Circuits

1. What is the voltage drop of a 480 V, 3φ motor that is drawing 52 A and is connected with No. 6 THHN copper conductors that are 250' long?

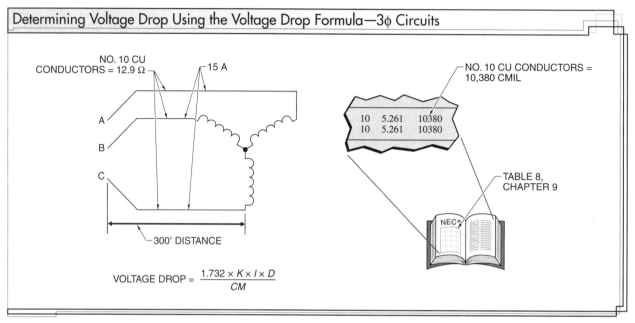

Figure 8-7. The square root of 3 (1.732) is the constant used in the voltage drop formula when determining the voltage drop of 3ϕ circuits.

Sizing Conductors for 3ϕ Circuits

The voltage drop formula variables for 3ϕ circuits can be rearranged to solve for the conductor size in circular mils. This formula will be used to find the correct conductor size for a circuit while keeping the recommended voltage drop of 3% or less. **See Figure 8-8.**

Ruud Lighting, Inc.
Voltage drop must be considered when designing electrical systems for large factories or warehouses because the circuits are more likely to have long runs.

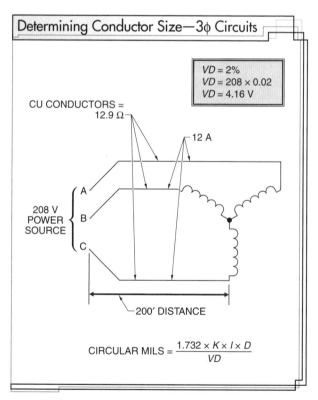

Figure 8-8. Variables from the voltage drop formula can be rearranged to determine the proper conductor size of 3ϕ circuits.

Once the conductor size is determined, it is important to check that the conductor can carry the load. Check Table 310.15(B)(16) to see if the conductor is allowed to carry the current found in the problem. To determine the proper conductor size for a 3ϕ circuit, apply the following formula:

$$CM = \frac{1.732 \times K \times I \times D}{VD}$$

where
CM = conductor size (in cmil) from Table 8 of chapter 9
1.732 = constant
K = direct current (in Ω), 12.9 for copper conductors or 21.2 for aluminum conductors
I = current of the load (in A)
D = one way distance (in ft)
VD = voltage drop (in V)

For example, what size of copper conductor is required to keep a 2% voltage drop for a 208 V, 3ϕ supply with a circuit that is 200′ away carrying 12 A?

$$CM = \frac{1.732 \times K \times I \times D}{VD}$$

$$CM = \frac{1.732 \times 12.9 \times 12 \times 200}{(208 \times 0.02)}$$

$$CM = \frac{1.732 \times 12.9 \times 12 \times 200}{4.16}$$

$$CM = \frac{53,623}{4.16}$$

$$CM = 12,890 \text{ cmil}$$

See Table 8 from chapter 9.

12,890 cmil = **No. 8 conductors at 16,510 cmil**

Per Table 310.15(B)(16), No. 8 conductors can carry up to 50 A.

Practice Question

Determining Conductor Size for 3ϕ Circuits

1. What size of copper conductors is needed to keep the voltage drop to 3% for a 240 V, 3ϕ circuit that is drawing a load of 17 A that is 150′ from the power source?

Determining Maximum Load for 3ϕ Circuits

The maximum load that can be placed on a circuit without passing the recommended 3% voltage drop can be determined using the voltage drop formula. The voltage drop formula variables can be rearranged to solve for the maximum load of a 3ϕ circuit. This formula will be used to find the maximum load that can be put on a circuit with a predetermined conductor size and distance while keeping the recommended voltage drop of 3% or less. **See Figure 8-9.** To determine the maximum load a 3ϕ circuit can carry, apply the following formula:

$$I = \frac{CM \times VD}{1.732 \times K \times D}$$

where
I = current of the load (in A)
CM = conductor size (in cmil) from Table 8 of chapter 9
VD = voltage drop (in V)
1.732 = constant
K = direct current (in Ω), 12.9 for copper conductors or 21.2 for aluminum conductors
D = one way distance (in ft)

For example, what is the maximum load that can be placed on a No. 6 aluminum conductor that is long enough to allow the load to be placed 250′ away from the 240 V, 3ϕ power source?

$$I = \frac{CM \times VD}{1.732 \times K \times D}$$

$$I = \frac{26,240 \times (240 \times 0.03)}{1.732 \times 21.2 \times 250}$$

$$I = \frac{26,240 \times 7.2}{1.732 \times 21.2 \times 250}$$

$$I = \frac{188,928}{9179.6}$$

$$I = \mathbf{20.58 \text{ A}}$$

Practice Question

Determining Maximum Load on a 3ϕ Circuit

1. What is the maximum load that can be placed on a circuit that is 300′ away from its source using No. 6 aluminum conductors on a 480 V, 3ϕ circuit?

TECH TIP

Conductors that are No. 8 and larger are stranded conductors. Conductors that are smaller than No. 8 may be stranded or solid. The measured area in circular mils from Table 8 is the same for both stranded and solid conductors.

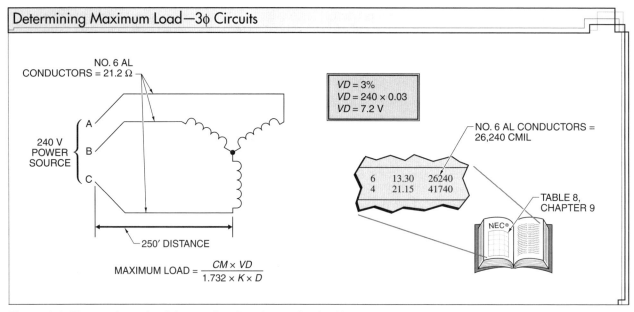

Figure 8-9. The maximum load that can be placed on a circuit without passing the recommended 3% voltage drop can be determined using the voltage drop formula.

Determining Maximum Distance from the Power Source to Load for 3ϕ Circuits

The maximum distance that is allowed for a circuit without passing the recommended 3% voltage drop can be determined using the voltage drop formula. The voltage drop formula variables can be rearranged to determine the maximum distance a 3ϕ load can be from a power source. This formula will be used with a predetermined conductor size, resistance, and current while keeping the recommended voltage drop of 3% or less. **See Figure 8-10.** To determine the maximum distance from the power source to the load for a 3ϕ circuit, apply the following formula:

$$D = \frac{CM \times VD}{1.732 \times K \times I}$$

where

D = one way distance (in ft)

CM = conductor size (in cmil) from Table 8 of chapter 9

VD = voltage drop (in V)

1.732 = constant

K = direct current (in Ω), 12.9 for copper conductors or 21.2 for aluminum conductors

I = current of the load (in A)

For example, what is the maximum distance from the power source a 25 A load can be placed using No. 4 copper conductors on a 480 V, 3ϕ circuit?

$$D = \frac{CM \times VD}{1.732 \times K \times I}$$

$$D = \frac{41{,}740 \times (480 \times 0.03)}{1.732 \times 12.9 \times 25}$$

$$D = \frac{41{,}740 \times 14.4}{1.732 \times 12.9 \times 25}$$

$$D = \frac{601{,}056}{558.57}$$

$$D = \mathbf{1076'}$$

Practice Question

Determining Maximum Distance between a 3ϕ Power Source and Load

1. What is the maximum distance that a 208 V, 3ϕ load that draws 15 A on No. 12 AWG copper conductors can be placed from the power supply?

DETERMINING POWER LOSS IN CIRCUITS

Not only is there a loss of voltage through conductors, there is a loss of power as well. Once the voltage drop is determined for a circuit, the loss of power can be determined by multiplying the voltage that was lost by the current of the circuit. The power loss is determined by using the power formula. **See Figure 8-11.** To determine the power loss in a 1ϕ circuit after voltage drop is determined, apply the following formula:

$P = E \times I$

where

P = power (in W)

E = voltage

I = current

TECH TIP

The current flowing through a wire must be kept below the wire's rated limit to maintain a safe system. Copper (Cu) is a better conductor (less resistant) than aluminum (Al) and may carry more current for any given AWG size. A large wire is a better conductor (less resistant) than a small wire and may also carry more current. High-temperature-rated insulation may also carry more current because a high temperature is required to break down the insulation material.

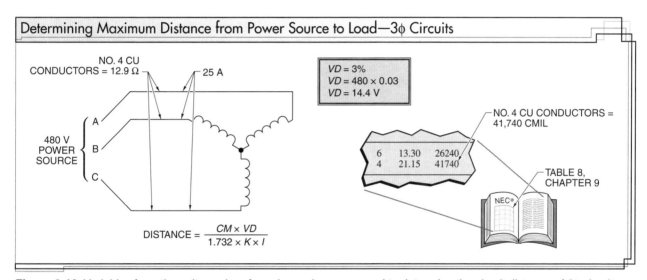

Figure 8-10. Variables from the voltage drop formula can be rearranged to determine the circuit distance of 3ϕ circuits.

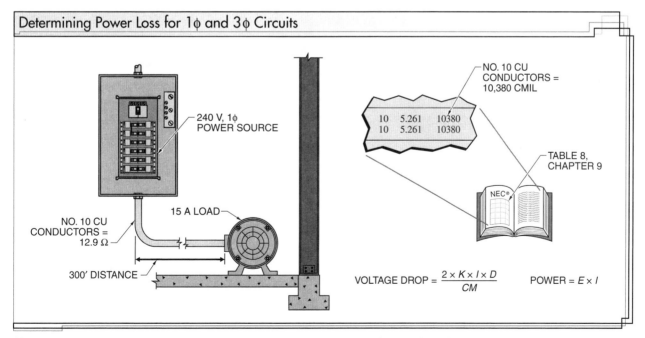

Figure 8-11. Voltage drop and current are used to determine the loss of power in a circuit.

For example, how much power is lost in a 15 A circuit that is 300′ from the source and is using No. 10 copper conductors?

First, determine the voltage drop:

$$VD = \frac{2 \times K \times I \times D}{CM}$$

$$VD = \frac{2 \times 12.9 \times 15 \times 300}{10,380}$$

$$VD = \frac{116,100}{10,380}$$

$$VD = 11.2 \text{ V}$$

Then, find the power loss:

$P = E \times I$
$P = 11.2 \times 15$
$P = \mathbf{168\ W}$

To determine the power loss in a 3ϕ circuit after voltage drop is determined, apply the following formula:

$P = E \times I \times 1.732$

where
P = power (in W)
E = voltage
I = current
1.732 = constant

For example, what is the power loss in a 3ϕ circuit that is 300′ from its source, carrying 15 A, and using No. 12 aluminum conductors?

First, find the voltage drop:

$$VD = \frac{1.732 \times K \times I \times D}{CM}$$

$$VD = \frac{1.732 \times 21.2 \times 15 \times 300}{6530}$$

$$VD = \frac{165,233}{6530}$$

$$VD = 25.3 \text{ V}$$

Then, find the power loss:

$P = E \times I \times 1.732$
$P = 25.3 \times 15 \times 1.732$
$P = \mathbf{657\ W}$

Practice Questions

Determining Power Loss for 1ϕ and 3ϕ Circuits

1. What is the power loss for a 120 V, 1ϕ load that is 200′ from the voltage source and draws 12 A on a No. 8 THHN copper conductor?

2. What is the power loss of a 480 V, 3ϕ motor that is drawing 52 A and is connected with No. 6 THHN copper conductors that are 250′ long?

VOLTAGE DROP FOR FEEDER AND BRANCH CIRCUITS COMBINED

When branch circuits are combined with feeders, the total voltage drop between them should be a maximum percentage of 5%. If the feeder has a 3% maximum voltage drop, then the branch circuit must not have more than a 2% maximum voltage drop. If the feeder has a 2% maximum voltage drop, then the branch circuit must not have more than a 3% maximum voltage drop. The total allowable voltage drop for a combined feeder and branch circuit varies depending on the voltage. **See Figure 8-12.**

Total Voltage Drop for Feeder and Branch Circuits Combined

VOLTAGE*	% OF VOLTAGE DROP	TOTAL VOLTAGE DROP*
120	5	6
208	5	10.4
240	5	12
480	5	24

* in V

Figure 8-12. As the power source increases, the total allowable voltage drop increases.

To perform a combined feeder and branch circuit calculation, there needs to be enough information to calculate the voltage drop for one of the two sides. Remember that the feeder or branch circuit part can have a voltage drop of less than 3% as long as the total combined voltage drop stays below 5%. **See Figure 8-13.**

For example, what size of branch circuit conductors is required for a 1ϕ circuit where the feeder has 30 A passing through No. 8 copper conductors, with a power supply of 240 V that is 100′ from the subpanel? The branch circuit load is 200′ from the subpanel.

$$VD = \frac{2 \times K \times I \times D}{CM}$$

$$VD = \frac{2 \times 12.9 \times 30 \times 100}{16,150}$$

$$VD = \frac{77,400}{16,150}$$

$$VD = 4.79 \text{ V}$$

$$VD\% = \frac{4.79}{240}$$

$$VD\% = \mathbf{2\%}$$

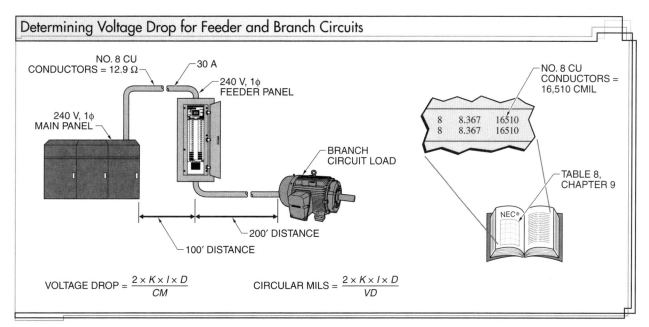

Figure 8-13. In a combined feeder and branch circuit calculation, the combined voltage drop percentage must not be more than 5%.

The voltage drop for the feeder is about 2%. That leaves an allowable 3% voltage drop for the branch circuit.

$$CM = \frac{2 \times K \times I \times D}{VD}$$

$$CM = \frac{2 \times 12.9 \times 30 \times 200}{(240 \times 0.03)}$$

$$CM = \frac{2 \times 12.9 \times 30 \times 200}{7.2}$$

$$CM = \frac{154,800}{7.2}$$

$$CM = 21,500$$

See Table 8 from chapter 9.

12,890 cmil = **No. 6 conductors at 26,240 cmil.**

Per Table 310.15(B)(16), No. 6 conductors can carry up to 65 A.

Practice Question

Determining Voltage Drop Calculations for Feeders and Branch Circuits

1. A feeder located 135′ away from a panel supplying 240 V is drawing 100 A. The branch circuit, located 150′ from the subpanel, is also drawing 100 A and is using No. 1/0 AWG copper conductors. What size of conductors is required for the feeders?

CHAPTER 8

Voltage Drop

Review Questions

Name _____   Date _____

_____ 1. The maximum load that can be placed on a 115 V, 1ϕ branch circuit using No. 12 THHN solid conductors that are 150′ long is ___ A.
   A. 5.00
   B. 5.82
   C. 6.45
   D. 8

_____ 2. A 60 A branch circuit located 150′ from a 240 V panel requires No. ___ aluminum THW conductors.
   A. 2
   B. 3
   C. 4
   D. 6

_____ 3. A 480 V, 3ϕ branch circuit should have a voltage drop of no more than ___ V.
   A. 12
   B. 14.4
   C. 24
   D. 25

_____ 4. The maximum distance that a 10 A, 120 V load can be placed from the power supply using No. 12 copper conductors is ___′.
   A. 91
   B. 45.5
   C. 126
   D. 182

_____ 5. The voltage drop on a branch circuit for an 11 A load that is located 175′ away from a 208 V, 3ϕ power source using No. 8 THWN aluminum conductors is ___ V.
   A. 2.8
   B. 4.5
   C. 4.28
   D. 5.34

6. A circuit that is 300′ in total length, carrying 5 A with a conductor resistance of 4 Ω per 1000′ will have a voltage drop of ___ V.
   A. 1.5
   B. 3
   C. 6
   D. 8

7. A maximum voltage drop of ___ V is recommended for a branch circuit supplying 240 V.
   A. 3.6
   B. 6.24
   C. 7.2
   D. 12

8. A 40 A, 240 V, 1φ noncontinuous load is located 100′ from the panel. If the voltage drop is 3%, then No. ___ copper conductors are the minimum size required for the load.
   A. 2
   B. 4
   C. 6
   D. 8

9. A 240 V branch circuit that uses No. 10 copper conductors supplying a load 200′ away from the power source will have a maximum load of ___ A.
   A. 14.5
   B. 16.7
   C. 20
   D. 30

10. A 2-wire circuit that carries 12 A located 300′ from a 240 V power source using No. 8 THHN copper conductors has a voltage drop of ___ V.
    A. 4.87
    B. 5.6
    C. 9.25
    D. 12

CHAPTER 9

Motors

Electric motors are the workhorses of today's manufacturing and processing plants. Electric motors can be very powerful and require protective devices to prevent them from overheating. A gasoline engine that has a load placed on it that is excessively large will slow down and stop. This protects the engine from damaging itself. An electric motor that has an excessively large load will continue to operate by drawing more current than it is designed to handle. This overload causes the electric motor to overheat and fail. Electric motors, along with their conductors and equipment, need protection against overloads, short circuits, and ground faults. The methods of sizing and protecting a motor are found in Article 430 of the National Electric Code® (NEC®).

OBJECTIVES

After completing this chapter, the learner will be able to do the following:

- Identify basic motor requirements as applied by the NEC®.
- Determine appropriate protection required for various types of motors.
- Size branch circuit and feeder conductors.
- Determine a motor's volt amp input.
- Calculate synchronous speed.
- Calculate motor torque and horsepower.
- Determine motor nameplate amperage.
- Identify common motor connections.
- Determine motor control circuit conductor protection.
- Explain the significance of locked-rotor code letters on motor nameplates.
- Size conductors for variable-speed drives.

MOTOR CALCULATIONS

Article 430 covers motor installation. The calculations that pertain to Article 430 include finding the motor full-load current, sizing the branch-circuit conductors, running overload protection, and short-circuit ground fault protection. *Motor full-load current (FLC)* is the current required by a motor to produce full-load torque at the motor's rated speed. Section 430.6(A)(1) states that the motor full-load current (FLC) tables are used in all the calculations except for overloads. Overload protection requires the use of the motor nameplate current, unless none is given. Multispeed motors have more than one current, which is listed on the motor nameplate. Motors are also considered to be of continuous duty unless otherwise mentioned.

Feeder conductors and feeder short-circuit and ground fault protection are also topics that appear on exams. Motors will be 1φ unless the question indicates otherwise. The motor nameplate and full-load current are needed to perform motor calculations. Motor full-load current is found on the following tables:

- 430.247—for DC motors
- 430.248—for 1φ AC motors
- 430.249—for 2φ AC motors
- 430.250—for 3φ AC motors

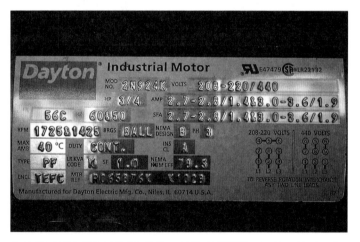

A motor nameplate contains data used for installation and troubleshooting purposes.

Each table lists the horsepower on the left-hand side and the voltage across the top. Table 430.250, which is for 3φ motors, uses this format as well, but it also lists different types of motors. Starting on the left side of the chart up to 2300 V, the table is for induction and wound rotor motors. The right side of the table is for synchronous motors at a 100% power factor. If a synchronous motor operates at a power factor of 90%, the chart current must be increased by 10%. If the motor operates at an 80% power factor, the chart current must be increased 25%.

Forms can be used to simplify the process of performing motor calculations. **See Figure 9-1.** Forms may be broken into six sections. The first four sections are used to determine motor full-load current, branch-circuit conductor size, motor standard and maximum overload protection, and branch-circuit ground fault and short-circuit protective devices. The last two sections are used to size conductor and feeder protection. **See Appendix.**

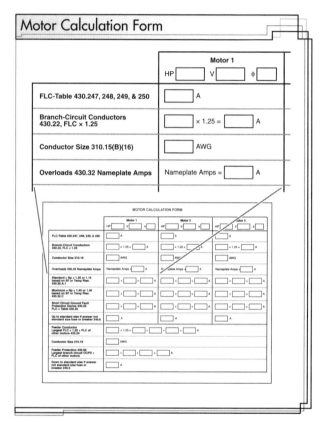

Figure 9-1. Using forms simplifies the process of performing motor calculations.

Determining Motor Full-Load Current

To determine motor FLC, first determine the appropriate table to use by determining the type of motor. For example, if the motor is 1φ, Table 430.248 would be used. Next, determine the voltage (V) and horsepower (HP). The FLC of the motor is found by following the appropriate HP row until it intersects with the appropriate V column. For example, a 10 HP, 230 V motor has an FLC of 50 A. **See Figure 9-2.**

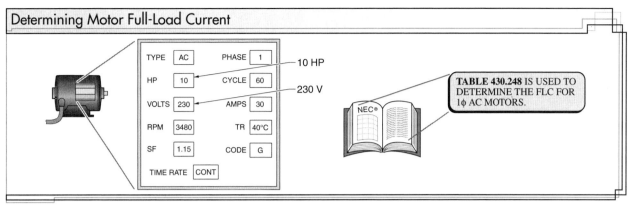

Figure 9-2. Per Table 430.248, the FLC for a 10 HP, 230 V motor is 50 A.

The tables used most often to calculate motor FLC are Table 430.248 for 1ϕ motors up to 10 HP and Table 430.250 for 3ϕ induction and synchronous motors. Table 430.250 consists of two parts. Induction motors with voltages from 115 V to 2300 V make up the first part, and synchronous motors with voltages from 230 V to 2300 V make up the second part. *Note:* The asterisk in the heading for synchronous motors indicates that synchronous motors with a power factor of 90% or 80% must have the FLC increased by 10% or 25%, respectively.

For example, what is the FLC of a 200 HP, 460 V, 3ϕ induction motor?
See Table 430.250 for 3ϕ motors.
FLC = **240 A**

Practice Questions

Determining Motor Full-Load Current
1. What is the FLC of a 5 HP, 208 V, 1ϕ induction motor?
2. What is the FLC of a 100 HP, 460 V, 3ϕ induction motor?
3. What is the FLC of a 100 HP, 230 V, 3ϕ synchronous motor with a power factor of 90%?

DETERMINING BRANCH-CIRCUIT CONDUCTOR SIZE

A *branch-circuit conductor* is a conductor that makes up the portion of an electrical circuit between the circuit breaker or fuse and the loads connected in the circuit. Branch-circuit conductor calculation is based on whether a motor is rated for continuous or other than continuous duty as determined by the motor nameplate.

Determining Conductor Size for Continuous-Duty Motors

To size single motor conductors, the FLC must be increased by 25% to find the ampacity according to Table 430.22. This is done by multiplying the FLC by 125% (1.25). For example a 50 HP, 460 V, 3ϕ motor has an FLC of 65 A according to Table 430.250. Multiplying 65 by 1.25 gives an ampacity of 81.25. Table 310.15(B)(16) is used to calculate conductor size. The temperature ratings for the conductors are listed across the top of the table. The conductor sizes are listed along the left side of the table.

To find the correct conductor size, follow the appropriate temperature rating down until the appropriate ampacity is found. The correct conductor size is found in the left column. For example, a 75°C THW conductor with an ampacity of 85 would require a 4 AWG conductor. If the ampacity equals a number between two numbers listed on the table, then the higher number is used. For example, a 75°C THW conductor with an ampacity of 81.25 falls between 65 and 85. The larger ampacity (85) is used. **See Figure 9-3.**

The voltage drop is calculated if the motor is a distance away from the source. The current on the motor nameplate is used for calculating voltage drop. A 3% voltage drop is allowed on branch circuits according to 210.19(A) Informational Note No. 4. The calculation that gives the largest conductor size should be used.

To find the current rating to size conductors for a continuous-duty motor, apply the following formula:
$I = FLC \times 125\%$
where
I = current (in A)
FLC = full-load current (in A)
125% = constant

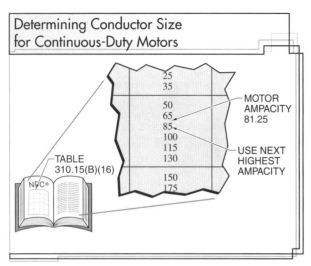

Figure 9-3. If the motor ampacity falls between two numbers listed in Table 310.15(B)(16), the larger ampacity is used.

For example, what is the conductor size for a 75°C rated conductor supplying power to a 20 HP, 208 V, 3ϕ continuous-duty motor?

$I = FLC \times 125\%$
$I = 59.4 \times 125\%$
$I = 59.4 \times 1.25$
$I = 74.25$ A

See Table 310.15(B)(16).
74.25 A = No. 4 AWG
Conductor size = **No. 4 AWG**

Practice Question

Determining Conductor Size for Continuous-Duty Motors

1. What size of 75°C rated copper conductor is required for a 100 HP, 460 V, 3ϕ induction motor?

Multispeed Motors. Per 430.22 (B), multispeed motor conductors on the line side of the motor controller shall be sized at 125% of the highest nameplate amperage of the different speeds. On the load side of the motor controller, each speed's nameplate amperage will be increased 125% to size the conductors.

Wye-Delta and Part Winding Motors. Per 430.22(C), wye-delta motors if used with a wye-delta motor starter require at least six conductors from the starter to the motor. These conductors, since there are six of them, are sized at 72% of the motor's FLC. Per 430.22 (D), part winding motors require six conductors, three for each half of the motor windings. These conductors are sized at 62.5% of the motor's FLC.

Determining Conductor Size for Other Than Continuous-Duty Motors

Motors are considered to be continuous duty unless otherwise marked. Subsection 430.22(E) and Table 430.22(E) refer to motors that are not continuous duty. Motors that are not continuous duty are short-time duty, intermittent duty, periodic duty, and varying duty motors. These motors are rated for time, such as 5 min, 15 min, 30 min, 60 min, and continuous duty. The basic rule for motor conductors is to size the conductor at 125% of the FLC. For other than continuous-duty motors, the current on the motor nameplate can be multiplied by a percentage lower than 125%. The percentage depends on the type of other than continuous motor. According to Table 430.22(E), a 50 A, short-time duty motor rated for 5 min can be sized at 110%. **See Figure 9-4.**

For example, find the conductor size required for a 3ϕ, 50 A, short-time duty motor rated for 15 min.

$I = $ amps on nameplate $\times 120\%$
$I = 50 \times 120\%$
$I = 50 \times 1.20$
$I = 60$ A

See Table 310.15(B)(16).
60 A = No. 6 AWG
Conductor size = **No. 6 AWG**

Practice Question

Determining Conductor Size for Other Than Continuous-Duty Motors

1. What size of THW copper conductor is required for a 100 A short-time duty motor rated for 15 min?

Determining Conductor Size for Wye-Start, Delta-Run Motors and Part-Winding Motors

Wye-start, delta-run motors and part-winding motors are used to reduce the starting current of the motors when they are started. Using either of these two types of motors will reduce the starting current as much as

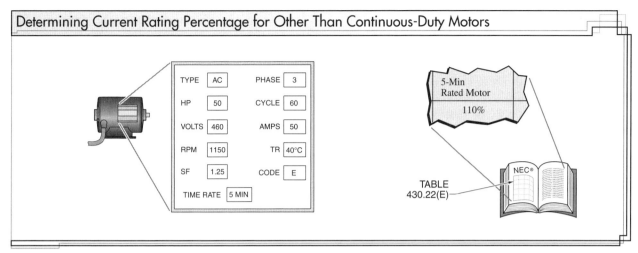

Figure 9-4. Table 430.22(E) is used to find nameplate current rating percentages.

33%. The conductors from the OCPD to the motor controller shall be sized at 125% for both motor types. Per 430.22(C) wye-start, delta-run motor conductors from the load side of the controller to the motor shall be sized at 72% of the motors FLC. Per 430.22(D) part-winding motor conductors from the load side of the controller to the motor shall be sized at 62.5% of the motors FLC.

Conductors for Small Motors

Per 430.22(G), conductors for small motors must not be less than a No. 14 AWG unless allowed per 430.22G(1) and (2). This subsection allows the use of No. 18 and No. 16 AWG copper conductors for small motors under certain circumstances. The conductors must be in multiwire cables or cords, or can be single conductors in enclosures or cabinets.

TECH TIP

Motors must be protected from short circuits, ground faults, and overloads. Fuses, circuit breakers, and overload relays are typically used to protect motors. Fuses and circuit breakers are used for ground-fault and short-circuit protection. Overload relays are used for overload protection. When properly selected, these devices provide a safety link in the motor circuits. They disconnect the circuits when there is a problem and are the first place to start when troubleshooting motors or motor circuits.

Per 430.22G(1), No. 18 AWG conductors can be used for motor circuits with an ampacity of no more than 5 A as long as the following conditions apply:
- The circuit is properly protected per 430.52.
- The circuit has Class 10 overload protection per 430.32.
- The circuit has proper overcurrent protection per 240.4(D)(1)(2).

No. 18 AWG conductors can be used for motor circuits with an ampacity of no more than 3.5 A as long as the following conditions apply:
- The circuit is properly protected per 430.52.
- The circuit has Class 20 overload protection per 430.32.
- The circuit has proper overcurrent protection per 240.4(D)(1)(2).

Per 430.22G(2), No. 16 AWG conductors can be used for motor circuits with an ampacity of no more than 8 A as long as the following conditions apply:
- The circuit is properly protected per 430.52.
- The circuit has Class 10 overload protection per 430.32.
- The circuit has proper overcurrent protection per 240.4(D)(2)(2).

No. 16 AWG conductors can be used for motor circuits with an ampacity of no more than 5.5 A as long as the following conditions apply:
- The circuit is properly protected per 430.52.
- The circuit has Class 20 overload protection per 430.32.
- The circuit has proper overcurrent protection per 240.4(D)(2)(2).

OVERLOAD CALCULATIONS FOR MOTORS

Motor overload devices (heaters) protect a motor and conductors from loads that are larger than the rating of the motor. *Single phasing* is the operation of a motor that is designed to operate on three phases but is only operating on two phases because one phase is lost. A motor draws 1.732 times its running current under this condition.

If the motor draws excessive current, it overheats before shutting down. Overload protective devices are used to prevent overheating but not for protecting against short circuits or ground faults. The current on the motor nameplate is used when sizing overload protective devices. Overload protective devices are sized for standard or maximum size. During the exam, the motor current is given or an actual motor nameplate may be shown.

Overload Devices

Overload devices are sized using the motor service factor (SF) and temperature rise (TR). *Service factor (SF)* is a number that represents the percentage of extra demand that can be placed on a motor for short time intervals without damaging the motor. The SF is given as a decimal such as 1.15. If the SF is 1.15 or greater, the motor overloads are sized at 125% of the motor nameplate.

Temperature rise (TR) is the amount of heat a motor generates without destroying the motor insulation. A common temperature rise is 40°C. If the motor has a temperature rise rating of 40°C or less, the overload is sized at 125% of the motor current. If the SF and TR are both outside the given limits, more than 40°C and less than 1.15 respectively, or if the motor nameplate does not have any SF or TR information, then the overloads shall be sized at 115%. Only one of the two factors must be within the limits to size the overloads at 125%. **See Figure 9-5.** Once the nameplate rating has been multiplied by the correct percentage, the answer shall not be rounded up or down to a standard size. Overloads come in many different sizes and the closest one shall be used.

To find the overload size for a motor, apply the following formula:

$$OS = I \times 125\%$$

where

OS = overload size (in A)

I = current (in A)

125% = constant

For example, what size of overload is required for a 100 HP, 480 V, 3φ motor when the current on the nameplate is 118 A? The motor has a service factor of 1.15 and a temperature rise of 50°C.

$$OS = I \times 125\%$$
$$OS = 118 \times 125\%$$
$$OS = 118 \times 1.25$$
$$OS = \mathbf{147.5\ A}$$

> **Practice Question**
>
> *Determining Motor Overload Protection Size*
>
> 1. What size of overload is required for a 5 HP, 208 V, 3φ induction motor with a service factor of 1.15? The current on the nameplate reads 15 A.
>
> 2. What is the required overload protection for a 25 HP, 460 V, 3φ motor with a temperature rise of 40°C? The current on the nameplate reads 30 A.
>
> 3. What is the required overload protection for a 25 HP, 460 V, 3φ motor with a temperature rise of 50°C? The current on the nameplate reads 30 A.

Higher-Size Overload Devices

Overload devices that are not large enough to start or carry the motor load are permitted to be increased as listed in 430.32(C) for motors of 1 HP or higher. The two factors used to determine the larger overloads are the SF and TR. If the SF is 1.15 or higher, the motor overloads are sized at 140% of the motor current. If the motor has a TR rating of 40°C or less, the motor overloads are sized at 140% of the motor current.

If the SF and TR are both outside the given limits, more than 40°C and less than 1.15 respectively, or if the motor nameplate does not have any SF or TR information, then the overloads shall be sized at 130%. Only one of the two factors must be within the limits to size the overloads at 140%. **See Figure 9-6.**

To find the higher-size overload for a motor, apply the following formula:

$$OS = A \times \%FLC$$

where

OS = overload size (in A)

I = current (in A)

$\%FLC$ = percentage of full-load current

Chapter 9—Motors

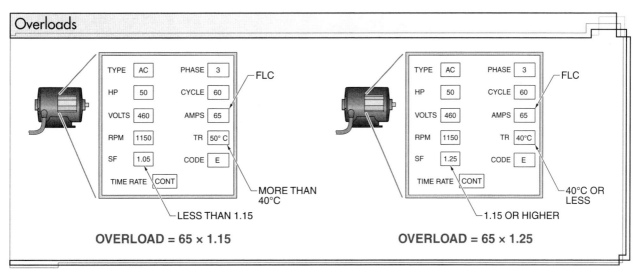

Figure 9-5. If the service factor is 1.15 or greater, the motor overloads are sized at 125% of the motor nameplate rating.

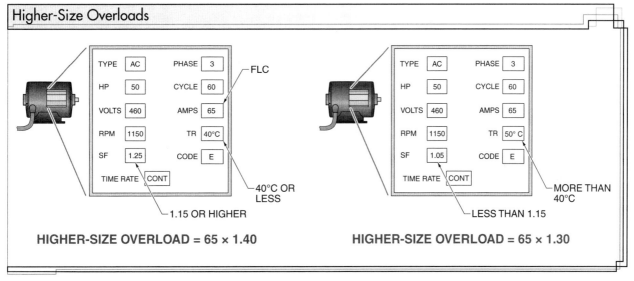

Figure 9-6. If a motor has a temperature rise rating of 40°C or less, the higher-size overload is sized at 140% of the motor's nameplate.

For example, what size of overload is required for a 100 HP, 480 V, 3ϕ motor with a temperature rise of 50°C? The current on the nameplate reads 118 A and the service factor is 1.15.

$OS = I \times \%FLC$
$OS = 118 \times 140\%$
$OS = 118 \times 1.40$
$OS = \mathbf{165.2\ A}$

Practice Question

Determining the Higher-Size Overload

1. What is the required overload protection for a 5 HP, 208 V, 3ϕ induction motor with a service factor of 1.15? The motor nameplate current is 15 A.

2. What is the required overload protection for a 25 HP, 460 V, 3ϕ motor with a temperature rise of 40°C? The current on the nameplate reads 30 A.

Integral Thermal Protectors

An integral thermal protector is an overload built into motor windings. Integral thermal protectors are listed in 430.32(A)(2). They are sometimes referred to as P-leads.

Integral thermal protectors are connected to the control circuit of the motor starter. On small motors, the integral thermal protectors carry the current of the motor when running. These integral thermal protectors require a higher percentage rating than other overloads. The current used for sizing integral thermal protectors is the FLC, which is found in Tables 430.248, 430.249, and 430.250. After the full-load current is determined, apply one of the following percentages:

- 0 A–9 A = 170% of the FLC of the motor.
- 9.1 A–20 A = 156% of the FLC of the motor.
- 20.1 A and up = 140% of the FLC of the motor.

To find the current required to size a thermal protector, apply the following formula:

$ITP = FLC \times \%FLC$

where

ITP = integral thermal protectors (in A)

FLC = full-load current (in A)

$\%FLC$ = percentage of full-load current

For example, what size of thermal protector is required for a 25 HP, 208 V, 3ϕ motor?

$ITP = FLC \times \%FLC$
$ITP = 74.8 \times 140$
$ITP = 74.8 \times 1.40$
$ITP = \mathbf{104.72\ A}$

Practice Question

Determining Integral Thermal Protector Size

1. What size of integral thermal protector is required for a 50 HP, 460 V, 3ϕ motor?

Number of Required Overloads

Section 430.37 lists the number of required overloads, location of the overloads, and the type of power supply for a particular motor. If the proper number of overloads is not installed, the motor may not be fully protected.

BRANCH SHORT-CIRCUIT/GROUND FAULT PROTECTION CALCULATION

According to 430.52(A), all motor branch circuits shall be protected in accordance with 430.52(B) and 430.52(C). Subsection 430.52(B) states that all motor short-circuit and ground fault protective devices shall be capable of carrying the starting current of the motor.

According to 430.52(C), protective devices shall be sized to not exceed the values calculated using the percentages found in Table 430.52. Table 430.52 lists four types of overcurrent protective devices. These overcurrent protective devices include nontime-delay fuses, dual-element (time-delay) fuses, instantaneous trip breakers, and inverse time circuit breakers. The standard rating for nontime-delay fuses is 300%, or 150% for wound rotor motors and direct current motors. The standard rating for dual-element fuses is 175% for all motors except wound-rotor motors and DC motors, which are rated at 150%. Instantaneous breakers can only be used if adjustable and part of a listed combination motor controller. This type of breaker is not commonly used on the exam. The standard rating is 800% except for design B motors, which are rated at 1100%, or for DC motors, which are rated at 250%. The standard rating for inverse time circuit breakers is 250%, except for wound rotor and DC motors, which are rated at 150%.

According to Exception 430.52(C)(1), Ex. 1, if a protective device value from Table 430.52 does not match a standard device rating from the list in 240.6(A), then it is acceptable to use the protective device that is the next size up.

Exception No. 2 refers to motors that will not start or stay running on the protective device sized at the next higher standard. Percentages are given for each type of protective device, and the next lower standard size must be used when applying this exception. This exception is based on the type of protective device used. The increases in percentages are as follows:

- A nontime-delay fuse increases from 300% to 400%.
- A time-delay fuse increases from 175% to 225%.
- An inverse time circuit breaker increases from 250% to 400% for breakers of 100 A or less and from 250% to 300% for breakers larger than 100 A. A fuse with the classification of 601-6000 shall be permitted to be increased to no more than 300%.

To find the current required to size a circuit breaker, apply the following formula:

$I = FLC \times \%FLC$

where

I = current (in A)

FLC = full-load current (in A)

$\%FLC$ = percentage of full-load current

For example, what size of inverse time circuit breaker is required for a 75 HP, 240 V, 3ϕ motor?

$I = FLC \times \%FLC$
$I = 192 \times 250$
$I = 192 \times 2.50$
$I = 480$ A

Use next higher standard size found in 240.6(A).

Inverse time circuit breaker size = **500 A**

To size for a time-delay fuse, multiply the FLC by 175% according to Table 430.52.

$I = FLC \times \%FLC$
$I = 192 \times 175$
$I = 192 \times 1.75$
$I = 336$ A

Use the next higher standard size found in 240.6(A).

Protective device size = **350 A**

Practice Question

Determining Branch/Short-Circuit Ground Fault Protection

1. What is the required short-circuit ground fault protective device for a 25 HP, 460 V, 3ϕ motor with a temperature rise of 40°C using a time-delay fuse?
2. What size of inverse time circuit breaker is required for a 100 HP, 460 V, 3ϕ induction motor?

FEEDER CONDUCTOR SIZING CALCULATION

A *feeder conductor* is a conductor that carries more than one load at a time. The feeder can supply a motor control center (MCC), other non-motor loads, or motors using the tap rules. The tap rules allow a smaller conductor to be tapped from a larger conductor. **See Figure 9-7.** The exam will have questions referring to motors of the same voltage and phase, and motors with a different voltage and phase.

Feeder Conductor Sizing for Motors of the Same Voltage

Feeder conductors supply power to more than one load. Section 430.24 states that the motor with the highest current shall determine the largest motor. The largest current is increased by 125%, and the sum of the other motor's FLC at 100% is added to provide the total load on the conductors. If non-motor loads are connected, then all noncontinuous loads are added in at 100% of their ratings. Continuous loads are added in at 125% of their ratings. To find the correct wire size, Table 310.15(B)(16) is used. Find the correct insulation type and move down the column until the calculated ampacity is determined. To find the current required to size feeder conductors, apply the following formula:

$I = LM \times \%FLC + M_1 + M_2$

where
I = current (in A)
LM = FLC of largest motor (in A)
$\%FLC$ = percentage of full-load current
M_1 = FLC of motor 1 (in A)
M_2 = FLC of motor 2 (in A)

For example, what is the current and THW wire size needed for the following motors on a feeder? Motor 1 is 25 HP, 480 V, 3ϕ. Motor 2 is a 50 HP, 480 V, 3ϕ. Motor 3 is 40 HP, 480 V, 3ϕ.

$I = LM \times \%FLC + M_1 + M_2$
$I = 65 \times 125 + 52 + 34$
$I = 65 \times 1.25 + 52 + 34$
$I = 81.25 + 52 + 34$
$I = \mathbf{167.25}$ **A**

See Table 310.15(B)(16).
Conductor size = **2/0 THW**

Practice Question

Determining Feeder Conductor Size for Motors of the Same Voltage

1. What is the required current needed to size a feeder for a 100 HP, 230 V, 3ϕ motor and a 50 HP, 230 V, 3ϕ motor?
2. What size of THW copper conductor is required for the feeder in question 1?

Power distribution equipment is used to measure, step down, and distribute all incoming power.

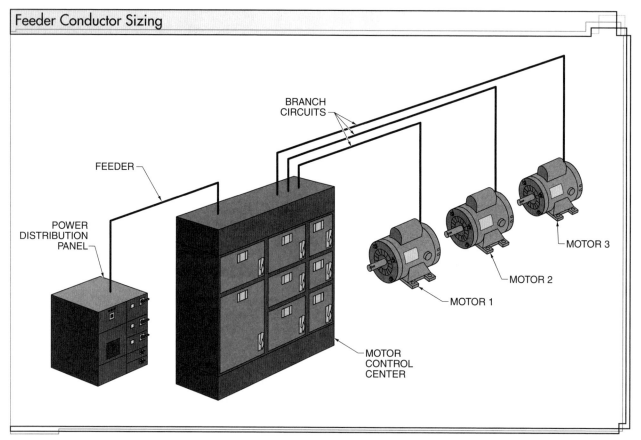

Figure 9-7. Feeder conductors carry more than one load at a time.

TECH TIP

In a 3ϕ motor, there are three stator windings for each pole. The windings are combined to provide the required number of poles. A common four-pole stator design has 36 slots, so each pole has nine windings. The nine windings are placed in three groups of three, with each group connected in series for each phase.

Feeders Supplying Three-Phase (3ϕ) and Single-Phase (1ϕ) Motors

Motor loads of 240/120 V, 1ϕ systems or 208/120 V, 3ϕ systems should be balanced as much as possible. On a 240/120 V, 1ϕ system, a 240 V motor can be placed across two ungrounded conductors. A 240 V motor requires both phases to self-balance. A 120 V motor can be connected to one of the two phases and the neutral. Two 120 V motors are placed on each phase to balance the current. If the motors are the same size, then one 120 V motor can be placed on phase A and the other 120 V motor can be placed on phase B. If the motors are of different sizes, they may have to be moved from one phase to another in order to be balanced.

A 3ϕ motor will self-balance and can be placed across all three ungrounded phases. Motors can be rated at 208 V, 1ϕ as well as 208 V, 3ϕ. Three 208 V, 1ϕ motors can be balanced by using phases A and B for the first motor, phases B and C for the second motor, and phases A and C for the third motor. Three 120 V, 1ϕ motors can be balanced by connecting across one of the ungrounded phases and the neutral. Each of the three motors must be connected to a different phase in order to be balanced. **See Figure 9-8.** This also helps balance the panel and the load on the transformer. If the motors are of different sizes, they may have to be moved from one phase to another in order to balance out. Once the motors are balanced, the phase with the highest current is used to size the conductors.

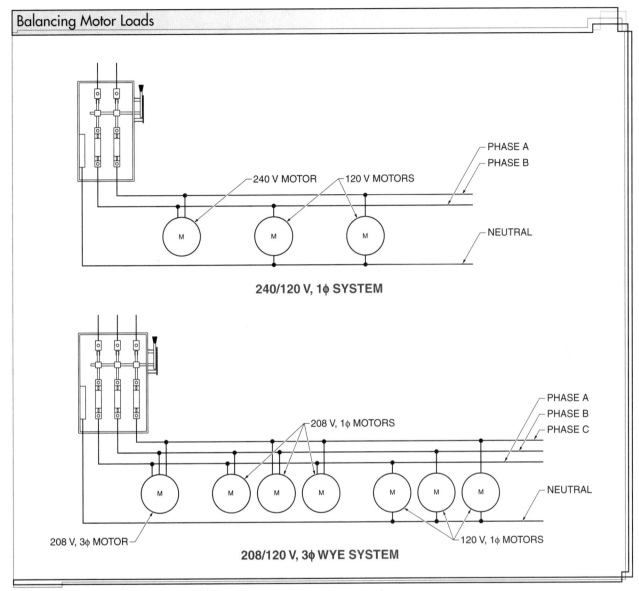

Figure 9-8. Motor loads should be balanced as much as possible across all phases.

Feeder Short-Circuit Ground Fault Protection Calculation

The short-circuit ground fault protective device protects the feeder conductors and allows the motors to start and run. Section 430.62 gives the procedure for sizing the short-circuit ground fault protective device.

The device is sized based on the largest branch-circuit protective device of the individual motors. When there are two branch-circuit protective devices of the same size, one device is chosen to be the largest. The FLC of the other motors on the feeder are added.

If the calculated current is not a standard size found in 240.6, the next lower size shall be used. To find the current required to size a short-circuit ground fault protective device, apply the following formula:

$$I = LM \times \%FLC + M_1 + M_2$$

where

I = current (in A)

LM = FLC of largest motor (in A)

$\%FLC$ = percentage of FLC

M_1 = FLC of motor 1 (in A)

M_2 = FLC of motor 2 (in A)

For example, what size of time-delay fuse is required for the feeder for the following motors?

Motor 1 is 25 HP, 480 V, 3ϕ. Motor 2 is 50 HP, 480 V, 3ϕ. Motor 3 is 40 HP, 480 V, 3ϕ.

$I = LM \times FLC\% + M_1 + M_2$
$I = 65 \times 1.75 + 52 + 34$
$I = 113.75 + 52 + 34$

According to 240.6, the next standard-size fuse higher than 113.75 A is a 125 A fuse.
$I = 125 + 52 + 34$
$I = 211$

According to 240.6(A), the next standard-size fuse lower than 211 A is a 200 A fuse.
size of fuse = **200 A**

Practice Question

Determining Short-Circuit Ground Fault Protection

1. What size of time-delay fuse is required to protect a feeder with a 100 HP motor and a 50 HP motor? Both motors are 3ϕ and are operating at 460 V.

MOTOR INPUT VOLT-AMP RATING

The *motor volt-amp (VA) rating* is the amount of power used to produce horsepower output. The output is never equal to the input due to efficiency and the power factor of the motor. The motor input VA is found by multiplying the input voltage times the input current (FLC).

Single-Phase (1ϕ) AC and DC Motor VA Calculation

For 1ϕ motors, the voltage is multiplied by the current the motor is drawing. To find the motor volt-amp rating for 1ϕ motors, apply the following formula:

$VA = E \times I$
where
VA = volt-amps
E = voltage of motor (in V)
I = FLC (in A)(Table 430.248)

For example, what is the VA input of a 5 HP, 120 V motor?

$VA = E \times I$
$VA = 120 \times 56$
$VA = $ **6720 VA**

Practice Question

Determining Volt-Amp Rating for Single-Phase (1ϕ) AC and DC Motors

1. What is the volt-amp input of a 2 HP, 200 V motor?

Three-Phase (3ϕ) AC Motor VA Calculation

For 3ϕ motors, the motor-operating voltage is multiplied by the current and the square root of 3 (1.732).

To find the motor volt-amp rating for 3ϕ motors, apply the following formula:

$VA = E \times I \times 1.732$
where
VA = volt-amps
E = voltage of motor (in V)
I = FLC (in A)(Table 430.250)
1.732 = square root of 3

For example, what is the volt-amp input of a 15 HP, 208 V, 3ϕ motor?

$VA = E \times I \times 1.732$
$VA = 208 \times 46.2 \times 1.732$
$VA = $ **16,644 VA**

Practice Question

Determining Volt-Amp Rating for Three-Phase (3ϕ) AC Motors

1. What is the VA input of a 5 HP, 230 V, 3ϕ motor?
2. What is the VA input of a 5 HP, 460 V, 3ϕ motor?

Kilovolt-Amps

Some questions on the exam will require the motor rating in kilovolt-amps (kVA). The motor rating in kilovolt-amps is found for 1ϕ motors by taking the total volt-amp rating and dividing it by 1000 (kilo = 1000). To find the motor rating in kilovolt amps for 3ϕ motors, the total volt-amp rating is multiplied by 1.732 and then divided by 1000.

To find the motor rating in kilovolt-amps for 1ϕ motors, apply the following formula:

$$kVA = \frac{E \times I}{1000}$$

where
kVA = kilovolt-amps
E = voltage of motor (in V)
I = FLC (in A)
1000 = constant

To find the motor rating in kilovolt-amps for 3φ motors, apply the following formula:

$$kVA = \frac{E \times I \times 1.732}{1000}$$

where
kVA = kilovolt-amps
E = voltage of motor (in V)
I = FLC (in A)
1.732 = square root of 3
1000 = constant

MOTOR SPEEDS

Motor speed is listed on the nameplate or measured using a tachometer. Motor speed can be calculated as synchronous speed or actual shaft speed.

Synchronous Speed

Synchronous speed is the speed in which the magnetic field travels around the windings of a motor. Synchronous speed is not the actual shaft speed. Two methods are used to determine motor synchronous speed. One method is to multiply the frequency of the AC voltage system by a constant number of 120 and divide by the number of poles per phase. The more poles, the slower the motor turns. The second method is to multiply the frequency by itself and divide by the number of paired poles.

To determine the synchronous speed, apply one of the following formulas:

$$\omega_S = \frac{120 \times F}{P}$$

where
ω_S = synchronous speed (in rpm)
120 = constant
F = frequency (in Hz)
P = number of poles

or

$$SS = \frac{F \times F}{PP}$$

where
SS = synchronous speed (in rpm)
F = frequency (in Hz)
PP = pair of poles

For example, what is the synchronous speed of a motor that operates at 60 Hz and has four poles per phase?

$$SS = \frac{120 \times F}{PP}$$

$$SS = \frac{120 \times 60}{4}$$

$$SS = \mathbf{1800 \text{ rpm}}$$

What is the synchronous speed of a motor that operates at 60 Hz and has two pairs of poles per phase?

$$SS = \frac{F \times F}{PP}$$

$$SS = \frac{60 \times 60}{2}$$

$$SS = \mathbf{1800 \text{ rpm}}$$

> **Practice Question**
>
> *Determining Synchronous Speed*
>
> 1. What is the synchronous speed of a 10 Hz motor that has six poles per phase?

Actual Shaft Speed

The actual shaft speed (in rpm) is found by subtracting the slip of a motor from the synchronous speed of a motor. The slip is given as a percentage of the synchronous speed. For example, the slip of a 1725 rpm motor (1800 rpm synchronous speed) is approximately 4% (1800 − 1725 = 75 ÷ 1800 = 0.041 = 4%). **See Figure 9-9.** The more slip a motor has, the more torque it delivers to operate the load. The slower a motor runs, the greater the amount of torque it creates.

To find the actual shaft speed, apply the following formula:

$$AS = SS - S$$

where
AS = actual shaft speed (in rpm)
SS = synchronous speed (in rpm)
S = slip (in %)

For example, what is the actual shaft speed of a four-pole motor operating on 60 Hz with a slip of 3%?

$$SS = \frac{120 \times F}{P}$$

$$SS = \frac{120 \times 60}{4}$$

$$SS = 1800$$

$S = SS \times 3\%$
$S = 1800 \times 0.03$
$S = 54$

$AS = SS - S$
$AS = 1800 - 54$
$AS = \mathbf{1746\ rpm}$

Practice Question

Determining Actual Shaft Speed

1. What is the actual shaft speed of a six-pole motor operating at 60 Hz with a 5% slip?

Actual Shaft Speed

Poles	Synchronous Speed	With 4% Slip	Actual Speed
2	3600	144	3456
4	1800	72	1728
6	1200	48	1152
8	900	36	864

* in rpm

Figure 9-9. Motor slip is listed as a percentage of the synchronous speed.

MOTOR TORQUE

Torque is the amount of twisting power that the shaft of a motor delivers. Torque is measured in pounds per feet (lb-ft). The amount of running torque is found by multiplying the constant 5252 by the horsepower of the motor and then dividing by the speed of the motor. The faster the motor, the lower the torque. The slower the motor, the more running torque the motor creates.

To find the running torque, apply the following formula:

$$RT = \frac{5252 \times HP}{rpm}$$

where
RT = running torque (in lb-ft)
5252 = constant
HP = horsepower
rpm = revolutions per minute

For example, what is the running torque of a 15 HP motor operating at 1725 rpm?

$$RT = \frac{5252 \times HP}{rpm}$$
$$RT = \frac{5252 \times 15}{1725}$$
$$RT = \frac{78,780}{1725}$$
$$RT = \mathbf{45.67\ lb\text{-}ft}$$

Practice Question

Determining Motor Torque

1. What is the running torque of a 10 HP motor operating at 1000 rpm?

Starting Torque

Starting torque is the torque required to start a motor with a load applied. Each motor delivers starting torque. To find the starting torque, multiply the running torque by a percentage based on the design of the motor. A common starting torque percentage is 150% to 225% of the running torque.

TECH TIP

All motors produce torque, but have different torque characteristics. Motors are classified by the National Electrical Manufacturers Association (NEMA) according to their electrical characteristics. Motor torque characteristics vary with the classification of the motor. Motors are classified as Class A through Class F. Classes B, C, and D are the most common classifications.

MOTOR HORSEPOWER

Horsepower is a unit of power equal to 746 W or 33,000 lb-ft per minute (550 lb-ft per second). A *watt* is a unit of measure equal to the power produced by a current of 1 A across a potential difference of 1 V. Electrical power is rated in horsepower (HP) or watts (W). A watt is 1/746 of 1 HP and is the base unit of electrical power. Motor power is rated in horsepower or watts. **See Figure 9-10.** To find the horsepower of an unmarked DC motor, the motor voltage, current, and efficiency must be known. For 1φ and 3φ AC motors, the power factor must be known, and for 3φ motors, the square root of 3 (1.732) is required.

To find the horsepower for DC motors, apply the following formula:

$$HP = \frac{E \times I \times Eff}{746}$$

where
HP = horsepower
E = voltage (in V)
I = current (in A)
Eff = efficiency
746 = constant to convert watts to horsepower

To find the horsepower for 1φ AC motors, apply the following formula:

$$HP = \frac{E \times I \times Eff \times PF}{746}$$

where
HP = horsepower
E = voltage (in V)
I = current (in A)
Eff = efficiency
PF = power factor
746 = constant to convert watts to horsepower

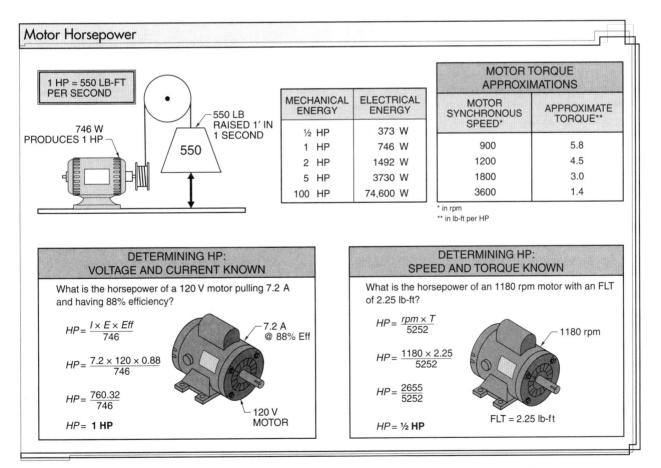

Figure 9-10. Motor voltage, amperage, and efficiency must be known in order to calculate the horsepower for an unmarked motor.

To find the horsepower for 3φ AC motors, apply the following formula:

$$HP = \frac{E \times I \times Eff \times PF \times 1.732}{746}$$

where
HP = horsepower
E = voltage (in V)
I = current (in A)
Eff = efficiency
PF = power factor
1.732 = square root of 3
746 = constant to convert watts to horsepower

For example, what is the horsepower of a 1φ motor that operates on 230 V, draws 50 A, and has an efficiency of 73% and a power factor of 80%?

$$HP = \frac{E \times I \times Eff \times PF}{746}$$

$$HP = \frac{230 \times 50 \times 0.73 \times 0.80}{746}$$

$$HP = \frac{6716}{746}$$

$$HP = \mathbf{9\ HP}$$

Practice Question

Determining Horsepower

1. What is the horsepower of a 3φ motor that operates on 460 V, draws 14 A, and has an efficiency of 80% and a power factor of 90%?

MOTOR NAMEPLATE CURRENT

Various equations are used to find the nameplate current of a motor. The equation used depends on the type of motor. The three different types of motors are DC, 1φ AC, and 3φ AC. To find the nameplate current for a DC motor, apply the following formula:

$$I = \frac{HP \times 746}{E \times Eff}$$

where
I = current (in A)
HP = horsepower
746 = constant to convert watts to horsepower
E = voltage (in V)
Eff = efficiency

To find the current for a 1φ AC motor, apply the following formula:

$$I = \frac{HP \times 746}{E \times PF \times Eff}$$

where
I = current (in A)
HP = horsepower
746 = constant to convert watts to horsepower
E = voltage (in V)
PF = power factor
Eff = efficiency

To find the current for a 3φ AC motor, apply the following formula:

$$I = \frac{HP \times 746}{E \times PF \times Eff \times 1.732}$$

where
I = current (in A)
HP = horsepower
746 = constant to convert watts to horsepower
E = voltage (in V)
PF = power factor
Eff = efficiency
1.732 = square root of 3

For example, what is the nameplate current of a 100 HP, 480 V, 3φ motor that operates at 87% efficiency and has an 85% power factor?

$$I = \frac{HP \times 746}{E \times PF \times Eff \times 1.732}$$

$$I = \frac{100 \times 746}{480 \times 0.85 \times 0.87 \times 1.732}$$

$$I = \frac{74,600}{614.79}$$

$$I = \mathbf{121.34\ A}$$

Practice Question

Determining Motor Nameplate Current

1. What is the nameplate current of a 2 HP, 120 V, 1φ motor that operates at 87% efficiency and has an 85% power factor?

MOTOR CONNECTIONS

Motors are one of the most common types of electrical devices. Knowing how to connect motors is a requirement for electricians. The exam will have questions concerning 1ϕ and 3ϕ motor connections. The NEC® will not be needed for these questions since they do not contain motor diagrams.

Single-Phase (1ϕ) Motor Connections

Single-phase motor connections are usually given on the exam in the form of a wiring diagram. The electrician will need to determine which motor is wired for high or low voltages. This is determined by looking at the run windings. If the run windings are connected in series, the motor is running on high voltage (240 V). If the run windings are connected in parallel, the motor is running on low voltage (120 V). **See Figure 9-11.**

Three-Phase (3ϕ) Motor Connections

Three-phase motors are available as wye-connected motors and delta-connected motors. Wye and delta describe the internal winding connections in the motor. Both motor types have 3, 6, 9, or 12 leads. The most common type is a wye-connected motor with 9 leads.

Windings connected in series are high voltage (480 V), and windings connected in parallel are low voltage (240 V). The high-voltage connection for wye and delta motors is the same. The low-voltage connection for wye and delta motors is different. **See Figure 9-12.**

MOTOR VOLTAGE DROP

Voltage drop is covered in 210.19(A) Informational Note No. 4 for branch circuits, and 215.2(A)(1)(b) Informational Note No. 2 for feeders. The recommended voltage drop is 3% for either the feeder or branch circuit and not more than 5% for both the feeder and branch circuit. Article 695 gives a different voltage drop for fire pumps. According to 695.7, the voltage source at the controller-line terminals shall not drop more than 15% below normal under motor starting conditions. The voltage source shall not drop more than 5% when the motor is running at 115% of its rated current. While this does not apply to general-duty motors, it does apply to fire-pump motors.

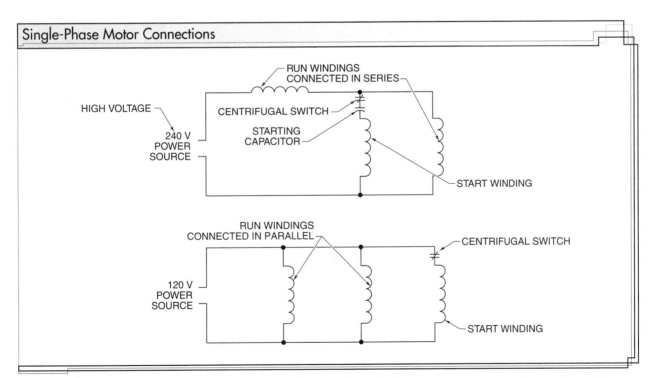

Figure 9-11. If the run windings are connected in series on a 1ϕ motor, the motor is running on high voltage. If the run windings are connected in parallel, the motor is running on low voltage.

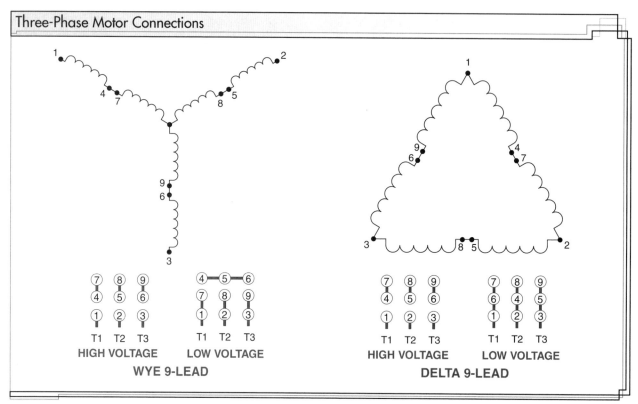

Figure 9-12. The connection of the windings of a 3φ motor are in series for high voltage and in parallel for low voltage.

MOTOR CONTROL CIRCUIT CONDUCTORS

Article 430 covers motors, motor circuits, and controllers. The correct overcurrent protective device is required based on the condition in which the conductors are used. Table 430.72(B) of the NEC® provides three conditions for protecting the control conductors.

Column A: Separate Protection Provided

The conductors are protected at their ampacity. The protection is limited to the ampacity of the conductor when the overcurrent protective device for the motor is not used for protecting the control conductors. Control conductors can tap off the load side of the overcurrent protective device (OCPD) if the device rating is not larger than that allowed in Column B or C of Table 430.72(B).

Column B: Conductors Within Enclosure

The conductors do not leave the enclosure and may be tapped off the load side of the OCPD. The OCPD may be tapped from the load side if the rating is not larger than that listed for the wire size. If the OCPD is larger, then it shall be sized out of Column A.

Column C: Conductors Extend Beyond Enclosure

The conductors leave the enclosure and travel to a remote start/stop station or other controlling device. Here, the control conductors can be protected by the motor OCPD as long as it does not exceed the rating in the table of Column C. If the OCPD is larger than that listed, the OCPD shall be sized out of Column A.

LOCKED-ROTOR CURRENT RATINGS

Locked rotor current (LRC) occurs when the rotor of a motor stops turning while the motor is connected to power. This happens when the motor first has power supplied to it. When the rotor is not turning and power is applied, the motor has a high inrush of current. Once the rotor starts to turn, the current draw is reduced. LRC may also occur if an excessively large load is placed on the motor causing it to stall. The current draw will be high until the overloads trip and remove the power to the control circuit, which can be used to stop the motor.

Locked-Rotor Code Letter

The LRC may be found using two different methods. The first method uses Table 430.7(B). The table gives code letters along with a range of kilovolt-amps per horsepower for a motor with a locked rotor. The maximum kilovolt-amp level per code letter is used when finding the maximum LRC. The code letter H indicates a range of 6.3 kVA to 7.09 kVA per horsepower. To find the locked rotor current, 7.09 kVA is used. **See Figure 9-13.**

To determine the LRC for a 1φ motor, apply the following formula:

$$LRC = \frac{kVA \times 1000 \times HP}{E}$$

where
LRC = locked rotor current (in A)
kVA = kilovolt-amps
1000 = constant
HP = horsepower
E = voltage (in V)

To determine the LRC for a 3φ motor, apply the following formula:

$$LRC = \frac{kVA \times 1000 \times HP}{E \times 1.732}$$

where
LRC = locked rotor current (in A)
kVA = kilovolt-amps
1000 = constant
HP = horsepower
E = voltage (in V)
1.732 = square root of 3

For example, what is the LRC of a 10 HP, 240 V, 1φ motor with a code letter of G?

$$LRC = \frac{kVA \times 1000 \times HP}{E}$$

$$LRC = \frac{6.29 \times 1000 \times 10}{240}$$

$$LRC = \frac{62,900}{240}$$

$$LRC = \mathbf{262.1\ A}$$

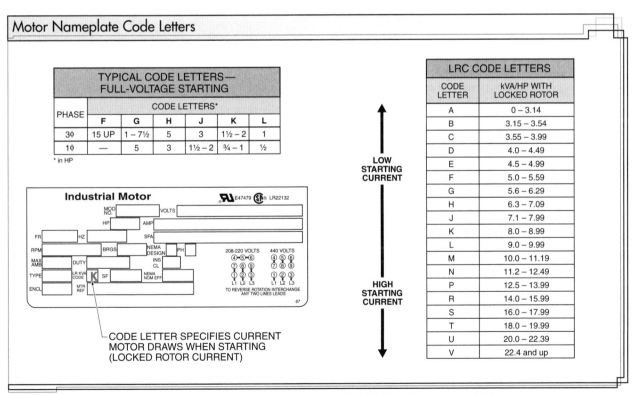

Figure 9-13. The locked rotor current (LRC) is found using the LRC code letters, which are found on a motor's nameplate.

What is the locked rotor current of a 50 HP, 480 V, 3ϕ motor with a code letter of J?

$$LRC = \frac{kVA \times 1000 \times HP}{E \times 1.732}$$

$$LRC = \frac{7.99 \times 1000 \times 50}{480 \times 1.732}$$

$$LRC = \frac{399,500}{831.36}$$

$$LRC = \textbf{480.54 A}$$

Practice Question

Determining Locked-Rotor Current Ratings

1. What is the locked rotor current of a 50 HP, 480 V, 3ϕ motor with a code letter of K?

Locked Rotor Tables

The second method used to find locked rotor current uses Tables 430.251(A) and 430.251(B). Table 430.251(A) is used for 1ϕ motors and Table 430.251(B) is used for 3ϕ motors. If the locked-rotor code letter is not known, then the horsepower and voltage can be used to provide a rating for the locked rotor current of the motor.

Part 430.110(C)(3) lists the six times rule for small motors not listed on the previous tables. The tables are based on about six times the full load current (FLC) of the motor as found in Tables 430.247, 430.248, 430.249, and 430.250. The six times method may be used in the field, but it is not accurate enough for the exam.

VARIABLE-FREQUENCY DRIVES

A *variable-frequency drive (VFD)* is an electronic unit designed to control the speed of a motor using solid-state components. A drive can handle motors that are smaller than what they are rated for, so there is a need to regulate the sizing of conductors and other devices. The NEC® addresses this topic in Part X of Subsection 430.

Variable-Frequency Drive Conductors

Section 430.122(A) is used to size conductors for VFD equipment and motors. The conductors supplying the VFD are sized at 125% of the conversion equipment input rating. The conductor to the motor must be sized for the motor installed, but the drive has to be supplied with conductors for the maximum load that the drive can handle. If an electrician replaced a motor with a drive having a larger horsepower, the conductors from the VFD to the motor would have to be changed, but not the conductors supplying the drive.

To determine the current rating used to size conductors from a VFD to a motor, apply the following formula:

$$I = \frac{VA}{E \times 1.732}$$

where
I = current (in A)
VA = volt-amps
E = voltage (in V)
1.732 = square root of 3

For example, what size of THW supply conductors is required for a 10 HP, 480 V, 3ϕ motor on a VFD that is rated at 25,000 VA?

$$I = \frac{VA}{E \times 1.732}$$

$$I = \frac{25,000}{480 \times 1.732}$$

$$I = \frac{25,000}{831.36}$$

$$I = 30$$

30 × 125% = 37.5 A
Table 310.15(B)(16); 37.5 A = **No. 8 THW**

Practice Question

Determining VFD to Motor Conductor Size

1. What size of THW supply conductors is required for a 15 HP, 480 V, 3ϕ motor on a VFD that is rated at 28,000 VA?

Bypass Devices

VFDs are sometimes removed from the circuit if the motor is going to be running at its rated speed. This is done with a bypass device. The bypass device removes the VFD from the circuit and diverts the current around the VFD. Per 430.122(B), the conductors for the bypass are sized according to 125% of the input rating of the conversion equipment or 125% of the motor FLC, whichever is larger.

To determine the current rating used to size a bypass device, apply the following formula:

$$I = \frac{VA}{E \times 1.732}$$

where

I = current (in A)
VA = volt-amps
E = voltage (in V)
1.732 = square root of 3

For example, what size of bypass conductors is required for a 10 HP, 480 V, 3ϕ motor on a VFD that is rated at 25,000 VA?

$$I = \frac{VA}{E \times 1.732}$$

$$I = \frac{25,000}{480 \times 1.732}$$

$$I = \frac{25,000}{831.36}$$

$$I = 30$$

30 × 125% = 37.5 A
Table 310.15(B)(16); 37.5 A = **No. 8 THW**

Practice Question

Determining Bypass Conductor Size

1. What size THW bypass conductors is required for a 15 HP, 480 V, 3ϕ motor on a VFD that is rated at 28,000 VA?

Overloads

VFDs that have overloads built into the equipment do not require additional overload equipment. VFDs that do not have overloads built into the equipment must have an overload device installed.

Overload devices are sized according to Part III of 430. If a circuit has a bypass device installed, it shall be sized based on Part III of Article 430. The overloads shall be sized based on the service factor or temperature rise of the motor and the motor nameplate current rating.

TECH TIP

The three main sections of a VFD are the converter, DC bus, and inverter. The converter receives incoming AC voltage and changes it to DC voltage. The DC bus filters the voltage and maintains the proper DC voltage level.

The DC bus also delivers DC to the inverter for conversion back to AC. The inverter controls the speed of a motor by controlling frequency and controls motor torque by controlling voltage sent to a motor.

CHAPTER 9
Motors
Review Questions

Name _____ Date _____

_____ 1. The ampacity for a feeder with a 50 HP, 480 V, 3φ electric motor, a 40 HP, 480 V, 3φ electric motor, and a 30 HP, 480 V, 3φ electric motor is ___ A.
  A. 120
  B. 157
  C. 173
  D. 186

_____ 2. The size of a dual-element fuse required for a 2 HP, 208 V, 1φ induction motor is ___ A.
  A. 15
  B. 20
  C. 25
  D. 30

_____ 3. A 50 HP, 480 V, 3φ induction motor has a nameplate rating of 60 A and an FLC of 65 A. The motor has a service factor of 1.1 and a temperature rise of 50°C. The required standard overload size for the motor is ___ A.
  A. 69
  B. 70
  C. 74.75
  D. 80

_____ 4. If a motor will not start with standard overloads sized for it, the maximum overload that can be used on a 50 HP, 480 V, 3φ motor with a service factor of 1.15 and a nameplate rating of 60 A is ___ A.
  A. 65
  B. 84
  C. 84.5
  D. 91

_____ 5. The input volt amp rating of a 5 HP, 230 V, 1φ induction motor is ___ VA.
  A. 3496
  B. 3730
  C. 6055
  D. 6440

_____ 6. Aluminum conductors sized No. ___ are required for a 1½ HP, 230 V, 1φ motor with a nameplate rating of 8 A.
   A. 8
   B. 10
   C. 12
   D. 14

_____ 7. The required current for sizing a 15 min, 5 HP, 240 V intermittent electric motor with a nameplate rating of 27 A and an FLC rating of 28 A is ___ A.
   A. 22.95
   B. 27
   C. 28
   D. 33.75

_____ 8. The smallest branch circuit conductor to one 5 HP, 208 V, 3φ motor and three 2 HP, 120 V, 1φ motors should have an ampacity of ___ A.
   A. 50
   B. 100
   C. 150
   D. 175

_____ 9. A 100 HP, 460 V, 3φ synchronous motor with a power factor rating of 80% has an ampacity of ___ A.
   A. 101
   B. 111
   C. 124
   D. 126

_____ 10. The maximum motor-running overload protection for a 3 HP, 230 V, 1φ motor is ___ A.
   A. 17
   B. 20
   C. 22.1
   D. 29.1

_____ 11. A feeder for an MCC has a 50 HP, a 200 HP, and a 100 HP motor connected to it. Each motor is 460 V and 3φ. The required amperage needed to size the feeder conductor is ___ A.
   A. 240
   B. 364
   C. 489
   D. 536

_____ 12. The input VA of a 5 HP, 230 V, 3φ induction motor is ___ VA.
   A. 3496
   B. 3730
   C. 6055
   D. 10,488

_____ **13.** The NEC® requires a ___ A dual-element fuse for a 2 HP, 208 V, 1φ motor.
        A. 20
        B. 25
        C. 30
        D. 35

_____ **14.** The FLC of a 200 HP, 460 V, 3φ synchronous electric motor with a power factor of 80% is ___ A.
        A. 201
        B. 221
        C. 251
        D. 302

_____ **15.** The locked rotor current for a 5 HP, 240 V, 1φ motor is ___ A.
        A. 28
        B. 92
        C. 168
        D. 186

CHAPTER 10

Digital Resources
ATPeResources.com/QuickLinks
Access Code: 637160

Conductor-Fill, Box-Fill, and Pull and Junction Box Sizing

The NEC® recognizes many different types of raceways for the protection and routing of electrical conductors. Raceway installations may include conduit, wireways, device or junction boxes, and pull boxes. Each of these items has a limit to the number of conductors it may contain. When these installations are overfilled, the chances of damaging the conductor's insulation increase. Damage to conductor insulation can be prevented by correctly sizing raceways, pull boxes, and junction boxes. Tables in chapter 9 of the NEC® are used to determine raceway-fill calculations. All tables in this chapter will refer to tables found in chapter 9 unless otherwise noted.

OBJECTIVES

After completing this chapter, the learner will be able to do the following:

- Apply conductor-fill requirements of Table 1 of chapter 9 and the nine notes that follow it.
- Identify the key data used in Table 4 of chapter 9 to determine allowable fill.
- Explain how Table 5 of chapter 9 is used to calculate conductor fill.
- Explain why the conductor fill for conduit nipples is higher than for conduits longer in length.
- Identify the fill requirements for wireways.
- Explain how to size wireways.
- Identify the five steps used to determine the additional number of conductors allowed in a conduit that already contains a conductor.
- List common items that are installed in boxes but that are not included in box-fill calculations.
- Size the length of a pull box or junction box for straight pulls, angle pulls, and U-pulls.

CONDUCTOR FILL

Conductor fill is the maximum percentage of cross-sectional area that can be occupied by conductors inside a raceway. As a general rule, raceways that are round in shape follow Table 1 in Chapter 9 of the NEC®. Table 1 shows that raceways with more than two conductors have a 40% fill. Raceways that are square or rectangular generally have a fill of 20%.

Conductor-fill requirements are found in Table 1. See the notes from the NEC® that go along with Table 1. Table 1 gives the maximum percentage of fill that is allowed for one, two, and over two conductors or cables installed in the conduit. **See Figure 10-1.** Under Table 1 are two Informational Notes containing information about common usage of the raceway as well as jamming problems that may occur under certain conditions.

- Informational Note No. 1 states that the fills in Table 1 are for common conditions, and the need to upsize the conduit might be required for length, number of bends, and so forth.
- Informational Note No. 2 states that when pulling in three conductors, jamming may occur, and a jam ratio should be calculated. The ratio is found by dividing the inside diameter of the raceway by the outside diameter of the conductors. If this ratio falls between 2.8 and 3.2, jamming is likely to occur.

There are notes that are associated with Table 1. These notes provide information on what needs to be counted toward conduit fill, as well as treatment of multiconductor cables with an outer jacket. These notes are as follows:

- **Note 1.** Use Informative Annex C for the maximum number of conductors or fixture wires that are all the same size based on cross-sectional area including the insulation.
- **Note 2.** The fill requirements are for complete sections of conduit not short pieces used for physical protection only.
- **Note 3.** Equipment grounding conductors count toward conduit fill.
- **Note 4.** A nipple is 24″ or less in length and can be filled to 60% of the total area.
- **Note 5.** The actual dimensions should be used for conductors that are not listed in the tables of chapter 9.
- **Note 6.** When there are conductors of different sizes, use Table 4 for conduit dimensions and Table 5 for sizing conductors.
- **Note 7.** When performing the fill calculations with conductors that are all the same size, the next higher whole number can be used if the calculated total is a decimal of 0.8 or greater.
- **Note 8.** Conductor sizes for bare conductors are found in Table 8.
- **Note 9.** Multiconductor cables are treated just like single conductors. Multiconductors that are oblong use the largest dimension for sizing. Multiconductor cables shall have an overall outer covering. If an outer covering is not present, then the conductors are treated as single conductors and are added together to calculate conduit fill.

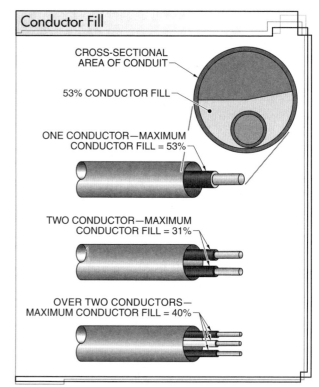

Figure 10-1. Conductor fill in a conduit is the maximum percentage of cross-sectional area that may be occupied by the conductors.

Conductor-Fill Tables for Same-Size Conductors

Informative Annex C includes 12 different tables for conduits recognized by the NEC®. The conductor fill for each type and size of conduit is calculated for the number of same-size conductors installed in it. These calculations include the conductor insulation. There are two tables for each type of conduit. The first table (Tables C-1 to C-12) is for standard building and fixture wire. The second table (Tables C-1A to C-12A) is for compact copper and aluminum conductors. Use these tables to find the maximum number of same-size conductors allowed in a specific type of conduit. **See Figure 10-2.**

Conductor Fill Tables—Informative Annex C

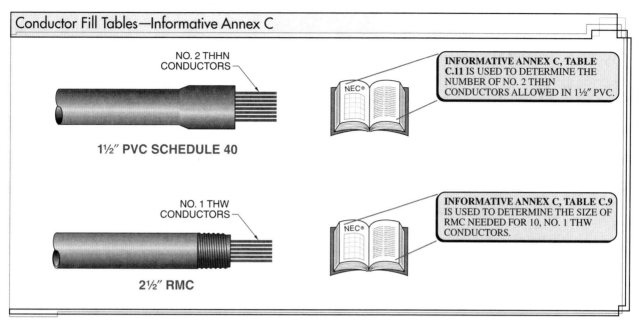

Figure 10-2. Tables from Informative Annex C are used to find the maximum number of same-size conductors allowed in a specific type of conduit, including the insulation.

For example, how many No. 2 THHN conductors can be installed in a 1½″ PVC schedule 40 conduit?

See Table C.11 in Informative Annex C.

No. 2 THHN in 1½″ PVC = **7 conductors**

For example, what size of rigid metal conduit is needed for ten No. 1 THW conductors?

See Table C.9 in Informative Annex C.

Size of rigid metal conduit = **2½″**

Practice Questions

Determining Conductor Fill for Same-Size Conductors

1. What size ENT conduit is required for ten No. 8 THHN conductors?
2. What size of rigid schedule 40 PVC conduit is needed for five No. 4/0 THW conductors?

Conduit Properties—Table 4 of Chapter 9. Table 4 lists the dimensions and percent area for the 12 conduits recognized by the NEC®. Each table is set up in the same manner with the conduit size listed on the left and the area on the right. The area of fill is listed under the number of conductors installed in the conduit. Each column on the right lists area values in both square inches and millimeters. On the exam, use square inches unless told otherwise. Answers will be in square inches for all conductor fill calculations in this book.

A separate table is used for ⅜″ flexible metal conduit because the fill is determined by the type of connector used, wire size, and insulation. The fill for ⅜″ flexible metal conduit is found in Table 348.22. Table 348.22 has the conductor sizes on the left side with the type of insulation listed across the top. There are two columns under each wire type. Each column specifies the type of connector used. The connector that screws into the conduit is an inside connector, and the type that clamps on the outside of the conduit is an outside connector. Table 348.22 allows an additional insulated, covered, or bare equipment grounding conductor of the same size in a ⅜″ flexible metal raceway. Use Table 4 of Chapter 9 to determine the allowable fill for various conduits. **See Figure 10-3.**

For example, what is the allowable fill for a 3″ schedule 80 PVC conduit with more than two conductors installed?

See Table 4 of Chapter 9.

Allowable fill = **2.577 sq in.**

Practice Questions

Determining Fill in Conduits Using Table 4

1. What is the allowable fill in square inches for more than two conductors in a 1½″ rigid metal conduit?
2. What is the allowable fill in square inches for two conductors in a 3″ EMT conduit?

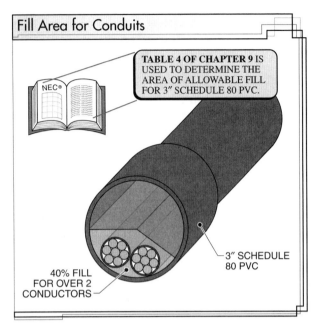

Figure 10-3. Table 4 of Chapter 9 is used to determine the allowable fill for various types of conduits.

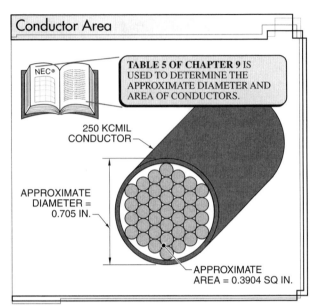

Figure 10-4. Table 5 of Chapter 9 is used to determine the cross-sectional area of specific conductors.

Conductor Dimensions—Table 5 of Chapter 9. Table 5 is used to determine the diameter and area for insulated conductors. Each table lists the insulation type on the left column. The second column lists the AWG size. The next two columns list approximate area, and the approximate diameter of the conductor.

Be aware of RHW, RHH, and RHW-2 conductors. These conductors are listed in two different tables. One table shows values with an outer cover, and the other table shows values without an outer cover. The asterisk next to the insulation type represents a conductor without an outer cover. Use Table 5 to determine the area of each specific conductor. **See Figure 10-4.**

For example, how much area in square inches does a 250 kcmil XHHW conductor take up in a conduit?

See Table 5.

250 kcmil XHHW conductor = 0.3904 sq in.

Area of conductor = **0.3904 sq in.**

Practice Questions

Determining the Cross-Sectional Area of Conductors

1. What is the area in square inches of a No. 10 THHN copper conductor?

2. What is the area in square inches of a No. 8 RHW conductor without an outer covering?

TECH TIP

All conductors (each phase, polarity, neutral, and grounded conductor) shall be the same insulation type. There are three basic temperature ratings for conductor insulations. Parallel conductors shall have the same temperature rating to ensure that the termination does not reach a temperature that exceeds the operating temperature of the insulation.

Compact Conductor Dimensions—Table 5A of Chapter 9. Table 5A is used to determine the diameter and area for compact copper and aluminum building wires. Compact building wire has a smaller outside diameter than normal building wire. The diameter of the conductor is made smaller by having the strands of the conductor shaped to fit together, eliminating the space between each strand of the conductor. The ampacity of these conductors is the same as standard conductors of the same AWG size. Compact conductors have limited sizes and insulations. Table 5A is only used on the exam when compact conductors are mentioned in the question. **See Figure 10-5.**

For example, how much area in square inches does a 350 kcmil THHW compact aluminum conductor take up in a conduit?

See Table 5A.

Area of conductor = **0.5281 sq in.**

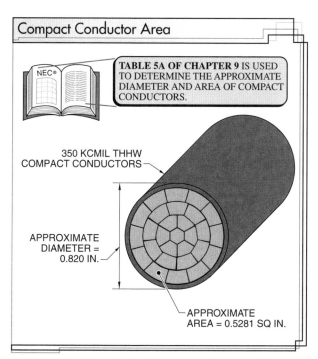

Figure 10-5. Table 5A of Chapter 9 is used to determine the diameter and area for compact copper and aluminum building wires only.

Practice Question

Determining the Cross-Sectional Area of Compact Conductors

1. What is the area in square inches of a 250 kcmil XHHW compact aluminum conductor?

Bare Conductor Properties—Table 8 of Chapter 9. Table 8 is used to determine the diameter and area for bare conductors. Bare conductors are used as equipment grounding conductors. Conductors that are No. 8 or smaller may be stranded or solid. Conductors larger than No. 8 conductors are stranded. Per 310.106(C), conductors that are No. 8 or larger installed in a raceway shall be stranded.

Sizing Conduit for Different-Size Conductors

When more than one conductor size or type of insulation is used, the size of each individual conductor needs to be determined to get a total for all conductors installed in the conduit. The area in square inches of each conductor is found in Table 5 or 5A. Once the total area of all conductors is determined, Table 4 is used to determine the proper size of conduit. Creating a table will help keep calculations organized. **See Figure 10-6.**

For example, what size EMT conduit is required to house all of the following conductors in a single conduit?
- four No. 6 THWN conductors
- four No. 4 XHHW conductors
- three No. 3 THHN conductors
- two No. 1 THW conductors

See Table 5 to determine the area of each conductor.

No. 6 THWN = 0.0507 sq in.
No. 4 XHHW = 0.0814 sq in.
No. 3 THHN = 0.0973 sq in.
No. 1 THW = 0.1901 sq in.

Multiply the area by the number of conductors:
0.0507 sq in. × 4 = 0.2028 sq in.
0.0814 sq in. × 4 = 0.3256 sq in.
0.0973 sq in. × 3 = 0.2919 sq in.
0.1901 sq in. × 2 = 0.3802 sq in.

Add the area of all the conductors:
0.2028 sq in. + 0.3256 sq in. + 0.2919 sq in. + 0.3802 sq in. = 1.2005 sq in.

See Table 4 (EMT).
The next highest area from 1.2005 for over two conductors is 1.342 sq in.
1.342 sq in. = **2″ EMT**

Practice Question

Determining Conduit Size for Conductors of Different Sizes

1. What size of rigid metal conduit (RMC) is required to house all of the following conductors in a single conduit?
- three No. 10 THHN conductors
- four No. 8 THHN conductors
- five No. 6 THHN conductors

The wall thickness of electrical metallic tubing (EMT) is approximately 40% less than that of rigid metal conduit (RMC).

Determining Conduit Size for Conductors of Different Sizes

INSULATION TYPE	SIZE OF CONDUCTOR	NUMBER OF CONDUCTORS	AREA IN SQ IN.	TOTAL AREA IN SQ IN.
THWN	No. 6	4	0.0507	0.2028
XHHW	No. 4	4	0.0814	0.3256
THHN	No. 3	3	0.0973	0.2919
THW	No. 1	2	0.1901	0.3802

Total Area in Sq In. **1.2005**

TABLE 4 IS USED TO DETERMINE THE SIZE OF EMT FOR CONDUCTORS MEASURING 1.2005 SQ IN.

Figure 10-6. Once the total area of all conductors is determined, Table 4 is used to determine the proper size of conduit.

Conductor Fill for Nipples

A *nipple* is a conduit raceway that is 24″ or shorter in length. Nipples are used to join enclosures that are mounted close together. Because of the short length of nipples, there is not as much heat produced by the current flowing in the conductors to damage conductor insulation. This is why nipples have a higher fill rating. As stated in Note 4 of Table 1, conduit nipples may be filled up to 60%. Table 4 has the 60% fill for nipples already calculated for each conduit. The tables in Informative Annex C do not include the 60% fill. The same calculation used to size raceways for different-size conductors shall be used to size nipple conduits. **See Figure 10-7.**

For example, what size of IMC conduit nipple is needed for ten No. 8 THW conductors and six No. 10 THW conductors?

See Table 5 to determine the area of each conductor.
No. 8 THW = 0.0437 sq in.
No. 10 THW = 0.0243 sq in.

Multiply the area by the number of conductors:
0.0437 sq in. × 10 = 0.4370 sq in.
0.0243 sq in. × 6 = 0.1458 sq in.

Add the area of all of the conductors:
0.4370 sq in. + 0.1458 sq in. = 0.5828 sq in.

See Table 4 Article 342 (IMC).
The next highest area from 0.5828 for over two conductors is 0.650 sq in.
0.650 sq in. = **1¼″ IMC**

Practice Questions

Determining Conduit Nipple Fill for Conductors of Various Sizes

1. What is the allowable fill for a 1″ EMT nipple?
2. What size of rigid metal conduit nipple would be required to house the following conductors in a single conduit?
   - three No. 10 THHN conductors
   - four No. 8 THHN conductors
   - five No. 6 THHN conductors

WIREWAYS

A *wireway* is a sheet metal or nonmetallic enclosure with a cover that opens to provide access to the conductors inside. Wireways are typically square and can range up to 10′ in length. The standard width and height for wireways are 2″ × 2″, 4″ × 4″, 6″ × 6″, 8″ × 8″, 10″ × 10″, and 12″ × 12″. Wireways may also be custom-built to any size. Wireways may be used to supplement wiring spaces for panels and disconnects. When used for this purpose, they are called auxiliary gutters. The conductors in auxiliary gutters cannot be longer than 30′ in length. Gutters are sized using Table 5.

TECH TIP

Each type of raceway has its own article in the NEC®. The sections entitled "Uses Permitted" and "Uses Not Permitted" should be reviewed for the particular raceway being considered. Some of these articles require the raceway to be listed. If that is the case, the raceway shall be installed in accordance with any instructions included in the listing or labeling per 110.3(B).

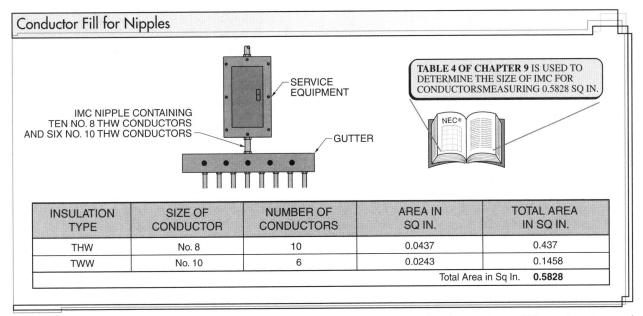

Figure 10-7. Conduit nipples may be filled up to 60%. Table 4 is used to determine the conductor fill for various types of conduit nipples.

Maximum Number of Same-Size Conductors in Wireways

The sum of the cross-sectional area of all of the contained conductors in a wireway shall not exceed 20%. This 20% fill includes all conductors, not just current-carrying conductors. In addition, wireways shall not contain conductors larger than what they are designed for. Requirements for auxiliary gutters are found in 366, requirements for metallic wireways in 376, and requirements for nonmetallic wireways in 378.

The NEC® does not provide wireway tables with the 20% fill calculated. To find the maximum fill allowed in a wireway, first find the total area of the wireway in square inches by multiplying the depth of the wireway times the height. A 6″ × 6″ standard size wireway is 36 sq in. The area (36 sq in.) is then multiplied by the 20% fill factor, which leaves 7.2 sq in. for conductor fill. Once the conductor fill is found, the number of same-size conductors can be determined by dividing the allowed conductor fill of the wireway by the area of the conductors being installed. **See Figure 10-8.**

For example, how many No. 3/0 THHN conductors can be installed in a 6″ × 6″ wireway?

Conductor fill = area of gutter × 20%
Conductor fill = (6″ × 6″) × 20%
Conductor fill = 36 sq in. × 0.20
Conductor fill = 7.2 sq in.

Number of same-size conductors = conductor fill/area of conductor.

See Table 5 to get the area of No. 3/0 THHN conductor.

Number of same-size conductors = area of wireway/area of conductors.
Number of same-size conductors = 7.2/0.2679
Number of same-size conductors = 26.87
Number of same-size conductors = **26**

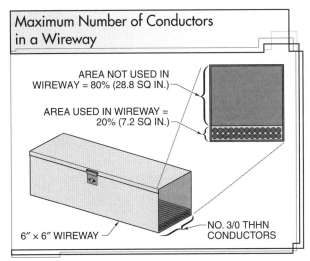

Figure 10-8. The number of same-size conductors can be determined by dividing the conductor fill of the wireway by the area of the conductors being installed.

> **Practice Question**
>
> *Determining the Maximum Number of Same-Size Conductors in a Wireway*
>
> 1. How many No. 1/0 THWN conductors can be installed in a 4″ × 4″ wireway?

Sizing Wireways

Sizing a wireway to hold a predetermined number of conductors is similar to sizing conduit. First, determine the total area of the conductors in square inches. This is found by using Table 5 for insulated conductors, Table 5A for compact conductors, or Table 8 for bare conductors. Take the total area of these conductors, divide by 20% or multiply by 5, and take the square root to find the appropriate size wireway. **See Figure 10-9.**

For example, what size of wireway is required for the following THNN conductors?

- four 500 kcmil THNN conductors
- four 300 kcmil THNN conductors
- three 250 kcmil THNN conductors

See Table 5 to determine the area of each conductor.
500 kcmil THHN = 0.7073 sq in.
300 kcmil THHN = 0.4608 sq in.
250 kcmil THHN = 0.3970 sq in.

Multiply the area by the number of conductors:
0.7073 sq in. × 4 = 2.8292 sq in.
0.4608 sq in. × 4 = 1.8432 sq in.
0.3970 sq in. × 3 = 1.1910 sq in.

Add the area of all the conductors:
2.8292 sq in. + 1.8432 sq in. + 1.1910 sq in. = 5.8634 sq in.

Divide the total area by 20% or multiply by 5 and take the square root:
Wireway size = $\sqrt{5.8634 \times 5}$
Wireway size = $\sqrt{29.317}$
Wireway size = 5.41

The next highest standard wireway size from 5.41″ is 6″ × 6″
Wireway size = **6″ × 6″**

> **Practice Question**
>
> *Determining Wireway Size*
>
> 1. What size of wireway is needed for the following conductors?
>   - three No. 1/0 THHN conductors
>   - four No. 2/0 THHN conductors
>   - five No. 3/0 THHN conductors

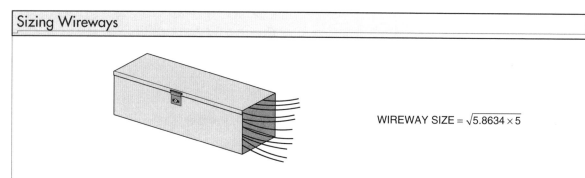

Sizing Wireways

WIREWAY SIZE = $\sqrt{5.8634 \times 5}$

INSULATION TYPE	SIZE OF CONDUCTOR	NUMBER OF CONDUCTORS	AREA IN SQ IN.	TOTAL AREA IN SQ IN.
THHN	500 KCMIL	4	0.7073	2.8292
THHN	300 KCMIL	4	0.4608	1.8432
THHN	250 KCMIL	3	0.3970	1.1910
			Total Area in Sq In.	5.8634

Figure 10-9. Sizing a wireway to hold a predetermined number of conductors is similar to sizing conduit.

INSTALLING CONDUCTORS INTO EXISTING RACEWAYS

It is common for electricians to install conductors into a conduit raceway system that already has conductors installed in them. If this is the case, the amount of space being used by the conductors must be determined in order to know the amount of space available to insert more conductors. **See Figure 10-10.** Apply the following steps to determine the number of conductors that may be added to a conduit:

1. Determine the space occupied by existing conductors by using Table 5.
2. Determine the allowable fill (typically 40%) based on conduit type and number of conductors already installed. See Table 4 for the type of conduit.
3. Subtract total conductor area from the allowable fill to find the remaining space left.
4. Divide the remaining space by the area in square inches of the new conductors being installed. The area for conductors is found in Table 5.

For example, how many No. 10 THHN conductors can be installed in a 1½″ EMT conduit that already has ten No. 8 THHN conductors installed in it?

See Table 5 to determine the area of the No. 8 THHN conductors.

No. 8 THHN = 0.0366 sq in.

Multiply the area by the number of conductors in the conduit:

0.0366 sq in. × 10 = 0.366 sq in.

See Table 4.

1½″ EMT with over 2 conductors = 0.814 sq in.

Subtract the area of the conductors in the conduit from the area allowed in the conduit to get the area that is left in the conduit.

Area left in conduit = 0.814 sq in. − 0.366 sq in.

Area left in conduit = 0.448 sq in.

Divide the remaining space available in the conduit by the area of the new conductors being installed.

No. 10 THHN conductors = 0.0211 sq in.

0.448/0.0211 = 21.23

Round down to 21.

Number of No. 10 THHN conductors allowed in conduit = **21 conductors**

BOX FILL

Box fill is the total volume of all conductors, devices, and fittings in a box. There are many types of boxes such as octagon, square, device, masonry, FS, and FD. Boxes shall be large enough to provide enough free space for all of the enclosed conductors. This free space is needed to dissipate the heat given off by the conductors.

The box size is determined by the box fill. Box fill includes all the parts attached to the box with a marked cubic-inch rating. This includes plaster rings, raised covers, and extension rings. If two boxes are joined, or "ganged" together, then the volumes in cubic inches of the boxes are added together to get the total volume of the box in cubic inches.

Requirements for boxes are found in Article 314. Boxes can be made of metal, plastic, or fiberglass. Plastic and fiberglass boxes usually have the size marked inside of them. Some even list the number of same-size conductors that can be installed. Sometimes barriers will be used to divide the box into sections. When this is the case, each barrier counts as ½ cubic inch for each metal barrier and 1 cubic inch for each plastic barrier unless the barrier is marked with its volume.

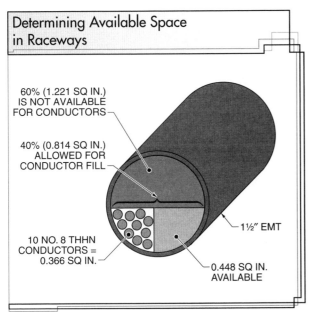

Figure 10-10. When adding more conductors to an existing conduit, the amount of space being used by the conductors must be determined in order to know the amount of space available to insert more conductors.

TECH TIP

Per 378.10(2) and (3), nonmetallic wireways may be used in areas subject to corrosive vapors where identified for the use and in wet locations where listed for the purpose.

Section 314.16 is used to determine the number of conductors in outlets, devices, and junction boxes. There are two types of box-fill calculations. One calculation is used for conductors of all the same size. The other calculation is for conductors of different sizes or with a device installed in them.

Before determining the number of conductors allowed in a box, it must be known what counts as a conductor. See 314.16 B for box-fill calculations.

Box-Fill Calculations—314.16(B)

Items in a box that are not conductors such as fittings, clamps, devices, and equipment grounding conductors reduce the amount of conductors allowed in a box. Understanding what counts as a conductor and what does not is important when determining box fill. Some common items that are installed inside a box that do not count as box fill include jumpers, pigtails, locknuts, wirenuts, and bushings. **See Figure 10-11.**

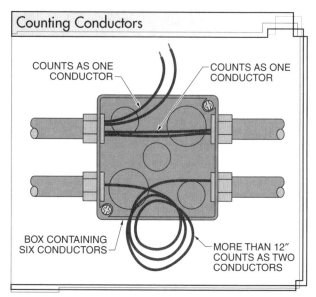

Figure 10-12. Conductors that start outside a box and enter a box to terminate and conductors that pass through the box count as one conductor. If a conductor has a loop or coil that is more than 12″ long, it is counted as two conductors.

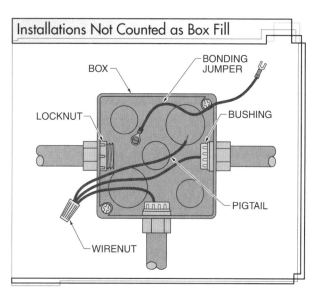

Figure 10-11. Installations that are not counted as part of the box fill include locknuts, wirenuts, pigtails, bushings, and bonding jumpers.

Conductors—314.16(B)(1). Each conductor that starts outside a box and enters a box to terminate or to be spliced counts as one conductor. Each conductor passing through a box counts as one conductor. If the conductor has a loop or coil that is more than 12″, it is counted as two conductors. **See Figure 10-12.** An exception allows four or fewer No. 16 or smaller fixture wires entering in through a domed cover from a fixture or fan to be omitted from the calculation. Conductor insulation is not a factor considered for box fill.

Clamps—314.16(B)(2). All field- or factory-installed clamps inside a box are counted as one conductor, even if they are of different types. Clamps are sized for the largest conductor in the box. Clamps installed outside the box are not counted. **See Figure 10-13.**

Support Fittings—314.16(B)(3). Support fittings count as one conductor for each type of fitting in the box. They are sized based on the largest conductor in the box. When one or more fixture studs or hickeys are installed inside a box, they are counted as one conductor.

Devices or Equipment—314.16(B)(4). Devices or equipment count as two conductors based on the largest-size conductor connected to the device. This includes switches and receptacles that are mounted on a strap or yoke connected to the box. If a device or utilization equipment is wider than a single gang box (more than 2″), it shall have a double-volume allowance for each gang required for mounting of the device.

Equipment Grounding Conductors—314.16(B)(5). Equipment grounding conductors count as one conductor for all the grounds in the box. The size is based on the largest grounding conductor in the box.

Isolated Grounds—314.16(B)(5). Article section 250.146(D) allows a second equipment grounding conductor for isolated ground systems. This second set of grounding conductors counts as a second ground and is sized based on the largest conductor of that group.

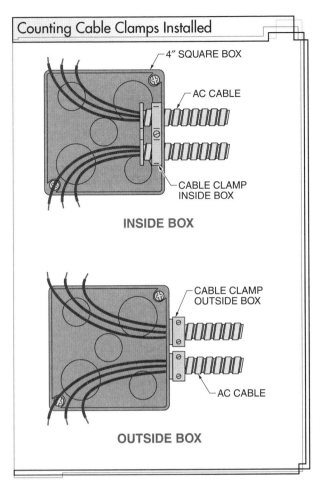

Figure 10-13. Clamps installed inside a box are counted as one conductor. Clamps installed outside a box are not counted.

Box Size and Fill for Same-Size Conductors

Table 314.16(A) can be used to find the total number of same-size conductors allowed in a box or to determine the required box size for a certain number of conductors. When determining conductor size for box-fill calculations, the AWG size in cubic inches is used. The type of insulation is not a factor. Table 314.16(A) lists the standard size of metal boxes recognized by the NEC® and the sizes of No. 18 to No. 6 conductors. This table is used to find the volume for boxes in cubic inches as well as the maximum number of conductors that are allowed if the conductors are all the same size. Keep in mind that the number of conductors allowed is reduced when additional features such as switches, straps, clamps, or equipment grounding conductors are installed in the box. These deductions are based on 314.16(B)(1) through 314.16(B)(5).

The box size and type of box is listed on the left side of Table 314.16(A). The middle column lists the minimum volume of the box in cubic inches. The size of the conductors is listed on the left side of the table. Under each AWG size is the number of conductors that are allowed in the box. These numbers do not take into account any additional installations in the box such as devices, clamps, or support fittings. **See Figure 10-14.**

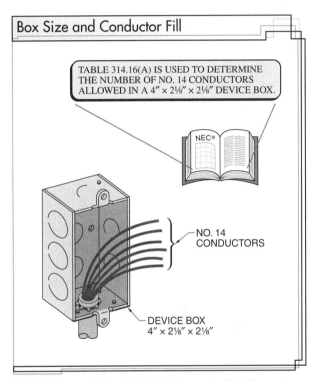

Figure 10-14. Table 314.16(A) is used to find the volume of boxes in cubic inches as well as the maximum number of conductors that are allowed if the conductors are all the same size.

For example, how many No. 14 conductors can be installed in a 4″ × 2⅛″ × 2⅛″ device box?
 See Table 314.16(A).
 Number of No. 14 conductors installed in a 4″ × 2⅛″ × 2⅛″ device box = **7 conductors**

For example, what is the depth of a 4″ square box containing nine No. 12 THW conductors?
 See Table 314.16(A).
 Depth of box = **1½″**

Practice Questions

Determining Box Size and Fill for Same-Size Conductors

1. What size of square junction box is required for six No. 12 THHN conductors and three No. 12 THW conductors?
2. How many No. 14 THHN conductors are allowed in a 4″ × 1½″ octagon box?

Box Size and Fill for Different-Size Conductors

When conductors of different sizes are installed in a box, Table 314.16(B) is used to determine the correct box size and fill. Table 314.16(B) provides the cubic inches for conductors No. 18 AWG to No. 6 AWG. The AWG sizes are on the left side of the tables, with the volume in cubic inches on the right. These numbers do not take into account any additional installations in the box such as clamps or support fittings. Once the total volume in cubic inches is found for all of the conductors, Table 314.16(A) is used to find the correct box size in cubic inches. Use the box size that just covers the cubic-inch requirement. To determine the correct box size for conductors of various sizes, follow the steps below:

1. Determine size and number of each conductor. **See Figure 10-15.**
2. Check for clamps and fittings.
3. Check for devices and determine the wire size that terminates to the devices.
4. Check for grounds.
5. Determine the total volume in cubic inches for each conductor, device, fitting, and equipment grounding conductor.
6. Use Table 314.16(A) to determine the type of box required.

For example, what size of square junction box is required for three No. 14 conductors spliced to three other No. 14 conductors and six No. 12 conductors passing through the box?
See Table 314.16(B).
No. 14 conductors = 2 cu in.
2 cu in. × 6 = 12 cu in.
No. 12 conductors = 2.25 cu in.
2.25 cu in. × 6 = 13.5 cu in.
Volume of all conductors = 12 cu in. + 13.5 cu in.
Volume of all conductors = 25.5 cu in.
See Table 314.16(A).
Minimum value of 25.5 cu in. = **4¹¹⁄₁₆″ × 1¼″**

TECH TIP

Pull boxes and junction boxes are points in the electrical system that provide access to the raceways entering and leaving the boxes. A pull box is a box used as a point to pull or feed electrical conductors into the raceway system. A junction box is a box in which splices, taps, or terminations are made.

Practice Question

Determining Box Size and Fill for Different-Size Conductors

1. What size of device box is required for a 14/3 type NM cable with a ground that terminates on a switch and a 12/2 NM cable with a ground that terminates on a receptacle? The box has internal clamps.

Plaster Rings and Raised Covers

Plaster rings and raised covers that are marked with the volume in cubic inches can be used as part of the total volume for box fill. If the plaster ring or raised cover does not have the volume marked in cubic inches, then it cannot be counted as part of the box assembly. The volume in cubic inches of the ring or raised cover is subtracted from the total volume in cubic inches of the conductors, devices, fittings, grounds, or clamps that are in the box to find the minimum size box. **See Figure 10-16.**

For example, what is the minimum volume in cubic inches required for a box that has a 4″ plaster ring on it that is rated for 2.75 cu in.? The box has four No. 14 THHN conductors passing through it, four No. 12 THHN conductors connected to a receptacle, one No. 14 ground wire, and one No. 12 ground wire. The box also has one No. 12 equipment bonding jumper in it.
See Table 314.16(B).
No. 14 conductors = 2 cu in.
2 cu in. × 4 = 8 cu in.
No. 12 conductors = 2.25 cu in.
2.25 cu in. × 4 = 9 cu in.
1 receptacle = 2.25 cu in.
2.25 cu in. × 2 = 4.5 cu in.
1 ground wire = 2.25 cu in.
8 cu in. + 9 cu in. + 4.5 cu in. + 2.25 cu in. = 23.75 cu in.
Total volume of all conductors = 23.75 cu in.
Subtract the 4″ (2.75) plaster ring from the total:
23.75 cu in. − 2.75 cu in. = 21 cu in.
Minimum volume in cu in. = **21 cu in.**

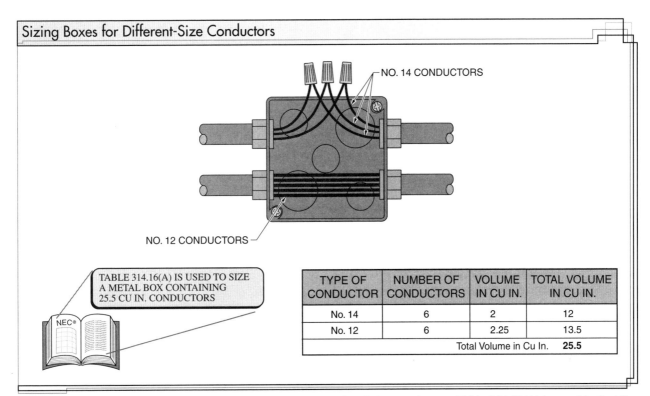

Figure 10-15. Once the total volume in cubic inches is found for all the conductors, Table 314.16(A) is used to find the correct box size.

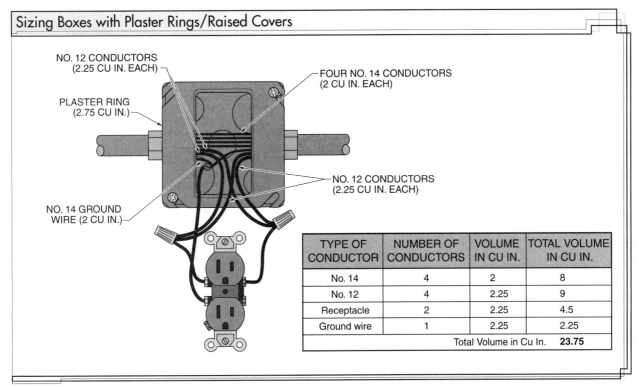

Figure 10-16. The volume listed on plaster rings or raised covers is subtracted from the total volume of the conductors, devices, fittings, grounds, or clamps to determine minimum box size.

Practice Question

Determining Box Size with Plaster Rings and Raised Covers

1. What is the depth of a 4″ square junction box containing the following conductors?
   - two No. 12 THW conductors passing through the box
   - two No. 12 THW conductors passing through the box
   - four No. 14 conductors that terminate to a receptacle
   - four No. 10 conductors that are spliced together
   - No. 10 ground wire
   - No. 14 ground wire
   - No. 12 ground wire

   The box has a raised cover that is marked 3.6 cu in.

SIZING PULL BOXES AND JUNCTION BOXES

A *pull box* is a box used as a point to route electrical conductors into a raceway system. A *junction box* is a box in which splices, taps, or terminations are made. Sizing these boxes depends on how the conductors enter and leave the enclosure. It also depends on the voltage rating of the conductors being used. The method used to determine a box size of 1000 V and under is different from the method used for a size of over 1000 V. Both methods contain calculations for straight, angle, and U-pulls. **See Figure 10-17.**

Sizing Pull Boxes and Junction Boxes—1000 Volts and Under

Boxes are sized for conductors that are No. 4 and larger. When No. 4 or larger conductors are in a raceway, the pull box is sized for the conduits entering the box. Requirements in 314.28(A)(1) are for straight pulls, while requirements in 314.28(A)(2) are for angle pulls, U-pulls, and conductors with splices.

Sizing Pull Boxes and Junction Boxes for Straight Pulls—1000 Volts and Under. A *straight pull* is a condition where conductors are pulled into a box and routed to a raceway on the opposite wall of the box. See 314.28(A)(1) for straight-pull requirements. The length of the box shall be eight times the largest conduit for conductors that enter one side of the box and leave the opposite side. Straight-pull questions on the exam will only ask for the length of the box and not the depth or width. **See Figure 10-18.**

For example, what is the minimum length for a box that houses a straight pull if there is a 3″ conduit entering one side and a 2½″ conduit leaving the opposite side?

Length of box = largest conduit × 8
Length of box = 3″ × 8
Length of box = **24″**

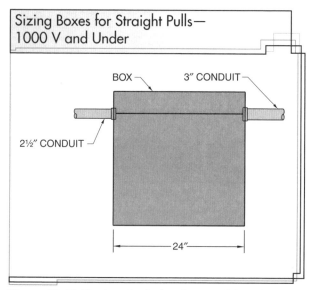

Figure 10-18. The length of a box for a straight pull is eight times the largest conduit entering the box.

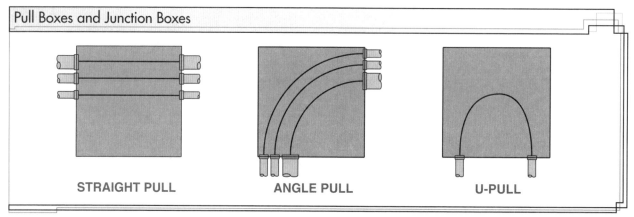

Figure 10-17. Pull boxes and junction boxes are sized for straight pulls, angle pulls, and U-pulls.

Practice Question

Determining Box Size for Straight Pulls—1000 Volts and Under

1. What is the minimum length of a box that houses a straight pull if there is a 3″ conduit entering one side and a 3½″ conduit leaving the opposite side?

Sizing Pull Boxes and Junction Boxes for Angle Pulls—1000 Volts and Under. An *angle pull* is a condition where conductors are pulled into a box and routed to a raceway entering an adjacent wall in the box. See 314.28(A)(2) for angle-pull requirements. The vertical and horizontal length of a box used for an angle pull shall be six times the largest conduit plus the trade size of any other conduits on the same side of the box. This determines the length of the box from the wall where the conductors enter to the opposite wall. If the box has multiple rows of conduits on one side, then each row is treated as a separate box. The length of the box is determined by the largest row. **See Figure 10-19.**

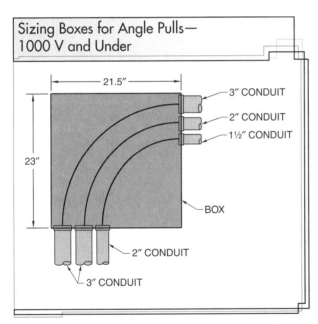

Figure 10-19. The length of a box for an angle pull is six times the largest conduit plus the size of any other conduits on the same side of the box.

For example, what size of box is required with one 3″ conduit, one 2″ conduit, and one 1½″ conduit on one side of the box, while two 3″ conduits and one 2″ conduit are on the adjacent side of the box?

Multiply the largest conduit by 6:
3″ × 6 = 18″
Add the sizes of the remaining conduits on the same side of the box:
18″ + 2″ + 1.5″ = 21.5″
Length of box = 21.5″

Multiply the largest conduit on the other side of the box by 6:
3″ × 6 = 18″
Add the sizes of the remaining conduits on the same side of the box:
18″ + 3″ + 2″ = 23″
Width of box = 23″
Box size = **21.5″ × 23″**

Practice Question

Determining Box Size for Angle Pulls—1000 Volts and Under

1. What size of box is required for an angle pull if there is a 3″ conduit, a 2″ conduit, and a 1″ conduit on the right side, while a 3″ conduit, a 2½″ conduit, and a 4″ conduit are on the bottom side of the box?

Sizing Pull Boxes and Junction Boxes for U-Pulls—1000 Volts and Under. A *U-pull* is a condition in which conductors are pulled into a box and routed to another raceway entering the same wall of the box. See 314.28(A)(2) for U-pull requirements. A box with a U-pull shall be sized at six times the largest conduit plus the trade size of the other conduits on the same side of the box. This determines the length of the box from the wall where the conductors enter to the opposite wall. U-pull questions on the exam will only ask for the length of the box and not the width. To make sure that the conductors' bending radius is not too small, there must be a distance of at least six times the largest conduit size between the two raceways with the same conductors. **See Figure 10-20.**

For example, what is the minimum length required for a U-pull enclosure that has two 4″ conduits on one side of the box?

Multiply the largest conduit by 6:
4″ × 6 = 24″
Add the sizes of the remaining conduits on the same side of the box:
24″ + 4″ = 28″
Length of box = **28″**

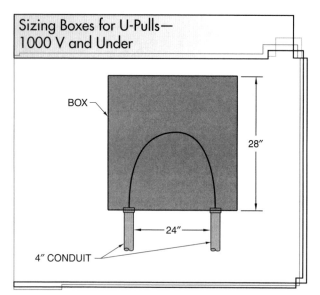

Figure 10-20. The length of a box for a U-pull is six times the largest conduit plus the trade size of the other conduits on the same side of the box.

For example, what is the distance between the conduits for a U-pull enclosure that has two 4″ conduits on one side of the box?

Multiply the largest raceway by 6:
4″ × 6 = 24″
Distance between conduits = **24″**

Practice Questions

Determining Box Size for U-Pulls—1000 Volts and Under

1. What is the length of a box using a U-pull that has two 3½″ conduits on one side of the box?
2. What is the distance between the conduits for a U-pull enclosure that has two 3½″ conduits on one side of the box?

Depth of Boxes with Removable Covers—1000 Volts and Under. The box depth is determined by the size of the conductors that enter into the box. The box should also be deep enough for the installation of conduits entering the side of the box. If a conduit comes through the back of the box (opposite the removable cover), it is sized per Table 312.6(A) for one conductor. This applies not only to boxes but to conduit bodies as well. **See Figure 10-21.**

For example, how deep shall a box be if 250 kcmil THHN conductors are entering through the back of the box?

See Table 312.6(A) for one terminal.
250 kcmil conductors = 4½″
Depth of box = **4½″**

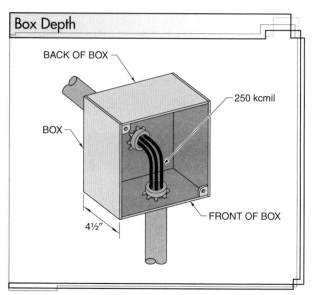

Figure 10-21. Use Table 312.6(A) to determine the wire-bending space and depth for a box when conduit enters through the back.

Practice Question

Determining Depth of Boxes with Removable Covers

1. How deep of a box is required for three 350 kcmil THW aluminum conductors?

Distance between Conduits in Pull Boxes and Junction Boxes that Contain the Same Conductors—1000 Volts and Under

When conduits contain the same conductors in an angle pull or U-pull, the distance between the two conduits shall be at least six times the largest raceway containing the same conductors. This distance is measured from one inside edge to the other inside edge of the conduits. **See Figure 10-22.**

For example, how far apart from each other must two 4″ conduits on an angle pull be placed?

Multiply the largest conduit by 6:
4″ × 6 = 24″
Distance between conduits = **24″**

Practice Question

Determining Distance between Conduits Containing the Same Conductors—1000 Volts and Under

1. How far apart from each other should two 3.5″ conduits be placed when connected to a box with a U-pull?

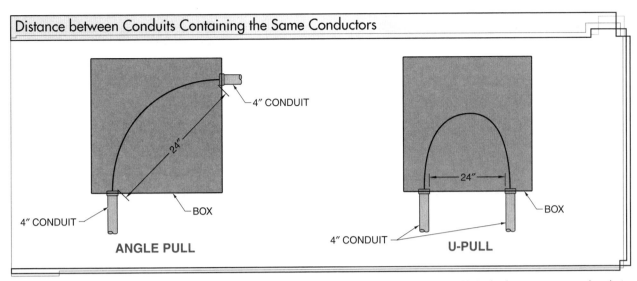

Figure 10-22. When conduits contain the same conductors in an angle pull or U-pull, multiply the largest raceway by six to determine the distance between the two conduits.

Handhole Enclosures—1000 Volts and Under. A *handhole* is a box, large enough to reach into but not large enough for a person to get into, that is placed in the ground and used for installing, operating, or maintaining equipment or wires. Requirements for handhole enclosures are found in Article 314.30. Handhole enclosures are not required to have a closed bottom, but since they are in the ground, they must be able to withstand any load that might be imposed on them. If the box does have a closed bottom, the conduit entries do not have to be connected mechanically to the box.

Sizing handhole enclosures for straight pulls or for angle pulls is the same as sizing for a pull box. Reference 314.28(A) for 1000 V and under and 314.71 for voltage over 1000 V.

If the enclosure does not have a bottom in it, the depth of the box to the removable cover is measured from the end of the conduits that are stubbed up in the handhole enclosure. The lid for the handhole enclosure should be fastened down to the box and tools must be used to open it unless it weighs more than 100 lb. The lid must also be marked to show what the box is for. If the box is made of metal, it shall be bonded to the equipment grounding conductor.

Sizing Pull Boxes and Junction Boxes—Over 1000 Volts

Requirements for sizing pull boxes over 1000 V are found in 314.71. The size of the box is determined by the diameter of the conductors (in inches). It must be determined if conductors or cables are lead-covered, shielded, or nonshielded cable. The rating of the conductors is used to determine if a circuit is over 1000 V, not the actual voltage flowing through the conductors. Some plants require the use of 5 kV cable for circuits below 1000 V.

Sizing Pull Boxes and Junction Boxes for Straight Pulls—Over 1000 Volts. See 314.71(A) for straight-pull requirements. Different multipliers are used for sizing a pull box for a straight pull that is over 1000 V. These multipliers are based on whether the cable is shielded or nonshielded. Per 310.10(E), all conductors rated over 2000 V shall be shielded. For conductors that are shielded or lead-covered, multiply the largest outside diameter of the conductor by 48. For conductors that are nonshielded, multiply the largest outside diameter of the conductor by 32. **See Figure 10-23.**

For example, what is the length of a box for 5 kV non-shielded cables that have an outside diameter of 1.25″?

Length of box = outside diameter of largest conductor × 32
Length of box = 1.25″ × 32
Length of box = **40″**

Practice Question

Determining Box Size for Straight Pulls—Over 1000 Volts

1. What size of box is required for a straight pull for 15 kV shielded cable that has an outside diameter of 1.375″?

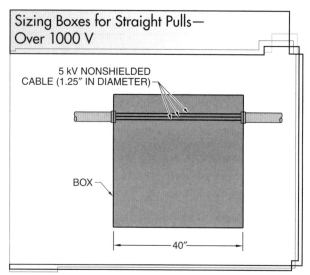

Figure 10-23. To determine the length of a box for a straight pull with conductors rated over 1000 V, multiply the largest outside diameter of the conductor by 48 for shielded or lead-covered conductors, or multiply by 32 for conductors that are nonshielded.

Sizing Pull Boxes and Junction Boxes for Angle Pulls and U-Pulls—Over 1000 Volts. See 314.71(B)(1) for angle-pull and U-pull requirements for boxes over 1000 V. For conductors that are shielded or lead-covered, multiply the largest outside diameter of the conductor by 36, and then add the outside diameter of all of the other conductors on the same side of the box. For conductors that are nonshielded, multiply the largest outside diameter of the conductor by 24, and then add the outside diameter of all of the other conductors on the same side of the box. This determines the length of the box from the wall where the conductors enter to the opposite wall. **See Figure 10-24.**

For example, what size of box is required for an angle pull containing three 15 kV shielded cables, each with an outside diameter of 1.45″?

   Multiply the largest conductor by 36:
   1.45″ × 36 = 52.2″
   Add the outside diameter of the remaining conduits on the same side of the box:
   52.2″ + 1.45″ + 1.45″ = 55.1″
   Length of box = **55.1″**

Practice Question

Determining Box Size for Angle Pulls and U-Pulls—Over 1000 Volts

1. What size of box is required for an angle pull containing three 5 kV cables, each with an outside diameter of 1.5″? The cables are shielded and installed in 5″ RMC.

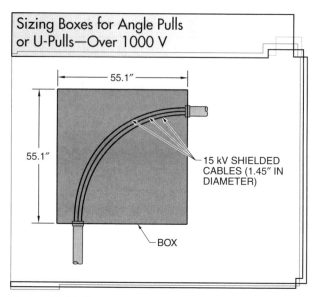

Figure 10-24. To determine the length of a box for an angle pull or U-pull with conductors rated over 1000 V, multiply the largest outside diameter of the conductor by 36 for shielded or lead-covered conductors, or multiply by 24 for conductors that are nonshielded. Then, add the outside diameter of all of the other conductors on the same side of the box.

Distance between Conduits in Pull Boxes and Junction Boxes—Over 1000 Volts. See 314.71(B)(2) for distance requirements between conduits for boxes over 1000 V. Conductor entry and exit points being too close to each other may cause the conductors to bend too tightly, causing damage to the insulation during installation. For conductors that are shielded or lead-covered, the distance between openings shall be 36 times the outside diameter of the same conductors entering and exiting the box. For conductors that are nonshielded, the distance is 24 times the outside diameter of the same conductor that enters and exits the box. **See Figure 10-25.**

TECH TIP

Frequently, code and licensing examinations focus on outlet and junction-box sizing requirements. Apprentices and students of the NEC® should pay close attention to 314.28 for pull- and junction-box sizing requirements. In addition, two tables are used to ensure that the box selected can safely handle the number of conductors that may be installed within it. Table 314.16(A) lists the maximum number of conductors permitted for various sizes and types of metal boxes. Table 314.16(B) contains the volume allowances per conductor size for use in calculating box fill where different-size conductors are present.

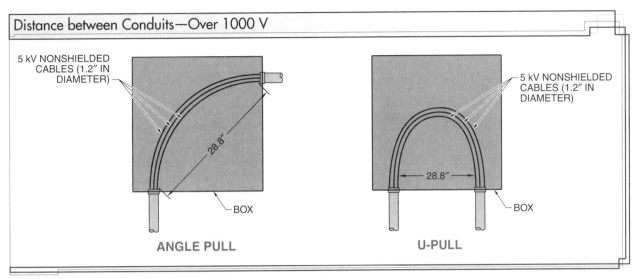

Figure 10-25. To determine the distance between conduits in a box over 1000 V, multiply the largest outside diameter of the conductor by 36 for shielded or lead-covered conductors, or multiply by 24 for conductors that are nonshielded.

For example, what is the distance between conduits for nonshielded 5 kV cables that are 1.2″ in diameter?

Multiply the outside diameter of the nonshielded conductor by 24:

1.2″ × 24 = 28.8″

Distance between conduits = **28.8″**

Practice Question

Determining Distance between Conduits in Boxes—Over 1000 Volts

1. What is the required distance between two 6″ PVC schedule 80 conduits that form a U-pull in a box that has three 15 kV shielded cables with an outside diameter of 1.3″?

Boxes are used in conduit runs for splicing conductors, allowing access to conductors during installation, and for branching off in a conduit run to supply additional circuits.

CHAPTER 10
Conductor-Fill, Box-Fill, and Pull and Junction Box Sizing

Review Questions

Name _____   Date _____

_____ 1. If a 1½" × 4" octagon outlet box contains a stud and a cable clamp, the number of wires must be reduced to ___ if using No. 14 conductors.
   A. four
   B. five
   C. six
   D. seven

_____ 2. A pull box containing a U-pull that has two, 3" schedule 80 PVC conduits entering from the bottom will have a vertical measurement of ___".
   A. 21
   B. 24
   C. 26
   D. 30

_____ 3. A box containing two No. 12 THHN grounding conductors and two No. 12 THW conductors will have a required volume of ___ cu in.
   A. 4.5
   B. 5
   C. 6.75
   D. 9

_____ 4. A device box that contains two switches, one receptacle, and three 14/2 NM (Romex) cables with grounds is required to be ___.
   A. 3" × 2" × 2¼"
   B. 3" × 2" × 2¾"
   C. 3" × 2" × 2"
   D. 4" × 2⅛" × 2⅛"

_____ 5. A 4" × 1½" square box containing two No. 14/2 Romex cables with ground, one duplex receptacle, and a 3.6 cu in. raised cover may contain ___ more No. 14/2 Romex cable(s) with grounds.
   A. one
   B. two
   C. three
   D. four

6. A pull box sized at ___ is required for an angle pull that has 4″, 2″, and 1.5″ conduits entering on the right side and 4″, 3″, and 3″ conduits leaving the bottom side of the box.
   A. 27.5″ × 30″
   B. 27.5″ × 32″
   C. 30″ × 30″
   D. 32″ × 30″

7. A 4″ × 4″ × 1½″ square junction box that has four Romex connectors inside may have up to ___ No. 10 conductors installed in it.
   A. six
   B. seven
   C. eight
   D. nine

8. A pull box containing a straight pull with 4″ EMT conduit entering on the right side and a 3″ EMT conduit leaving the left side will have a length of ___″.
   A. 24
   B. 27
   C. 32
   D. 35

9. A 2″ PVC schedule 80 raceway that is 24′ long containing eight conductors will have an allowable fill of ___ sq in.
   A. 0.891
   B. 1.150
   C. 1.913
   D. 2.874

10. EMT conduit sized at ___″ is required for ten No. 2 THW conductors.
    A. 1¼
    B. 1½
    C. 2
    D. 2½

11. A 2″ PVC schedule 80 conduit may contain up to ___ No. 6 XHHW-2 conductors.
    A. 18
    B. 19
    C. 20
    D. 24

12. An EMT conduit nipple containing fourteen No. 8 THHN conductors and seven No. 10 THHN conductors is sized at ___″.
    A. 1
    B. 1¼
    C. 1½
    D. 2

13. A ⅜" flexible metal conduit with inside fittings may contain up to ___ No. 12 THHN aluminum conductors.
    A. two
    B. three
    C. four
    D. seven

14. The internal diameter of a 3" PVC schedule 80 conduit is ___".
    A. 2.469
    B. 2.95
    C. 3.042
    D. 3.068

15. The area of a No. 2 THWN copper conductor that is to be installed in 2" RMC is ___ sq in.
    A. 0.0824
    B. 0.1158
    C. 0.1562
    D. 0.384

CHAPTER 11

Dwellings

Digital Resources
ATPeResources.com/QuickLinks
Access Code: 637160

A dwelling, according to Article 100, is a single unit providing complete and independent living facilities for one or more persons, including permanent provisions for living, sleeping, cooking, and sanitation. Dwellings can be one-family, two-family, or multifamily apartments. The NEC® provides a standard calculation method and an optional calculation method to determine the load for one-family and multifamily dwellings. Forms designed specifically for determining the standard calculation method and the optional calculation method standardize and simplify the process. An electrician taking a master electrician exam will have to know how to complete all parts of the dwelling calculations in order to find the total service load and the size of service conductors.

OBJECTIVES

After completing this chapter, the learner will be able to do the following:

- Describe the steps involved in calculating the general lighting load for one-family dwellings.
- Determine the demand load for dryers in a dwelling.
- Use Table 220.55 to determine one-family cooking-equipment loads.
- Identify how to size service conductors for one-family dwellings.
- Describe the two main areas that make up the optional calculation method for one-family dwellings.
- Identify the main difference between the standard calculation method for multifamily dwellings and for one-family dwellings.
- Determine the service load for mobile and manufactured homes.
- Identify the demand factors used to make load calculations for mobile home parks.

STANDARD CALCULATION METHOD FOR ONE-FAMILY DWELLINGS

The standard calculation method for one-family dwellings contains six individual calculations that are performed before the minimum-size service or feeder conductors required for the calculated load can be determined. The six calculations include general lighting, fixed appliances, dryer, cooking equipment, heating or air conditioning, and largest motor calculations. **See Figure 11-1.** Standard calculation forms can be used to keep track of each load. **See Appendix.**

Figure 11-1. Forms designed specifically for determining the standard calculation for one-family dwellings contain six individual calculations that are performed before the size of the service or feeder conductors can be determined.

General Lighting for One-Family Dwellings

The general lighting section includes all of the lighting and general-purpose receptacles as well as the small-appliance and laundry circuits that are required in a dwelling. Areas such as attics, open porches, garages, and unused or unfinished spaces not adaptable for future use are not considered in the general lighting section. Section 220.12 and Table 220.12 provide information needed to calculate the general lighting load.

If the dwelling is designed for energy codes that have been adopted by the local authority having jurisdiction, then the energy code figures can be used. If the local energy codes are used, then three criteria must be met: a power-monitoring system is installed in the dwelling, alarms are built into the power-monitoring system, and the derating factors from 220.42 have not been used for the lighting load.

There are two steps used in calculating the general lighting load. The first step is to calculate the area of the building or structure in square feet. The area of a building is found by multiplying the width by the length. This is done by using the outside dimensions of the building, dwelling unit, or other area involved. If the building has a basement or a second story that can be used for living space, it must also be calculated into the total square footage of the dwelling, even if the space is not finished at the time. **See Figure 11-2.**

The second step is to find the volt-amp value per square foot from Table 220.12 for the applicable occupancy and to multiply that value by the calculated floor area. Table 220.12 provides the volt-amps (VA) per square foot for many common types of buildings including dwellings. Table 220.12 lists the unit lighting load for dwellings as 3 VA per square foot.

For example, what is the lighting load for a dwelling that is 60′ × 30′?

Area = 60′ × 30′
Area = 1800 sq ft
See Table 220.12 for load in volt-amp per square foot for dwelling units.

Lighting load = volt-amps per square foot × area
Lighting load = 3 × 1800
Lighting load = **5400 VA**

Practice Questions

Determining Lighting Load for One-Family Dwellings—Standard Calculation Method

1. What is the general lighting load per square foot for dwellings?
2. What is the general lighting load for a 3000 sq ft home?
3. What is the general lighting load for a 3600 sq ft home that has a 300 sq ft garage included in the total square feet?

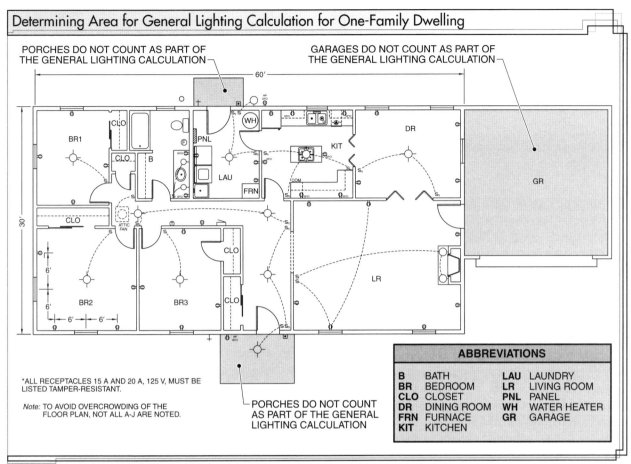

Figure 11-2. Areas such as attics, open porches, garages, and unused or unfinished spaces not adaptable for future use are not considered in the general lighting section.

Minimum Number of Branch Circuits Required for One-Family Dwellings. Section 210.11 contains provisions for determining the minimum number of branch circuits, including the requirements for small-appliance branch circuits, laundry branch circuits, bathroom branch circuits, and garage receptacle circuits.

Per 210.11(B), the required branch circuits shall be evenly divided across the branch-circuit panelboard. In addition, branch-circuit overcurrent devices and circuits need only be provided for connected loads. Any future loads need not be considered. See 210.11(A through C).

To find the required number of circuits for the lights and general-purpose receptacles, first determine the current for the lighting and receptacle circuits. To find the current, multiply 3 VA/sq ft by the square footage of the dwelling. This is the total load in volt-amps. Divide the total volt-amps by the voltage, which is usually 120 V. This determines the current for the light and receptacle circuits. The current is then divided by the size of the overcurrent protective device, which is usually a 15 A or 20 A circuit breaker, to determine the required number of circuits. If the answer comes out to be a decimal, always round up to a whole number.

For example, how many 20 A, 120 V circuit breakers are required for an 1800 sq ft dwelling?

$$\text{Number of circuit breakers} = \frac{\frac{3 \text{ W/sq ft} \times \text{area of dwelling}}{\text{voltage}}}{\text{circuit breaker rating}}$$

$$\text{Number of circuit breakers} = \frac{\frac{3 \times 1800}{120}}{20}$$

$$\text{Number of circuit breakers} = \frac{\frac{5400}{120}}{20}$$

$$\text{Number of circuit breakers} = \frac{45}{20}$$

Number of circuit breakers = 2.25

Round up to 3

Number of circuit breakers = **3**

174 MASTER ELECTRICIAN'S EXAM WORKBOOK

Practice Questions

Determining the Minimum Number of Branch Circuits for One-Family Dwellings—Standard Calculation Method

1. How many 20 A, 120 V circuit breakers are required for the lighting and general purpose receptacles in a 4000 sq ft dwelling?

2. How many 15 A, 120 V circuit breakers are required for the lighting and general purpose receptacles in a 4000 sq ft dwelling?

Small-Appliance and Laundry Circuits for One-Family Dwellings. Per 210.11(C)(1), at least two 20 A circuits for small appliances shall be used to supply the kitchen, dining room, breakfast room, and pantry, as well as similar areas. Per 210.11(C)(2), a dwelling is required to have at least one laundry circuit. This circuit includes all the receptacles in the laundry area along with the washing machine. **See Figure 11-3.** The receptacles in the garage shall have a circuit that only contains garage receptacles and no other outlets inside the dwelling. However, the garage circuit may supply readily accessible outdoor receptacle outlets. All of these circuits shall be installed using No. 12 AWG copper or No. 10 aluminum conductors to meet the 20 A requirement.

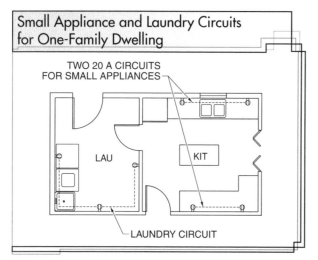

Figure 11-3. One-family dwellings should have at least two 20 A circuits for small appliances. The laundry circuit may include the washing machine and remaining receptacles in the laundry area.

General Lighting Demand Factors for One-Family Dwellings. The general lighting load for one-family dwellings consists of three separate loads that are calculated individually and added together, after which a demand factor may be applied. The first portion of the general lighting load is calculated from the total area of the dwelling. The second portion of the general lighting load consists of the calculation for small-appliance branch circuits. Each small-appliance circuit shall be sized at 1500 VA according to 220.52(A). The last portion of the general lighting load consists of the laundry load. Each laundry circuit shall be sized at 1500 VA according to 220.52(B). When these three sections are added together, the demand load for the general lighting section can be found.

Table 220.42 requires that the first 3000 VA of the general lighting load be calculated at 100%. Loads between 3001 VA and 120,000 VA are calculated at 35%, and any remaining load over 120,000 VA is calculated at 25%. The neutral load for the general lighting load is the same as the load of the ungrounded conductors. **See Figure 11-4.**

For example, what is the general lighting load for an 1800 sq ft dwelling with the minimum number of small-appliance and laundry circuits?

General lighting = 1800 sq ft × 3
General lighting = 5400 VA

Small appliances = 1500 VA × 2
Small appliances = 3000

Laundry = 1500 VA × 1
Laundry = 1500 VA

Total VA = general lighting + small appliances + laundry
Total VA = 5400 + 3000 + 1500
Total VA = 9900

First, the 3000 VA load is calculated at 100%.
3000 × 1.00 = 3000

Next, the load from 3001 VA to 120,000 VA is calculated at 35%.
9900 − 3000 = 6900
6900 × 0.35 = 2415 VA
3000 + 2415 = 5415 VA
The general lighting load including demand factors = **5415 VA**

Practice Questions

Determining General Lighting Load for One-Family Dwellings—Standard Calculation Method

1. What is the general lighting demand load for a 3000 sq ft dwelling with the minimum number of small-appliance and laundry circuits?

2. What is the general lighting demand load for a 2400 sq ft home that has three small-appliance circuits and one laundry circuit?

General Lighting for One-Family Dwellings—Standard Calculation Method

1. GENERAL LIGHTING: *Table 220.12*

__1800__ sq ft × 3 VA =	__5400__ VA	
Small appliances: *220.52(A)*		
__1500__ VA × __2__ circuits =	__3000__ VA	
Laundry *220.52(B)*		
__1500__ VA × 1 =	__1500__ VA	
	__9900__ VA	

Applying Demand Factors: *Table 220.42*

		PHASES	NEUTRAL
First 3000 VA × 100% =	3000 VA		
Next __6900__ VA × 35% =	__2415__ VA		
Remaining _____ VA × 25% =	_____ VA		
Total	__5415__ VA	__5415__ VA	__5415__ VA

Figure 11-4. Demand factors for dwelling units are applied to all general lighting loads of 3001 VA and greater.

Fixed Appliances for One-Family Dwellings

Per 220.53, four or more appliances that are fastened-in-place in a one-family dwelling are calculated at 75% of the total load for all four appliances. This demand factor applies to the nameplate rating of the appliances such as electric water heaters, garbage disposals, well pumps, spas, central heat blower motors, microwaves, and dishwashers. This demand factor does not apply to electric ranges, clothes dryers, space-heating equipment, or air conditioning equipment, which are served by the same feeder or service in a one-family, two-family, or multifamily dwelling. If there are fewer than four appliances, then no demand factor is applied. This is true for the neutral load as well. The 75% demand factor is only applied to the neutral when there are four or more loads rated at 120 V. **See Figure 11-5.**

For example, what is the fixed-appliance load for a 240 V water heater rated at 4500 W, a 120 V garbage disposal rated at 1000 W, a 120 V microwave rated at 1500 W, and a 120 V dishwasher rated at 1500 W?

Demand load = (4500 + 1000 + 1500 + 1500) × 75%
Demand load = 8500 × 0.75
Demand load = **6375 W**

Practice Question

Determining Fixed-Appliance Load for One-Family Dwellings—Standard Calculation Method

1. What is the fixed-appliance load for a 240 V water heater rated at 5000 W, a 120 V dishwasher rated at 1500 W, and a 120 V garbage disposal rated at 1000 W?

TECH TIP

The total calculated load is rarely placed on the electrical system of a one-family or multifamily dwelling because all of the electrical loads are not on at the same time. This is why demand factors may be applied.

Dryer Load for One-Family Dwellings

The load for household clothes dryers in dwelling units is required by 220.54 to be 5000 VA or the nameplate rating, whichever is greater. Table 220.54 provides demand factors for household electric clothes dryers. It does not allow a reduction in the load for the first four dryers supplied by a feeder or service. Where the number of dryers exceeds four, the load is reduced by the applicable demand factor found in Table 220.54. A calculation is required to find the demand for 12 through 23 dryers and another calculation is required for 24 through 42 dryers.

The calculation for the demand percentage for 12 through 23 dryers requires two steps. The first step involves subtracting 11 from the total number of dryers. The result of this subtraction is then subtracted from 47 to obtain the demand percentage. For example, the demand percentage for 12 dryers would be 47 − (12 − 11) = 46%.

The calculation for the demand percentage for 24 through 42 dryers requires three steps. The first step requires the subtraction of the number 23 from the total number of dryers. This number is then multiplied by 0.5, and the result of that operation is subtracted from 35. For example, the demand percentage for 25 dryers would be 35 − [0.5 × (25 − 23)] = 35 − [0.5 × 2] = 34%. Section 220.61(B)(1) allows the neutral load to be based on 70% of the ungrounded conductor load. **See Figure 11-6.**

Fixed Appliances for One-Family Dwellings—Standard Calculation Method

2. FIXED APPLIANCES: 220.53

					PHASES		NEUTRAL
Dishwasher =	1500 VA						(Four or more 120 V Loads × 75%)
Disposer =	1000 VA						
Compactor =	___ VA						
Water Heater =	4500 VA						
Microwave =	1500 VA						
___ =	___ VA						
___ =	___ VA						
Total	8500 VA × 75% =	6375 VA			6375 VA		4000 VA

Figure 11-5. Fixed appliances in a one-family dwelling are calculated at 75% of the total load only if there are four or more appliances.

Dryer for One-Family Dwellings—Standard Calculation Method

3. DRYER: 220.54; Table 220.54

		PHASES	NEUTRAL
5000 VA × 100 % =		5000 VA	5000 VA × 70% = 3500 VA

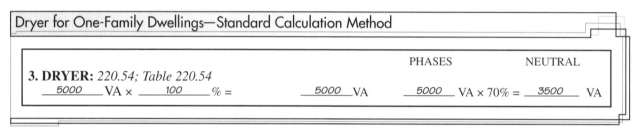

Figure 11-6. The total load for household dryers is calculated at 5000 VA or the nameplate rating, whichever is greater.

Per 220.54 the load for household dryers is 5000 VA. If the dryer nameplate is higher than 5000 VA, then the nameplate rating is used.

For example, what is the dryer load for a dryer rated at 4500 W?
   See 220.54.
   Use the larger rating between 4500 W and 5000 W.
   Dryer load = **5000 W**

For example, what is the dryer load for a dryer rated at 5.5 kW?
   See 220.54.
   Use the larger rating between 5.5 kW (5500 W) and 5000 W.
   Dryer load = **5.5 kW**

For example, what is the dryer load for five dryers rated at 5 kW?
   Total rating = number of appliances × rating
   Total rating = 5 × 5 kW
   Total rating = 25 kW

   See Table 220.54.
   Demand factor for five dryers = 85%
   Dryer load = 25 × 0.85
   Dryer load = **21.25 kW**

Practice Questions

Determining Dryer Load for One-Family Dwellings—Standard Calculation Method

1. What is the dryer load for two dryers rated at 4000 W?
2. What is the dryer load for 13 dryers rated at 5 kW?
3. What is the neutral dryer load for a dryer rated at 4500 W?

Single-Phase Dryers on Three-Phase Services for One-Family Dwellings. Where two or more 1ϕ dryers are supplied by a 3ϕ, 4-wire feeder or service, the total load must be calculated on the basis of twice the maximum number connected between any two phases. This calculation is generally required for multifamily dwellings with 3ϕ electrical service.

For example, a multifamily dwelling with nine dryers rated at 4500 VA each would be calculated by first increasing the load for each dryer to 5000 VA as required by 220.54. The dryers would be connected as follows:
- 3 dryers A phase to B phase
- 3 dryers B phase to C phase
- 3 dryers A phase to C phase

The maximum number of dryers connected on any one phase is six. The NEC® requires that the calculation be based on twice the maximum number of dryers connected on one phase, which would be 12. Multiplying 12 by 5000 VA would provide a load of 60,000 VA. Table 220.54 provides a demand factor of 46% for 12 dryers. The demand load on a 3ϕ service for the dryers is 60,000 VA × 46% = 27,600 VA.

There may be situations where the number of dryers will not distribute equally around the three phases. Again, connect the dryers to phases A and B, B and C, and A and C. Take the phase with the highest number of dryers connected to it and multiply by 2 to determine the number of dryers to use. A simple method to find the total number of dryers is to take the total number of appliances and divide by 3. If there is a remainder, round up to a whole number. Multiply the whole number by 2. This is the number of appliances that is used for Table 220.54. Section 220.61(B)(1) allows the neutral load to be based on 70% of the ungrounded conductor load.

For example, what is the dryer load for 20 dryers rated at 12 kW on a 208/120 wye service?

$$\text{Number of appliances per phase} = \left(\frac{20}{12}\right) = 6.67$$

Number of appliances per phase = 6.67
Round up to 7
Number of appliances per phase = 7 × 2
Number of appliances per phase = 14

Demand factor = 47 − (14 − 11)
Demand factor = 44

Demand load = appliance rating × number of appliances × demand factor
Demand load = 6 × 14 × 0.44
Demand load = **36.96 kW**

Cooking Equipment for One-Family Dwellings

Cooking equipment includes electric ranges, counter-mounted cooking units, and wall-mounted ovens. Requirements for ranges are found in Article 220.55. Table 220.55 is used to calculate the service demand for household cooking equipment. This table is used for cooking equipment rated over 1.75 kW up to 12 kW.

Table 220.55 has four columns. The first column lists the number of appliances being used. The second column (Column A) lists the demand percentage for appliances over 1.75 to less than 3.5 kW. The third column (Column B) lists the demand percentage for appliances 3.5 kW to 8.75 kW. The fourth column (Column C) lists the actual demand load in kilowatts for the number of appliances over 8.75 kW to 12 kW. Per 220.61(B)(1), the neutral may have an additional demand factor of 70%.

For example, what is the cooking-equipment load for one range rated at 12 kW?
See Table 220.55, Column C.
Demand = **8 kW**

> **Practice Question**
>
> *Determining Cooking Equipment Load for One-Family Dwellings—Standard Calculation Method*
> 1. What is the cooking-equipment load for four ranges rated at 11 kW?

Table 220.55, Note 1. Note 1 provides a calculation method for finding the Column C maximum demand for appliances that exceed 12 kW. The largest rating permitted by this note is 27 kW. Appliances rated larger than 27 kW are considered to be commercial cooking appliances and are not subject to the demand factors found in Table 220.55.

Note 1 is applied where one appliance is rated over 12 kW through 27 kW. It also applies when two or more appliances rated over 12 kW through 27 kW are supplied by the same feeder or service conductors. All of the appliances must have the same rating to apply Note 1. The rule states that the Column C maximum demand must be increased by 5% for every kilowatt or major fraction of a kilowatt that the appliance kilowatt rating exceeds 12 kW. A major fraction is 0.5 or greater. **See Figure 11-7.**

Increased Demand for Cooking Equipment Exceeding 12 kW

EQUIPMENT RATING IN kW	PERCENT INCREASE
12.0 – 12.49	0%
12.5 – 13.49	5%
13.5 – 14.49	10%
14.5 – 15.49	15%
15.5 – 16.49	20%
16.5 – 17.49	25%
17.5 – 18.49	30%
18.5 – 19.49	35%
19.5 – 20.49	40%
20.5 – 21.49	45%
21.5 – 22.49	50%
22.5 – 23.49	55%
23.5 – 24.49	60%
24.5 – 25.49	65%
25.5 – 26.49	70%
26.5 – 27.0	75%

Figure 11-7. The maximum demand in Column C must be increased by 5% for every kilowatt or major fraction of a kilowatt that the appliance rating exceeds 12 kW.

For example, what is the cooking-equipment load for one electric range rated at 15 kW?

    15 kW is 3 kW more than 12
    3 × 5% = 15%
    100% + 15% = 115%
    See Table 220.55, Column C.
    Demand factor for one range = 8 kW
    Demand load = 8 kW × 1.15
    Demand load = **9.2 kW**

Practice Question

Determining Cooking-Equipment Load for One-Family Dwellings—Table 220.55, Note 1

1. What is the cooking-equipment load for five electric ranges rated at 15 kW?

Table 220.55, Note 2. This note applies to household cooking appliances over 8¾ kW through 27 kW, where the appliance ratings are not of equal values. This note requires a calculation to find the average rating of the appliances supplied by the same feeder or service conductors. Before calculating the average value, all appliances rated less than 12 kW but more than 8¾ kW must be increased to 12 kW. Add all ranges together using 12 kW for all ranges rated less than 12 kW. Divide the total kilowatts by the number of ranges to obtain the average kilowatt rating of the ranges. Once the average kilowatt rating is determined, the rules are the same as those found in Note 1. The Column C maximum demand is increased by 5% for each kilowatt or major fraction of a kilowatt that the average rating exceeds 12 kW.

For example, what is the cooking-equipment load for the following electric ranges?
- 1 range at 9 kW
- 1 range at 13 kW
- 2 ranges at 12 kW
- 1 range at 15 kW

    Round up all appliances to 12 kW that are under 12 kW.
    Find the average of all appliances.
    Average rating of appliances =
$$\frac{12 + 13 + 12 + 12 + 15}{5}$$
    Average rating of appliances = 12.8 kW

12.8 kW is 0.8 kW more than 12
100% + 5% = 105%
See Table 220.55, Column C.
Demand factor for 5 ranges = 20 kW
Demand load = 20 kW × 1.05
Demand load = **21 kW**

Practice Question

Determining Cooking Equipment Load for One-Family Dwellings—Table 220.55, Note 2

1. What is the cooking-equipment load for the following electric ranges?
   - 1 range at 12 kW
   - 1 range at 14 kW
   - 1 range at 10 kW
   - 1 range at 11 kW

Table 220.55, Note 3. This note provides alternate methods for calculating appliances rated over the 1¾ kW through 8¾ kW rating. The alternate demands are found in Column A and Column B of Table 220.55. To use Note 3, the appliances must be separated into groups based on the nameplate rating. The first group applies to any appliances rated over 1¾ kW but less than 3½ kW and uses the demand percentage found in Column A for the number of appliances. The second group applies to any appliances rated 3½ kW through 8¾ kW and uses the demand percentage found in Column B for the number of appliances. After calculations are done for each group, add Column A and B to determine the total demand. Any appliances with a rating that exceeds 8¾ kW are calculated using Column C and Notes 1 and 2 where applicable. Column C is then added to any appliances from columns A and B to get the total rating in kW.

After the demand is calculated using Note 3, the resulting load must be compared to the demand load found in Column C. The smaller demand load is used. For instance, the demand load for five 3 kW counter-mounted cooking units using Column A would be 5 × 3 kW = 15 kW × 62% = 9.3 kW. The maximum demand for five appliances in Column C is 20 kW. In this case, the lower value of 9.3 kW calculated using Column A is permitted to be used for the feeder or service-conductor load. **See Figure 11-8.** However, this is not always the case.

For instance, the demand load for a 1φ feeder supplying ten 8 kW ranges could be determined using Column B. The demand would be 27.2 kW (8 kW × 10 = 80 kW; 80 kW × 34% = 27.2 kW). The maximum demand found in Column C for 10 appliances is 25 kW. The phrase "Maximum Demand" found in the heading for Column C means that the values determined using Column C are the maximum values that must be used. In the example of the ten 8 kW ranges, the 25 kW value found in Column C is a lower value than the 27.2 kW value calculated using Column B. In this case, 25 kW is the maximum demand required.

For example, what is the cooking equipment load for a counter-mounted cooking unit rated at 3 kW?

See Table 220.55, Column A.
1 appliance = 80%
Demand load for Column A = 3 kW × 80%
Demand load for Column A = 2.4 kW

See Table 220.55, Column C.
1 appliance = 8 kW
Demand load for Column C = 8 kW

Use the smaller of the two columns.
Demand load = **2.4 kW**

For example, what is the cooking-equipment load for two wall-mounted ovens rated at 5 kW each?
Total rating = number of appliances × rating
Total rating = 2 × 5 kW
Total rating = 10 kW

See Table 220.55, Column B.
2 appliances = 65%
Demand load for Column B = 10 × 65%
Demand load for Column B = 6.5 kW

See Table 220.55, Column C.
2 appliances = 11 kW
Demand load for Column C = 11 kW

Use the smaller of the two columns.
Demand load = **6.5 kW**

MayTag
Table 220.55 provides the demand factors and loads for cooking appliances rated between 1.75 kW and 12 kW.

Cooking Equipment for One-Family Dwellings—Standard Calculation Method

4. COOKING EQUIPMENT: *Table 220.55; Notes*

Col A __15,000__ VA × __62__ % =	__9300__ VA			
Col B _____ VA × _____ % =	_____ VA	PHASES		NEUTRAL
Col C _____ VA × _____ % =	_____ VA			
Total	__9300__ VA	__9300__ VA × 70% = __6510__ VA

Figure 11-8. Per Note 3, nameplate ratings between 1¾ kW and 8¾ kW for cooking equipment are added and then multiplied by the appropriate demand from Column A or B.

For example, what is the cooking-equipment load for two counter-mounted cooking units rated at 3000 W and two ovens rated at 5000 W?

First, find the load for the counter-mounted cooking units.
Total rating (Column A) = number of appliances × rating
Total rating = 2 × 3000 W
Total rating = 6000 W

See Table 220.55, Column A.
2 appliances = 75%
Demand load for Column A = 6000 W × 75%
Demand load for Column A = 4500 W

Then, find the load for the ovens.
Total rating (Column B) = number of appliances × rating
Total rating = 2 × 5000 W
Total rating = 10,000 W

See Table 220.55, Column B.
2 appliances = 65%
Demand load for Column B = 10,000 W × 65%
Demand load for Column B = 6500 W

Demand load for Column A and B = 4500 W + 6500 W
Demand load for Column A and B = 11,000 W

See Table 220.55, Column C.
4 appliances = 17 kW
Demand load for Column C = 17,000 W

Smaller of the two loads = 11,000 W
Demand load = **11,000 W**

Practice Questions

Determining Cooking-Equipment Load for One-Family Dwellings—Table 220.55, Note 3

1. What is the cooking-equipment load for three counter-mounted cooking units rated at 2.5 kW?
2. What is the cooking-equipment load for 10 ovens rated at 4000 W?
3. What is the cooking-equipment load for five counter-mounted cooking units rated at 7500 W and five electric ranges rated at 3000 W?

Table 220.55, Note 4. Note 4 is used to calculate the demand loads for branch-circuit conductors only. The note cannot be used to determine the loads for feeder and service conductors. Note 4 has the following three rules:

- Rule 1 allows the demand factors of Table 220.55 to be used to determine the load on the branch-circuit conductors supplying a single range.
- Rule 2 applies to a branch circuit that supplies a single counter-mounted cooking unit or a single oven. In this case, the load on the branch-circuit conductors must be based on the nameplate rating of the appliance.
- Rule 3 allows the nameplate ratings of a counter-mounted cooking unit and not more than two wall-mounted ovens to be added together and treated as one range for the purpose of calculating the branch-circuit load. The appliances must be supplied by the same branch circuit and must be located in the same room. The demand factors found in Table 220.55 may be applied to the total rating to determine the load on the branch-circuit conductors.

Section 220.61 does not allow the use of the 70% demand for the neutral conductor for branch-circuit conductors, but 210.19(A)(3), Ex. 2 does allow a reduction of not less than 70% of the branch-circuit rating. The derated neutral conductor cannot be smaller than 10 AWG copper.

For example, what is the branch-circuit load for a range rated at 12 kW?
Apply Rule 1.
See Table 220.55, Column C.
1 appliance = 8 kW
Branch-circuit demand = **8 kW**

For example, what is the branch-circuit load for a counter-mounted cooking unit rated at 3 kW?
Apply Rule 2.
Nameplate rating = 3 kW
Branch-circuit demand = **3 kW**

For example, what is the branch-circuit load for a counter-mounted cooking unit rated at 3 kW and 2 ovens rated at 4 kW all on the same circuit?
Apply Rule 3.
3 kW + 4 kW + 4 kW = 11 kW
See Table 220.55, Column C.
1 appliance = 8 kW
Demand load from Column C = **8 kW**

TECH TIP

Examples of dwelling calculations can be found in the back of the NEC® in Informative Annex D and can be used when answering load calculation problems on the exam.

Practice Questions

Determining Cooking-Equipment Branch-Circuit Load for One-Family Dwellings—Table 220.55, Note 4

1. What is the branch-circuit load for an electric range rated at 10 kW?
2. What is the branch-circuit load for a counter-mounted cooking unit rated at 5 kW?
3. What is the branch-circuit load for a counter-mounted cooking unit rated at 4 kW and two ovens rated at 5 kW?

Table 220.55, Note 5. This note allows the demand factors found in Table 220.55 to be applied to household cooking appliances rated over 1¾ kW where they are used in an instructional program. An example is a high school classroom used to teach cooking. If the ranges used are typical household cooking appliances, the demand factors found in Table 220.55 may be applied. The demand factors of Table 220.55 would not be applied, however, to that portion of the load that consists of commercial electric cooking appliances used in the cafeteria kitchen.

For example, what is the service demand for 10 ranges rated at 11 kW used for educational purposes in a school?

See Table 220.55, Column C.
Service demand = **25 kW**

Practice Questions

Determining Cooking-Equipment Load for One-Family Dwellings—Table 220.55, Note 5

1. What is the cooking-equipment load for a range rated for 13 kW?
2. What is the cooking-equipment load for six ranges rated at 15 kW?
3. What is the cooking-equipment load for three counter-mounted cooking units rated at 3 kW and three ovens rated at 6.5 kW?

Heating or Air Conditioning (A/C) Loads for One-Family Dwellings

Heating or A/C loads often represent the largest loads in one-family dwellings. Fortunately, these are noncoincidental loads. A *noncoincidental load* is a load that is not normally on at the same time as another load. Section 220.60 permits the smaller of the two noncoincidental loads to be omitted from the load calculation, provided it is unlikely that the loads will be in use simultaneously. Heating and A/C loads are both calculated at 100% of the nameplate rating. The one with the largest load is used in determining the total feeder or service demand. **See Figure 11-9.**

For example, what is the heating or A/C load when the heating load is 10 kW and the A/C load is 12 kW?

See 220.60.
Use 100% of the nameplate rating of the largest load.
Heating or A/C load = **12 kW**

Practice Question

Determining Heating or A/C Loads for One-Family Dwellings—Standard Calculation Method

1. What is the heating or A/C load when the heating load is 10 kW and the A/C load is 8.5 kW?

Largest Motor for One-Family Dwellings

Section 220.50 refers to the following four code sections for determining motor loads:

- 430.24, Several Motors or a Motor(s) and Other Loads
- 430.25, Multimotor and Combination-Load Equipment
- 430.26, Feeder Demand Factor
- 440.6, Hermetic Refrigerant Motor Compressors

Section 430.24 requires conductors supplying several motors to have an ampacity based on the full-load table currents of all motors supplied plus 25% of the highest-rated motor. Sections 430.25 and 430.26 apply to special motor applications that are not generally found in dwelling-unit applications. Section 440.6 covers hermetic refrigerant motor compressors. It does not cover the blower motors found in air-handling units of split air conditioning units, which are covered by Article 430. **See Figure 11-10.**

For example, what is the largest motor load for a dwelling with a 120 V blower rated at 660 VA, a 120 V attic fan rated at 1300 VA, and a 120 V vent motor rated at 400 VA?

Largest motor = 1300 VA
Largest motor load = 1300 × 25%
Largest motor load = **235 VA**

Practice Question

Determining Largest Motor Load for One-Family Dwellings

1. What is the demand load if the largest motor is a 1500 VA pool pump?

Heating or A/C for One-Family Dwellings—Standard Calculation Method

5. HEATING or A/C: *220.60*

			PHASES	NEUTRAL
Heating unit = __10,000__ VA × 100% =	__10,000__ VA			
A/C unit = __12,000__ VA × 100% =	__12,000__ VA			
Heat pump = _____ VA × 100% =	_____ VA			
Largest Load	__12,000__ VA	__12,000__ VA	__—__ VA

Figure 11-9. The largest load between the heating and A/C load is used. The largest load is rated at 100%.

Largest Motor for One-Family Dwellings—Standard Calculation Method

6. LARGEST MOTOR: *220.14(C)*

		PHASES	NEUTRAL
φ __1300__ VA × 25% =	__325__ VA	__325__ VA	
N __1300__ VA × 25% =	__325__ VA		__325__ VA

Figure 11-10. The largest motor load is multiplied by 25%.

Sizing Service Conductors for One-Family Dwellings

After the six loads (general lighting, fixed appliances, dryer, cooking equipment, heating or A/C, and largest motor) are individually calculated, the results of the individual calculations are added to determine the maximum load. The maximum load is divided by the voltage of the service, usually 240 V. This will determine the required load in A for the house. The required load is used to size the service equipment. **See Figure 11-11.** The standard sizes for overcurrent protective devices are located in 240.6.

To size the service conductors for a 120/240 V, 1φ dwelling, use 310.15(B)(7) for services rated 100 A to 400 A. An adjustment factor of 0.83 for the service equipment ampere rating shall be permitted to be used to determine the size of the ungrounded conductors. Any temperature correction for raceway fill must be applied. Per Article 230.79 C, the smallest service allowed for a dwelling is 100 A. Table 310.15(B)(16) is used for services over 400 A and applies to individual units of one-family, two-family, or multi-family dwellings.

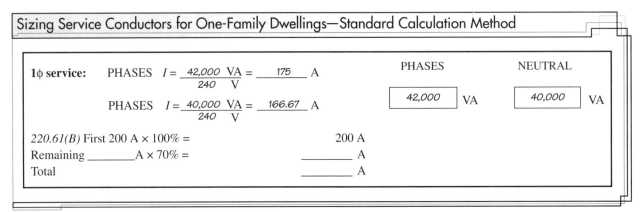

Figure 11-11. Once the total current is determined for the phase and neutral loads, 310.15(B)(7) is used to size the service conductors rated for 100 A to 400 A.

For example, what size of conductors is required for a 120/240 V, one-family dwelling with a total load of 42,000 VA?

See 310.15(B)(7).

General load = $\dfrac{42{,}000}{240}$

General load = 175 A

$175 \times 0.83 = 145$ A

See Table 310.15(B)(16), Column C.

145 A = **No. 1/0 copper conductors**

Practice Question

Determining Service Conductor Size for One-Family Dwellings

1. What size of copper conductors is required for a 225 A service for a single-family dwelling on a 120/240 V service?

OPTIONAL CALCULATION METHOD FOR ONE-FAMILY DWELLINGS

There is an optional method for calculating the loads for one-family dwellings. This method provides a more accurate amperage requirement than the standard house calculation method. Per 230.79, the service size shall not be less than the calculated load as found by referring to Article 220. The two areas that make up the optional dwelling calculation are heating or air conditioning (A/C) and general loads. **See Figure 11-12.** Optional calculation forms can be used to keep track of each load. **See Appendix.**

Heating or Air Conditioning (A/C) Loads for One-Family Dwellings—Optional Calculation

Section 220.82(C) lists six heating and A/C load selections that are considered in the optional calculation for one-family dwellings. The largest of these selections shall be used in determining the total feeder or service load. **See Figure 11-13.** The six loads are as follows:

- A/C and cooling equipment at 100% of nameplate ratings
- A/C equipment including heat pump compressors (at 100% of nameplate ratings, no supplemental electric heating)
- central electric space-heating equipment, including integral supplemental heat for heat pumps (at 100% of heat pump compressor and 65% of supplemental electric heating)
- less than four separately controlled electric space-heating units (at 65% of nameplate ratings)
- four or more separately controlled electric space-heating units (at 40% of nameplate ratings)
- electric thermal storage loads (at 100% of nameplate ratings) where the usual load is expected to be continuous at the full nameplate value

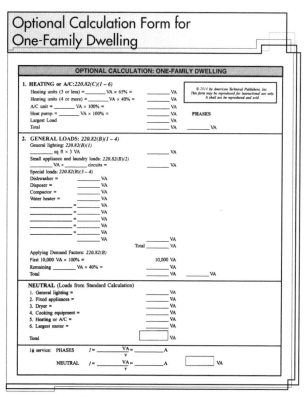

Figure 11-12. The two main areas that make up the optional calculation for one-family dwellings are heating or air conditioning loads and general loads.

TECH TIP

An experienced electrician should be familiar with both the standard and optional calculations for one-family dwellings. Always use the standard calculation method on the exam unless the question states otherwise. The optional calculation is less complex and easier to use but does not typically answer all the demand load questions on the exam. Only use the optional calculation method if the question requires it.

Heating or A/C for One-Family Dwellings—Optional Calculation Method

1. HEATING or A/C: *220.82(C)(1 – 6)*

Heating units (3 or less) = __10,000__ VA × 65% =	__6500__ VA		
Heating units (4 or more) = _____ VA × 40% =	_____ VA		
A/C unit = __6000__ VA × 100% =	__6000__ VA		
Heat pump = _____ VA × 100% =	_____ VA	PHASES	
Largest Load	__6500__ VA		
Total	__6500__ VA	__6500__ VA	

Figure 11-13. After applying demand factors, the largest of the heating and A/C load shall be used in determining the total feeder or service load.

For example, what is the heating and A/C load for a dwelling that has a central furnace rated at 10,000 W and an air conditioner rated at 6000 W? Use the optional calculation method.

Apply the demand factors to each load in order to find the largest load.

Demand factor for heating = 65%
Heating load = 10,000 × 65%
Heating load = 6500 W

Demand factor for AC = 100%
A/C load = 6000 × 100%
A/C load = 6000 W

Heating and A/C load = **6500 W**

Practice Question

Determining Heating and A/C Load for One-Family Dwellings—Optional Calculation Method

1. What is the heating and A/C load for a dwelling that has electric heat rated at 10 kW and an air conditioner rated at 8.5 kW? The heat is controlled by three thermostats.

General Loads for One-Family Dwellings—Optional Calculation

Per Section 220.82(B), general calculations shall be 100% of the first 10 kVA of all other loads, plus 40% of the remainder of all other loads shall be included. Four loads shall be added together to constitute the "general loads" portion of the optional calculation. The other loads include: (1) 3 VA per square foot for the general lighting and general-use receptacle load (as with the standard calculation method, the floor area is calculated using the outside dimensions of the dwelling, not including open porches, garages, or unfinished space), (2) 1500 VA for each 2-wire, 20 A small-appliance and laundry branch circuit, (3) the nameplate rating of all fastened-in-place, permanently connected appliances (ranges, wall-mounted ovens, water heaters, clothes dryers, and counter-mounted cooking units); and (4) the nameplate rating (in either amps or kilovolt-amps) of all motors not otherwise included in the other loads of (3). **See Figure 11-14.**

For example, what is the general load for a 1500 sq ft one-family dwelling with the following loads?
- 2 small-appliance circuits
- 1 laundry circuit
- 12 kW range
- 4 kW dryer
- 1.5 kW disposal
- 4 kW water heater
- 1.2 kW microwave

Apply demand factors to each load.
1500 sq ft × 3 = 4500 VA
2 small-appliance ciructs and one laundry
  circuit × 1500 VA = 4500 VA
12 kW range = 12,000 VA
4 kW dryer = 4000 VA
1.5 kW disposal = 1500 VA
4 kW water heater = 4000 VA
1.2 kW microwave = 1200 VA

Determine the total load.
Total load = 4500 + 4500 + 12,000 + 4000 + 1500 + 4000 + 1200
Total load = 31,700 VA

The first 10,000 VA of the load is rated at 100%.
10,000 × 1.00 = 10,000 VA
31,700 – 10,000 = 21,700
Remaining 21,700 is rated at 40%.
21,700 × 0.40 = 8680 VA
General load = 10,000 + 8,680
General load = **18,680 VA**

General Loads for One-Family Dwellings—Optional Calculation Method

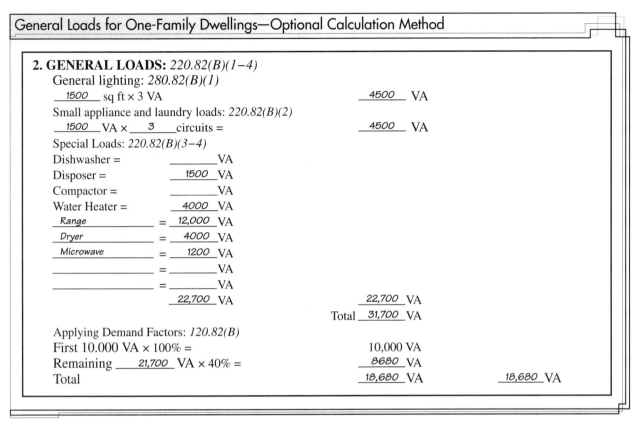

Figure 11-14. The four loads that make up the "general loads" section for the one-family optional calculation method are general lighting, small appliances, laundry, and special loads.

Practice Question

Determining the General Load for One-Family Dwellings—Optional Calculation Method

1. What is the general load for a 3000 sq ft dwelling with a range rated at 12 kW, a dryer rated at 4500 W, a heater rated at 4000 W, a dishwasher rated at 1500 W, a microwave rated at 1000 W, and a blower motor rated at 800 W?

Sizing Service Conductors for One-Family Dwellings—Optional Calculation

Once the heating or A/C load is determined, and the general load demand is calculated for the first 10 kVA at 100% and the rest at 40%, then the demand loads are added together to get the total demand load for the dwelling. The total demand load is used to size the service.

Sizing the service conductors after using the optional dwelling calculation method is the same as for using the standard calculation method. The total load in volt-amps is divided by the voltage of the service, usually 240 V. This will determine the required current for the house. The required current is used to size the service equipment. To size service conductors of 100 A to 400 A, use 83% of the service equipment size per 310.15(B)(7). Then use Table 310.15(B)(16) to size the service conductors. If over 400 A, use Table 310.15(B)(16) without the 83% applied to the service equipment rating.

For example, what size of conductors is required for a 120/240 V, single-family dwelling with a total load of 38,300 VA and a total neutral load of 40,000 VA?

See 310.15(B)(7).
38,300 / 240 volts = 160 amps
See 240.6
175 A is a standard size.
175 × 0.83 = 145 A
See 310.15(B)(16)
75°C column = 1/0 AWG Copper
Service conductor size = **No. 1/0 copper conductors**

Practice Question

Determining Service Conductor Size for One-Family Dwellings—Optional Calculation Method

1. What size of service entrance conductors is required for a single-family dwelling with a 120/240 V single-phase service that has a total calculated load of 46,350 VA.

STANDARD CALCULATION METHOD FOR MULTIFAMILY DWELLINGS

The standard calculation method for multifamily dwellings uses the same six loads as used in the standard calculation method for one-family dwellings. The main difference is that some of the individual loads are calculated and multiplied by the total number of units in the multifamily dwelling to arrive at the total connected load. The areas that require this will be general lighting, fixed appliances, and heating or air conditioning. **See Figure 11-15.**

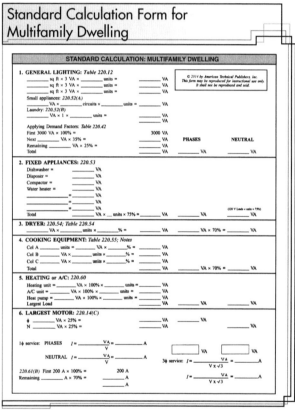

Figure 11-15. The main difference with multifamily dwellings compared to one-family dwellings is that some of the individual loads are multiplied by the total number of units in the multifamily dwelling.

As with the one-family dwelling calculations, both standard and optional methods are used to apply demand factors. If any additional house loads are present, they are added at 100% to the total dwelling load. House loads include hallway lighting, parking lot lighting, security, fire alarms, and common laundry facilities. Standard calculation forms for multifamily dwellings can be used to keep track of each load. **See Appendix.**

General Lighting for Multifamily Dwellings

The general lighting load for multifamily dwellings consists of the same three loads that were used in the one-family dwelling standard calculation. The first portion of the general lighting load is calculated from the total square footage of the dwelling. Table 220.12 lists the unit load per square foot for dwelling units at 3 VA. For this portion of the lighting load, the total square footage of the dwelling, excluding open porches, garages, and unused or unfinished spaces, is multiplied by 3 VA to determine the calculated load for one dwelling. This value is then multiplied by the number of dwelling units in the multifamily dwelling to determine the first portion of the general lighting load.

For multifamily dwellings in which the square footage of the individual units is different, the square footage of each unit is calculated separately and added together to get the total square footage of the multifamily dwelling. This value, multiplied by 3 VA, provides the first portion of the total general lighting load. Section 220.14(J) indicates that general-use receptacles in multifamily dwellings as well as one-family dwellings are considered outlets for general illumination and are included in the 3 VA per-square-foot calculation.

As with the one-family dwellings, each unit shall have two small-appliance circuits and one laundry circuit sized at 1500 VA per circuit. **See Figure 11-16.** If there is a common laundry facility in the building, the laundry circuit can be ignored. The common laundry facility will be rated at 100% for each washing machine and electric dryer, and the load is placed on the main panel of the building.

The demand factors found in 220.42 apply to multifamily dwellings as well as one-family units. The connected load is derated as follows:
- The first 3000 VA of the load is rated at 100%.
- Loads from 3001 VA to 120,000 VA have a demand factor of 35%.
- Loads 120,000 VA and up have a demand factor of 25%.

For example, what is the general lighting load for an apartment building with 10 units if the square footage of each unit is 1000 sq ft?

General lighting = 1000 sq ft × 3 VA × 10
General lighting = 30,000 VA

Small appliances = 1500 VA × 2 × 10
Small appliances = 30,000

Laundry = 1500 VA × 1 × 10
Laundry = 15,000 VA

General Lighting for Multifamily Dwellings—Standard Calculation Method

1. GENERAL LIGHTING: *Table 220.12*

_1000_ sq ft × 3 VA × _10_ units = _30,000_ VA
_____ sq ft × 3 VA × _____ units = _____ VA
_____ sq ft × 3 VA × _____ units = _____ VA

Small appliances: *220.52(A)*
_1500_ VA × _2_ circuits × _10_ units = _30,000_ VA

Laundry *220.52(B)*
_1500_ VA × 1 × _10_ units = _15,000_ VA
                                _75,000_ VA

Applying Demand Factors: *Table 220.42.*
First 3000 VA × 100% =   3000 VA
Next _72,000_ VA × 35% =   _25,200_ VA   PHASES   NEUTRAL
Remaining _____ VA × 25% =   _____ VA
Total   _28,200_ VA   _28,200_ VA   _28,200_ VA

Figure 11-16. Both one-family and multifamily dwellings shall have two small-appliance circuits and one laundry circuit sized at 1500 VA per circuit.

Total VA = general lighting + small appliances + laundry
Total VA = 30,000 + 30,000 + 15,000
Total VA = 75,000

The first 3000 VA of the load is rated at 100%.
3000 × 1.00 = 3000

The remaining part of the load, from 3001 VA to 120,000 VA, is rated at 35%.
75,000 − 3000 = 72,000
72,000 × 0.35 = 25,200 VA

3000 + 25,200 = 28,200 VA

The general lighting load, including demand factors = **28,200 VA**

Practice Question

Determining the General Lighting Load for Multifamily Dwellings—Standard Calculation Method

1. What is the general lighting load for a 3000 sq ft multifamily apartment building that has 30 units with a minimum number of small-appliance and laundry circuits?

Fixed Appliances for Multifamily Dwellings

Each fixed appliance in multifamily dwellings is calculated using the nameplate rating. For four or more appliances, the nameplate ratings are added together and multiplied by the 75% demand factor specified in 220.53. As with one-family dwellings, this demand factor applies to the nameplate rating of the appliance but does not include electric ranges, clothes dryers, space-heating equipment, or A/C equipment served by the same feeder or service.

In multifamily dwellings, the total number of fixed appliances includes all of the fixed appliances in the entire dwelling and not just the fixed appliances in each apartment. The neutral load applies to appliances that are rated for 120 V only. **See Figure 11-17.**

For example, what is the fixed-appliance load and the neutral load for a 10-unit apartment building that has in each unit a 240 V water heater rated at 4500 W, a 120 V garbage disposal rated at 1000 W, a 120 V microwave rated at 1500 W, and a 120 V dishwasher rated at 1500 W?

Fixed-appliance load = (4500 + 1000 + 1500 + 1500) × 10 × 75%
Fixed-appliance load = 8500 × 10 × 0.75
Fixed-appliance load = **63,750 W**

Neutral fixed-appliance load = (1000 + 1500 + 1500) × 10 × 75%
Neutral fixed-appliance load = 4000 × 10 × 0.75
Neutral fixed-appliance load = **30,000 W**

Practice Question

Determining Fixed-Appliance Load for Multifamily Dwellings—Standard Calculation Method

1. What is the fixed-appliance load for a 30-unit apartment building that has in each unit a 120 V microwave rated at 1000 W, a 120 V blower motor rated at 750 W, and a 120 V garbage disposal rated at 500 W?

Fixed Appliances for Multifamily Dwellings—Standard Calculation Method

2. FIXED APPLIANCES: *220.53*

				PHASES	NEUTRAL
Dishwaher =	_1500_ VA				
Disposer =	_1000_ VA				
Compactor =	_____ VA				
Water Heater =	_4500_ VA				
Microwave =	_1500_ VA				(Four or more 120 V Loads × 75%)
_____ =	_____ VA				
_____ =	_____ VA				
Total	_8500_ VA × _10_ units × 75% =	_63,750_ VA	_63,750_ VA	_40,000_ VA

Figure 11-17. In multifamily dwellings, the total number of fixed appliances includes all of the fixed appliances in the building, not just the fixed appliances in each apartment.

Dryer Load for Multifamily Dwellings

The total load for household electric clothes dryers is calculated at 5000 VA or the nameplate rating, whichever results in the greater value. Table 220.54 lists demand factors that are applied to clothes dryers. The first four clothes dryers are calculated at 100%. Subsequent numbers of dryers are calculated with decreasing demand-factor percentages according to Table 220.54. **See Figure 11-18.** If the dryers are not located in the individual apartments, but in a laundry room, they are treated as a commercial system and sized at 100% of their nameplate rating. Derating is not allowed if over four units are being used.

For example, what is the dryer load for a 10-unit apartment building that has an electric dryer in each unit rated at 4500 W?

Total rating = number of appliances × rating
Total rating = 10 × 5000 W
Total rating = 50,000 W

See Table 220.54.
Demand factor for 10 dryers = 50%
Demand load = 50,000 × 0.50
Demand load = **25,000 W**

> **Practice Question**
>
> *Determining Dryer Load for Multifamily Dwellings—Standard Calculation Method*
>
> 1. What is the dryer load for a 30-unit apartment building that has an electric dryer rated at 4500 W in each unit?

Cooking Equipment for Multifamily Dwellings

Household electric ranges rated in excess of 1¾ kW shall be calculated per Table 220.55. Column C values are used less frequently for multifamily dwellings since the size of the individual ranges is generally smaller in these occupancies than in one-family dwellings.

Note 3 is commonly used in multifamily dwellings with the smaller, individual cooking units. As with the one-family calculations, if the cooking appliances are of different ratings that fall under both Columns A and B, the demand factors for each column shall be applied to the appliance for that column and then all of the results are added together. Check Column C for the total number of cooking appliances, and whichever is the smallest is the demand load.

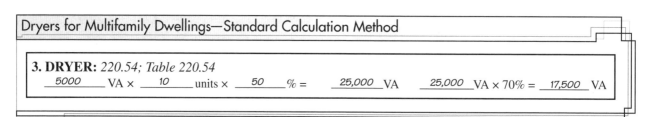

Dryers for Multifamily Dwellings—Standard Calculation Method

3. DRYER: *220.54; Table 220.54*

_5000_ VA × _10_ units × _50_ % = _25,000_ VA   _25,000_ VA × 70% = _17,500_ VA

Figure 11-18. Demand factors from Table 220.54 are applied to clothes dryers in multifamily dwellings.

Note 4, which applies to branch-circuit loads, can also be used frequently with multifamily dwelling units. Note 4 permits the load to be calculated per Table 220.55. For single wall-mounted ovens or single counter-mounted cooking units, the branch-circuit load shall be the nameplate rating of the appliance. For installations from a common branch circuit in which a single counter-mounted cooking unit is installed with not more than two wall-mounted ovens, Note 4 permits all of the cooking appliances to be added together and treated like a single range, provided all of the cooking appliances are located in the same room. **See Figure 11-19.**

For example, what is the cooking-equipment load for 10 counter-mounted cooking units rated at 3000 W and 10 ovens rated at 5000 W?

First, find the load for the counter-mounted cooking units.
Total rating (Column A) = number of appliances × rating
Total rating = 10 × 3000
Total rating = 30,000 W

See Table 220.55, Column A.
10 appliances = 49%
Demand load for Column A = 30,000 × 49%
Demand load for Column A = 14,700 W

Then, find the load for the ovens.
Total rating (Column B) = number of appliances × rating
Total rating = 10 × 5000
Total rating = 50,000 W

See Table 220.55, Column B.
10 appliances = 34%
Demand load for Column B = 50,000 × 34%
Demand load for Column A = 17,000 W

Demand load for Column A and B = 14,700 + 17,000
Demand load for Column A and B = 31,700 W

See Table 220.55, Column C.
20 appliances = 35 kW
Demand load for Column C = 35,000 W

Smaller of the two loads = 31,700 W
Cooking-equipment load = **31,700 W**

> *Practice Question*
>
> *Determining Cooking-Equipment Load for Multifamily Dwellings—Standard Calculation Method*
>
> 1. What is the cooking-equipment load for a 30-unit apartment building with an electric range rated at 10 kW in each unit?

Heating or Air Conditioning (A/C) Load for Multifamily Dwellings

As with one-family dwellings, the heating or A/C load for multifamily dwellings is often the largest load in the multifamily dwelling calculation. Section 220.60 permits the smaller of the two noncoincidental load calculations to be omitted from the multifamily dwelling load calculation, provided it is unlikely that the loads will be in use simultaneously.

Heating and A/C loads are both calculated at 100% of the nameplate rating. These loads are multiplied by the number of units that contain the loads. The calculation with the largest load is used in determining the total feeder or service demand for the multifamily dwelling. If the individual units contain heating or A/C loads of different values, the values are added together and the total loads are compared. The largest load is used in the multifamily dwelling calculation. **See Figure 11-20.**

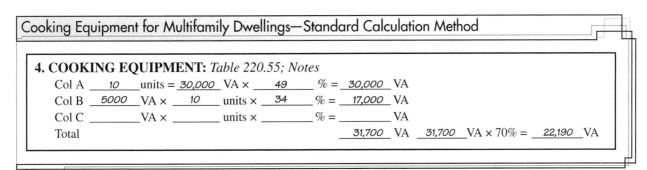

Figure 11-19. Household electric ranges rated over 1¾ kW shall be calculated per Table 220.55.

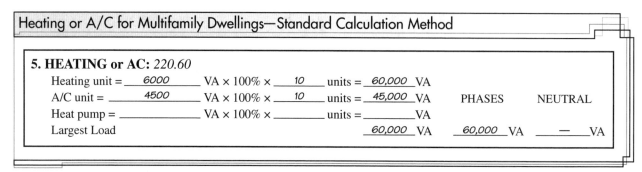

Figure 11-20. In multifamily dwellings, the heating and A/C loads are multiplied by the number of units that contain the loads. The calculation with the largest load is used to determine the total feeder or service demand for the multifamily dwelling.

For example, what is the heating or A/C load for a 10-unit apartment building when the heating load in each apartment is 6000 W and the A/C load in each apartment is 4500 W?

See 220.60.

Heating and A/C load = 100% of the nameplate rating of the largest load × number of units

Heating and A/C load = 6000 × 10

Heating and A/C load = **60,000 W**

Practice Question

Determining Heating or A/C Load for Multifamily Dwellings—Standard Calculation Method

1. What is the heating or A/C load for a 30-unit apartment building when the heating load in each apartment is 7500 W and the A/C load in each apartment is 8000 W?

Largest Motor for Multifamily Dwellings

Despite the fact that each unit in a multifamily dwelling may have the same motor load, 220.50 requires by its reference to 430.24 that only 25% of the largest motor in the entire multifamily dwelling be added to the standard calculation for multifamily dwellings. Once the largest motor in the multifamily dwelling is found, add 25% to the service demand. If there are multiple largest motors of the same size, only use one of the motors.

OPTIONAL CALCULATION METHOD FOR MULTIFAMILY DWELLINGS

In addition to the standard calculation, there is a multifamily dwelling optional calculation that can be performed if three conditions are present. First, 220.84(A)(1) requires that no dwelling within the multifamily structure be supplied by more than one feeder. The second condition is that each dwelling shall be equipped with electric cooking equipment. If the individual dwelling units have natural gas cooking equipment, the optional calculation method cannot be used. The last condition is that each of the individual dwelling units shall have either electric space-heating equipment, A/C, or both loads present. See 220.84(A)(3). Optional calculation forms for multifamily dwellings can be used to keep track of each load. **See Appendix.**

When performing the optional multifamily dwelling calculation method, apply the following rules:

1. Multiply the total area of each unit by 3 VA/sq ft.
2. Each unit must have two small-appliance circuits rated at 1500 VA and a laundry circuit rated at 1500 VA, unless there is a common laundry facility.
3. The nameplate ratings of all appliances including electric cooking equipment are used in determining the load. The nameplate rating of dryers is used if they are not connected to the laundry circuit.
4. Use the nameplate rating of all permanently connected motors.
5. Use the largest load between the electric heat and air conditioning.

Once the calculation for each area is completed, the calculations are added together to get the total load for each unit. Multiply the total load for each apartment by the number of units. If units are of different sizes or have different items in them, they must be added together. This total is then multiplied by a percentage from Table 220.84 to determine the total demand for the dwelling. If any additional house loads are present, they are added at 100% to the total dwelling load. House loads include hallway lighting, parking lot lighting, security, fire alarms, and common laundry facilities. **See Figure 11-21.**

Chapter 11—Dwellings

OPTIONAL CALCULATION: MULTIFAMILY DWELLING

1. HEATING or A/C: *220.82(C)(1–6)*

Heating unit = _____ VA × 100% × _____ units = _____ VA
A/C unit = __6000__ VA × 100% = __20__ units = __120,000__ VA
Heat pump = _____ VA × 100% = _____ units = _____ VA
Largest Load __120,000__ VA
Total

PHASES
__120,000__ VA

©2017 by American Technical Publishers, Inc. This form may be reproduced for instructional use only. It shall not be reproduced and sold.

2. GENERAL LOADS: *220.82(B)(1 – 4)*

General lighting: *280.82(B)(1)*

__1000__ sq ft × 3 VA × __20__ units = __60,000__ VA __60,000__ VA
_____ sq ft × 3 VA × _____ units = _____ VA
_____ sq ft × 3 VA × _____ units = _____ VA
_____ sq ft × 3 VA × _____ units = _____ VA

Small appliance and laundry loads: *220.82(B)(2)*
__1500__ VA × __3__ circuits × __20__ units = __90,000__ VA __90,000__ VA

Special Loads: *220.82(B)(3 –4)*
Dishwasher = __950__ VA
Disposer = __500__ VA
Compactor = _____ VA
Water Heater = __4500__ VA
Microwave = __1000__ VA
Range = __8000__ VA
_____ = _____ VA
_____ = _____ VA
_____ = _____ VA

Total __14,950__ VA × __20__ units = __299,000__ VA __299,000__ VA

Total Connected Load __569,000__ VA

Applying Demand Factors: *Table 220.84*
__569,000__ VA × __38__ % = __216,220__ VA

__216,220__ VA

NEUTRAL (Loads from Standard Calculation)
1. General Lighting = _____ VA
2. Fixed appliances = _____ VA
3. Dryer = _____ VA
4. Cooking equipment = _____ VA
5. Heating or A/C = _____ VA
6. Largest motor = _____ VA

Total _____ VA

1ϕ service: PHASES $I = \dfrac{216,220 \text{ VA}}{240 \text{ V}} =$ __901__ A

3ϕ service: $I = \dfrac{\text{VA}}{V \times \sqrt{3}} =$ _____ A

NEUTRAL $I = \dfrac{\text{VA}}{V} =$ _____ A

$I = \dfrac{\text{VA}}{V \times \sqrt{3}} =$ _____ A

220.84; First 200 A × 100% = 200A
Remaining _____ A × 70% _____ A
_____ A

Figure 11-21. Optional calculation forms for multifamily dwellings can be used to document each load.

For example, use the optional calculation method for multifamily dwellings to calculate the service size for an apartment building with 20 units. The service supplied to the dwelling is 120/240 V. Each unit is 1000 sq ft and contains the following loads:
- 2 small-appliance circuits
- 1 laundry circuit
- 8 kW electric range
- 950 W dishwasher
- 1000 W microwave
- 500 W disposal
- 4500 W water heater
- 6000 W A/C

1000 sq ft × 3 = 3000 VA
2 small-appliance circuits × 1500 VA = 3000 VA
Laundry circuit × 1500 VA = 1500 VA
8 kW range = 8000 VA
950 W dishwasher = 950 VA
500 W disposal = 500 VA
4500 W water heater = 4500 VA
1000 W microwave = 1000 VA
6000 W A/C = 6000 VA

Determine the total load.
Total load = (3000 + 3000 + 1500 + 8000 + 950 + 500 + 4500 + 1000 + 6000) × number of units
Total load = 28,450 × 20
Total load = 569,000 VA

See Table 220.84.
18 – 20 units = 38%

$$\text{Service size} = \frac{\text{total load} \times \text{demand factor}}{\text{service size}}$$

$$\text{Service size} = \frac{569{,}000 \times 0.38}{240}$$

Service size = **901 A**

Practice Question

Determining Service Size for Multifamily Dwellings—Optional Calculation Method

1. Use the optional calculation method for multifamily dwellings to determine the service size for an apartment building with 28 units. The dwelling has a common laundry room with 10 washers rated at 1200 W and 10 dryers rated at 4500 W. The service supplied to the dwelling is 120/240 V. Each unit is 1100 sq ft and contains the following loads:
   - 2 small-appliance circuits
   - 1 laundry circuit
   - 8000 W electric range
   - 1200 W microwave
   - 800 W disposal
   - 4500 W water heater
   - 6000 W air conditioner

CALCULATIONS FOR FARM DWELLINGS AND OTHER FARM LOADS

Farms load can be made up of both dwelling loads and other farm loads. Other farm loads are loads that are used for the operation of the farm and are not dwelling units. The calculation method used to determine the total demand load for the farm depends on the number of structures and the size of the loads on the farm.

Subsection 220.102(A) is used for determining the load calculations for a farm dwelling unit. The feeder and service load calculations for a farm dwelling unit is determined per 220 Parts III and IV, which are the same calculations used for a typical dwelling unit. If the farm dwelling has electric heat and the farm has electric grain-drying equipment, then Part III can be used for determining feeder and service load calculations.

Other farm loads are calculated per 220.102(B). Table 220.102 is used to determine the demand factor for 1φ or 3φ other-than-dwelling-unit farm loads that operate on voltages up to 240 V. A demand factor of 100% is applied to loads that operate at the same time, but not less than 125% FLC of the largest motor and not less than the first 60 A of the load. A demand factor of 50% is applied to the next 60 A of all the other loads. A demand factor of 25% is applied to the rest of the loads.

The total farm load is determined by applying article 220.103 and Table 220.103. Per Table 220.103, the largest load has a demand factor of 100%. The second largest load has a demand factor of 75%. The third largest load has a demand factor of 65%. The remaining loads have a demand factor of 50%.

If the farm dwelling unit and the other farm loads are supplied by the same service, the dwelling load,

TECH TIP

A mobile home is a transportable factory-assembled structure, or structures, constructed on a permanent chassis for use as a dwelling. A mobile home is not constructed on a permanent foundation but is connected to the required utilities. The term "mobile home" includes manufactured homes. Mobile homes not intended for use as dwelling units (e.g., on-site construction offices) are not required to have the number or capacity of circuits required by Article 550, unless the service is 120/240 V, 3-wire.

which is calculated from a standard or optional dwelling calculation, is added to the total of the other farm loads. This total determines the service demand, which is used to determine the size of the service conductors.

For example, what is the service demand for a farm that has a workshop rated at 115 A and a grain storage bin rated at 50 A?

Workshop:
See Table 220.102.
The first 60 A is rated at 100%.
60 × 1.00 = 60
The next 60 A is rated at 50%.
115 − 60 = 55
55 × 0.50 = 27.5
60 + 27.5 = 87.5
Total demand load for workshop = 87.5 A

Grain storage bin:
See Table 220.102
The first 60 A is rated at 100%.
50 × 1.00 = 50
Total demand load for grain storage bin = 50 A
See Table 220.103.
Largest load is rated at 100%.
87.5 × 1.00 = 87.5
Second largest load is rated at 75%.
50 × 0.75 = 37.5
Add the two demand loads together.
Service demand = 87.5 + 37.5
Service demand = **125 A**

Practice Question

Determining Service Demand for Farms

1. What is the service demand for a farm that has a grain elevator rated at 150 A and a horse stable rated at 120 A?

STANDARD CALCULATION METHOD FOR MOBILE HOMES AND MANUFACTURED HOMES

Demand load calculations for mobile homes are similar to those for one-family dwellings. The size of the service is based on many of the same conditions as a standard home. Standard calculation forms for mobile and manufactured homes can be used to keep track of each load. **See Appendix.** The load of many of the sections is calculated in amps. Convert the volt-amps into amps by dividing the volt-amps by the voltage of the appliance.

Once the load for each section is calculated, the total amperage is obtained by adding the current of each section. The loads must be balanced between the two phases as closely as possible to provide a reasonable service size. For example, once the lighting load is determined, the load will be divided by 240 V and distributed equally between the two phases. The 120 V loads will be placed on the phase that better balances the system.

Lighting, Small-Appliance, and Laundry Loads for Mobile and Manufactured Homes

Per 550.18(A)(1), the general lighting load for a mobile or manufactured home is found by multiplying the square feet of the home (outside dimensions) by the volt-amps per square foot that is given in Table 220.12. As with other dwellings, the lighting and receptacle loads require 3 VA for each square foot.

Per 550.18(A)(2), each small-appliance circuit is calculated at 1500 VA. Unlike a typical one-family dwelling, there is not a two-circuit requirement.

Per 550.18(A)(3), each mobile home is required to have one laundry circuit. This circuit is calculated at 1500 VA. Once each area is calculated above, the total connected load is found by adding them together.

Calculating Total Demand Load for Mobile and Manufactured Homes

The same derating factors that apply to standard one-family dwellings apply here as well. Per 550.18(A)(5), the first 3000 VA is rated at 100%, the remaining volt-amp is rated at 35%. The total volt-amps is divided by the supply voltage, which is usually 240 V, to obtain the current of each supply phase.

Sizing Service Load for Mobile and Manufactured Homes

Other items such as cooking equipment, clothes dryers, water heaters, disposals, and microwaves as well as any other item that might be installed are covered in 550.18(B). Once each area is calculated, the total current is obtained by adding the current of each area.

Per 550.18(B)(2), use the nameplate rating for electric heaters, A/C units, and blower motors. Use the largest load between the heating and A/C. If the blower motor is used for both heating and A/C, it must be counted separately. If there is no A/C, use 15 A per phase for the A/C load for a 40 A service.

Per 550.18(B)(3), the largest motor shall have a demand factor of 25%. Use the largest motor from 550.18(B)(2).

Per 550.18(B)(4), use the nameplate rating for fixed appliances. This includes disposals, dishwashers, water heaters, clothes dryers, wall-mounted ovens, and counter-mounted cooking units. Where there are more than three of these loads, apply a 75% demand factor to the total connected load.

If there is a free-standing range, a demand factor is applied from 550.18(B)(5). The demand factor is determined by the rating of the range.

Per 550.18(B)(6), if other circuits are provided that are not factory-installed, estimate the load size of each additional circuit. **See Figure 11-22.**

For example, what is the service size required for a 840 sq ft mobile home with the following loads?
- 1 small-appliance circuit
- 1 laundry circuit
- 240 V range = 8000 VA
- 240 V heater = 1200 VA
- 120 V blower motor = 300 VA
- 120 V garbage disposal = 600 VA

Apply demand factors to each load.
840 sq ft × 3 = 2520 VA
1 small-appliance circuit × 1500 VA = 1500 VA
Laundry circuit × 1500 VA = 1500 VA

Determine the total load for the general lighting.
Total load = 2520 + 1500 + 1500
Total load = 5520

Determine total current per phase.
See 50.18(A)(5).
The first 3000 VA of the load is rated at 100%
3000 × 1.00 = 3000
The remaining volt-amps of the load is rated at 35%
5520 − 3000 = 2520
2520 × 35% = 882

$$\text{Total current} = \frac{3000 + 882}{240}$$

Total current = 16.2 per phase (Phase A and B)

Apply demand factors to the rest of the circuits.

$$\text{Heater} = \frac{1200}{240}$$

Heater = 5 A per phase (Phase A and B)

$$\text{Blower motor} = \frac{300}{120}$$

Blower motor = 2.5 A for one phase (Phase A)

$$\text{Largest motor} = \frac{600 \times 0.25}{120}$$

Largest motor = 1.25 A for one phase (Phase A)

$$\text{Garbage disposal} = \frac{600}{120}$$

Garbage disposal = 5 A for one phase (Phase B)

$$\text{Range} = \frac{8000 \times 0.80}{240}$$

Range = 26.67 A per phase (Phase A and B)

Phase A = 51
Phase B = 52.87

The phase with the higher load will be used to set the service size.

Phase B = **60 A service**

Practice Question

Determining Service Size for Mobile and Manufactured Homes

1. What is the service size required for a 1120 sq ft mobile home with the following loads?
   - two small-appliance circuits
   - 240 V range rated = 6000 VA
   - 240 V heater = 1500 VA
   - 120 V blower motor = 500 VA
   - 120 V dishwasher = 500 VA

CALCULATIONS FOR MOBILE HOME PARKS

Mobile home parks can accommodate mobile homes of different sizes. Most of the lots are sized based on 550.31(1), which allows 16,000 VA per lot. The 16,000 VA load is multiplied by the number of lots in the park to determine the total load in volt-amps for the park. Once the total load in volt-amps is determined, a demand factor is applied from Table 550.31. The service load for the park is determined by dividing the final load in volt-amps, including the demand factor, by the voltage. The voltage is usually 240 V.

STANDARD CALCULATION: MOBILE AND MANUFACTURED HOMES

1. GENERAL LIGHTING: *Table 550.18(A)(1)*

    __840__ sq ft × 3 VA =   __2520__ VA

    Small appliances: *550.18(A)(2)*

    __1500__ VA × __1__ circuits =   __1500__ VA

    Laundry *550.18(A)(2)*

    __1500__ VA × 1 =   __1500__ VA

                                                __5520__ VA

©2017 by American Technical Publishers, Inc. This form may be reproduced for instructional use only. It shall not be reproduced and sold.

Applying Demand Factors: *Part 550.18(A)(5).*
First 3000 VA × 100% =   3000 VA
Next __2520__ VA × 35% =   __882__ VA   PHASE A   PHASE B
Remaining _____ VA × 25% =   VA
Total   __3882__ VA ÷ __240__ V = __16.2__ A   __16.2__ A

2. Motors and Heaters: *550.18(B)(2)*

    Heating unit = __1200__ VA × 100% =   __1200__ VA
    A/C unit = _____ VA × 100% =   _____ VA
    Largest Load _____   __1200__ VA ÷ __240__ V = __5__ A   __5__ A
    Blower Motor = __300__ VA × 100% =   __300__ VA ÷ __120__ V = __2.5__ A   _____ A

3. LARGEST MOTOR: *550.18(B)(3)*

    ϕ __300__ VA × 25% =   __75__ VA ÷ __120__ V = __0.63__ A   _____ A

4. FIXED APPLIANCES: *550.18(B)(4)*

    Dishwasher =   _____ VA
    Disposer =   __600__ VA
    Compactor =   _____ VA
    Water Heater =   _____ VA
    _____ = _____ VA
    _____ = _____ VA
    _____ = _____ VA
    Total   __600__ VA × __100__ % __600__ VA ÷ __120__ V = _____ A   __5__ A

5. RANGE: *550.18(B)(5)*

    Range = __8000__ VA × __80__ % =   __6400__ VA ÷ __240__ V = __26.67__ A   __26.67__ A

                                                                     | **51** | A | **52.87** | A

Figure 11-22. When calculating the loads for mobile and manufactured homes, the loads must be balanced between the two phases as close as possible to provide the required service size.

For example, what is the minimum current rating for a 30-lot mobile home park?

    See *550.31(1).*
    Each lot is rated at 16,000 VA.
    Total VA = rated VA × number of lots
    Total VA = 16,000 × 30
    Total VA = 480,000 VA

See *Table 550.31.*
Demand for 30 lots = 24%
480,000 × 0.24 = 115,200 VA

$$\text{Current rating} = \frac{115{,}200 \text{ VA}}{240 \text{ V}}$$

Current rating = **480 A**

Practice Question

Determining Service Size for Mobile Home Parks

1. What is the service size required for a mobile home park that has 45 lots?
2. What is the service size required for a mobile home park that has 10 lots?

TECH TIP

Per 550.18(C) the optional calculation for determining lighting and appliance loads for one-family dwellings shall be permitted to be used on mobile homes as well. Only use the optional calculation for mobile homes on the exam when the question specifies.

CHAPTER 11
Dwellings
Review Questions

Name _____   Date _____

_____ 1. An 1800 sq ft house is required to have ___ 20 A circuit breakers.
   A. three
   B. five
   C. six
   D. seven

_____ 2. If the heating load for a one-family dwelling is 12,500 W at 240 V and the air-conditioning is rated at 10,000 W, the service demand for the heat or air-conditioning load is ___ W.
   A. 7000
   B. 10,000
   C. 12,500
   D. 15,625

_____ 3. The service demand for the following fixed appliances in a one-family dwelling is ___ VA.
   • Microwave at 1250 VA
   • Dishwasher at 1400 VA
   • Trash compactor at 960 VA
   • Overhead garage door motor at 960 VA

   A. 3428
   B. 3620
   C. 4570
   D. 6093

_____ 4. If a 15′ × 20′ garage is added on to an 1800 sq ft dwelling, the general lighting load for the dwelling would be ___ VA.
   A. 4500
   B. 5000
   C. 5400
   D. 6300

_____ 5. A 4500 sq ft house that has a 500 sq ft garage is required to have ___ 15 A circuits.
   A. five
   B. six
   C. seven
   D. eight

197

6. A 40′ × 70′ one-family dwelling that has three small appliance circuits, a laundry circuit, and fifteen extra 75 W recessed fixtures installed in the family room will have a service demand of ___ VA.
   A. 3990
   B. 6990
   C. 11,400
   D. 14,400

7. A dwelling has electric heating with one control rated at 25 kW and an A/C load rated at 17.5 kW. Using the optional dwelling calculation, the amount of load required for the service is ___ kW.
   A. 10
   B. 16.25
   C. 17.5
   D. 25

8. The general lighting load for a 4000 sq ft house is ___ W.
   A. 6150
   B. 7725
   C. 9600
   D. 12,000

9. The branch circuit demand for a 4500 W electric oven in a one-family dwelling is ___ W.
   A. 2520
   B. 3600
   C. 4500
   D. 8000

10. The service demand for two 5000 W wall-mounted ovens used on separate circuits in a one-family dwelling is ___ kW.
    A. 5
    B. 6.5
    C. 8
    D. 10

11. The service demand for thirty 12 kW ranges in an apartment building is ___ kW.
    A. 15
    B. 30
    C. 45
    D. 60

12. The service demand for a dwelling containing a 16.25 kW electric range is ___ W.
    A. 6720
    B. 8000
    C. 9600
    D. 16,250

13. The neutral demand for a 12 kW electric range that operates on 240 V is ___ kW.
   A. 5.6
   B. 8
   C. 8.4
   D. 12

14. The service demand for forty 5000 W electric dryers in an apartment building is ___ W.
   A. 37,100
   B. 53,000
   C. 100,000
   D. 200,000

15. The neutral service demand for four 4500 W electric clothes dryers in a four-unit apartment building is ___ kW.
   A. 12.6
   B. 14
   C. 18
   D. 20

Digital Resources
ATPeResources.com/QuickLinks
Access Code: 637160

CHAPTER 12

Nondwelling Occupancies

Most nondwelling occupancies are buildings such as schools, stores, restaurants, and office buildings. Equipment found in a dwelling is typically found in a nondwelling occupancy. The difference is in how and when the equipment is used. For example, the lights in a dwelling are usually not on for 8 hr to 10 hr at a time. In a nondwelling such as a store, the lights will be on the entire time the store is open for business, so the demand applied to the lighting circuits is different. The same basic format that was used for dwelling calculations will be used for nondwelling occupancies. Specific parts of Article 220 are used to calculate loads.

OBJECTIVES

After completing this chapter, the learner will be able to do the following:

- Determine the lighting load for nondwelling occupancies.
- Know how to identify a continuous load.
- Determine the required number of lighting circuits for nondwelling occupancies.
- Calculate the load for show window lighting.
- Calculate the load for track lighting.
- Determine the service size for a school using the optional school calculation method.
- Describe the major difference between the optional restaurant calculation method and the optional school calculation method.

LIGHTING LOADS

The lighting load for nondwelling occupancies is determined by multiplying the area of the building by the volt amps per square foot. This is called the connected load. Demand factors can be applied to the connected load if applicable. Table 220.12 gives the volt-amp requirement for various types of buildings. As the table shows, there are different volt-amp requirements for different types of buildings. This is also true for the different areas inside a building.

For example, a restaurant is required to have at least 2 VA per square foot for the main dining areas, but the hallways are rated at 0.5 VA per square foot and storage areas at 0.25 VA per square foot. **See Figure 12-1.** When figuring lighting load calculations, each lighting section is calculated separately and then added together to get the final lighting load for the building. Receptacle loads are not calculated as part of the lighting load for nondwelling occupancies.

Like dwelling loads, if the nondwelling occupancy is designed for energy codes that have been adopted by the local authority having jurisdiction, then the energy code figures can be used. If the local energy codes are used, the three criteria must be met: a power-monitoring system is installed in the nondwelling occupancy, alarms are built into the power-monitoring system, and the derating factors from 220.42 have not been used for the lighting load.

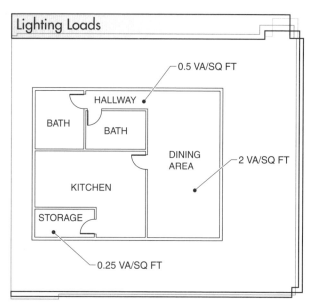

Figure 12-1. The volt-amp rating may be different in specific locations throughout a building.

Continuous Lighting Loads

In most cases the lighting load for nondwelling occupancies will be continuous. The NEC® defines a continuous load as one that is on for 3 hr or more at one time. **See Figure 12-2.** Per 210.19 and 210.20, conductors and overcurrent protective devices (OCPDs) used for continuous loads are sized at 125% of the connected load. This limits the amount of current to 80% of the rating for conductors and overcurrent protective devices.

Figure 12-2. A lighting load that is on for 3 hr or more at one time, such as in an office building, is considered to be a continuous load.

For example, what is the continuous lighting load for an assembly hall that is 6500 sq ft?
See Table 220.12.
Assembly hall = 1 VA/sq ft
OCPD = 125%
Lighting load = area × VA/sq ft × 125%
Lighting load = 6500 × 1 × 1.25
Lighting load = **8125 VA**

Practice Question

Determining Continuous Lighting Loads

1. What is the continuous lighting load for a school that has 12,500 sq ft of space used for classrooms?

Demand Factors

Demand factors for lighting are not applied to most commercial buildings. However, Table 220.42 does allow for some demand factors for the lighting load for hospitals, hotels, motels, and warehouses. Once the lighting load is calculated for these buildings, the demand factor as shown in Table 220.42 can be applied. Demand factors shall not be applied in the areas of these buildings where the entire lighting load will be used at the same time as in ballrooms, operating rooms, and dining rooms.

For example, what is the general lighting load for a 45,000 sq ft warehouse?

See Table 220.12.
Warehouse = ¼ (0.25) VA/sq ft
Lighting load = area × VA/sq ft × 125%
Lighting load = 45,000 × 0.25 × 1.25
Lighting load = 14,063

See Table 220.42.
The first 12,500 VA is calculated at 100%.
12,500 × 1.00 = 12,500
The remainder of the load is calculated at 50%.
14,063 − 12,500 = 1563
1563 × 0.50 = 782 VA
12,500 + 782 = **13,282 VA**

Practice Question

Determining Lighting Loads with Demand Factors

1. What is the lighting load for a 37,000 sq ft hotel? There are no cooking appliances in the rooms.

Required Number of Lighting Circuits

The required number of lighting circuits is calculated for the connected load. The connected load is the load in volt amps before any demand factors are applied. The connected load is the area inside a building in square feet multiplied by the required volt amps per square foot. This product is increased by 25% due to the fact that it is a continuous load and the lights will be ON for more than 3 hours at a time. Determine the required current for the lighting load by dividing the connected load by the voltage of the lighting load. The voltage for the lighting load is usually 120 V or 277 V, but it may also be 208 V, 240 V, or 480 V. To determine the number of circuits, divide the load by the size of the circuit breaker (CB). CB sizes for general lighting loads are typically 15 A or 20 A. If the answer is not a whole number, round up to the next whole number.

For example, how many circuit breakers are required for the lighting circuits in a 45,000 sq ft warehouse? The voltage for the warehouse is 120/240 V, and 20 A CBs are used.

$$\text{Number of circuit breakers} = \frac{0.25 \text{ W/sq ft} \times \text{area of building} \times 125\%}{\frac{\text{voltage}}{\text{circuit breaker rating}}}$$

$$\text{Number of circuit breakers} = \frac{\frac{0.25 \times 45{,}000 \times 1.25}{120}}{20}$$

$$\text{Number of circuit breakers} = \frac{\frac{14{,}063}{120}}{20}$$

$$\text{Number of circuit breakers} = \frac{117}{20}$$

Number of circuit breakers = 5.82
Round up to 6
Number of circuit breakers = **6**

Practice Question

Determining the Required Amount of Lighting Circuits

1. How many 15 A circuits are required for a 24,000 sq ft lodge room with a service supplying 120/240 V?

Nondwelling Occupancy Lighting

Nondwelling occupancy lighting typically includes fluorescent, metal halide, and high- and low-pressure sodium luminaires. Metal halide and high- and low-pressure sodium fixtures are listed as high-intensity discharge (HID). **See Figure 12-3.** Per 220.18(B), the number of HID luminaires on a circuit is determined by the current of each unit.

Show Window Lighting

Show window lighting is lighting used to enhance the window of a store that displays merchandise for sale. These windows usually use higher amounts of lighting to better show products. Per 220.43(A), show windows are calculated at 200 VA per linear foot as measured along the base of the window. **See Figure 12-4.** As with general lighting loads, the show window loads are continuous loads that are on for 3 hr or more at a time. This requires the connected load to be calculated at 125%. The NEC® also requires that the branch circuit be calculated per 220.14G(1) and 220.14G(2).

High-Intensity Discharge (HID) Lighting

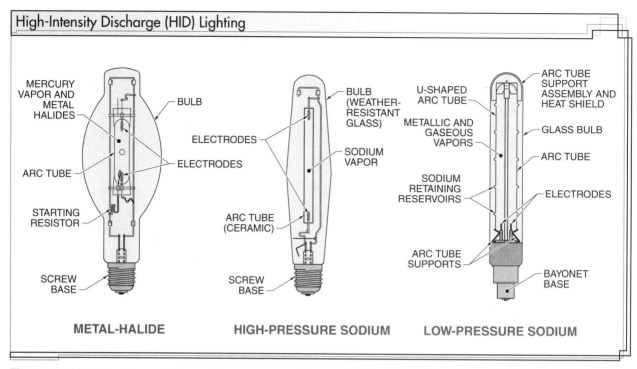

Figure 12-3. Metal halide and high- and low-pressure sodium are lighting fixtures listed as high-intensity discharge (HID).

Show Window Lighting

Figure 12-4. The lighting load for a show window is calculated by multiplying the length of the window at the base by 200 VA.

For example, what is the lighting load for 45′ of show window operating at 120 V?

Lighting load = VA/ft × length of window × 125%

Lighting load = 200 × 45 × 1.25

Lighting load = **11,250 VA**

Practice Question

Determining Show Window Lighting Load

1. What is the lighting load for a 30′ long show window?

Track Lighting

Track lighting is lighting used to illuminate an area using adjustable spotlight fixtures that are fastened to a current-carrying metal track. Track lighting may be used in commercial areas to spotlight a product or picture. Per 220.43(B), when track lighting is used in dwellings or in hotels and motels, the load for the track lighting is part of the general lighting load. When used in commercial applications, track lighting is calculated at 150 VA per linear foot. **See Figure 12-5.** As with the show window lighting, if the track lighting is on for 3 hr or more, it is a continuous load and will be calculated at 125%.

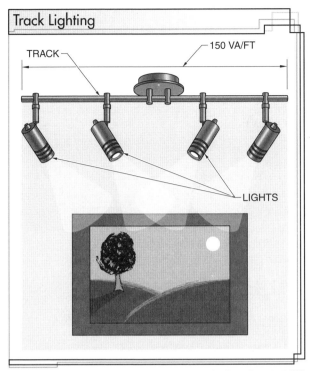

Figure 12-5. Track lighting loads are calculated at 150 VA per linear foot.

For example, what is the lighting load for 25′ of track lighting used in a store?

Lighting load = VA/ft × length of track lighting × 125%
Lighting load = 150 × 25 × 1.25
Lighting load = **4687.5 VA**

Practice Question

Determining Track Lighting Load

1. What is the lighting load for 15′ of track lighting that is on for 5 hr at a time?

Heavy-Duty Lampholders

A *lampholder* is a device that functions as a support for a lamp for the purpose of making electrical contact with the lamp. Per 210.21(A), lampholders shall be of the heavy-duty type when connected to a branch circuit in excess of 20 A. Heavy-duty and medium lampholders shall have a rating of not less than 660 W. All other lampholders shall have a rating of not less than 750 W.

Per 220.14(E), heavy-duty lampholders are calculated at 600 VA per fixture. When these lampholders are on for 3 hr or more, they are calculated at 125% of the connected load.

For example, what is the service demand for 10 heavy-duty lampholders that are on for 6 hr at a time?

Lighting load = VA/ft × number of lampholders × 125%
Lighting load = 600 × 10 × 1.25
Lighting load = **7500 VA**

Signs

Per 600.5, all commercial buildings that have grade-level access shall be required to have at least one 20 A circuit supplied for a sign. Per 220.14(F), each required sign circuit is calculated at 1200 VA. Where signs are on for 3 hr or more, they are calculated at 125% of the connected load.

RECEPTACLE LOADS

In dwelling units, hotels, and motels, the receptacle load is included in the unit load for lighting as found in Table 220.12. In commercial locations, receptacles are not counted as part of the lighting load. Per 220.14(I), receptacles in commercial locations are calculated at 180 VA for each single or multiple outlet on one yoke or strap. The yoke or strap openings can be for single, duplex, or triple receptacles. **See Figure 12-6.**

Figure 12-6. Each strap-supporting or yoke-supporting receptacle equals 180 VA. It does not matter how many receptacles are connected to each strap or yoke.

Per 220.14(K)(2), add an additional 1 VA/sq ft to the lighting load of banks and office buildings when the actual number of receptacles is unknown. When the actual number of receptacles is known in a bank or office building, there are two calculations used to determine the receptacle load per 220.14(K). The largest value of the two calculations is used for the receptacle load. The load may be derated based on Table 220.44 if the number of receptacles is known.

For example, what is the receptacle load for a 12,500 sq ft office building that has 100 duplex receptacles?

First calculation:
Receptacle load = VA/ft × number of receptacles
Receptacle load = 180 × 100
Receptacle load = 18,000 VA

See Table 220.44.
The first 10,000 VA is calculated at 100%.
10,000 × 1.00 = 10,000
The remainder of the load is calculated at 50%.
18,000 − 10,000 = 8000
8000 × 0.50 = 4000 VA
10,000 + 4000 = 14,000 VA

Second calculation:
Receptacle load = area × VA/sq ft
Receptacle load = 12,500 × 1
Receptacle load = 12,500

Calculation 1 = 14,000 VA
Calculation 2 = 12,500 VA

Largest calculation is 14,000 VA.
Receptacle load = **14,000 VA**

Practice Question

Determining Receptacle Load—Nondwelling
1. What is the receptacle load for a bank that has 75 duplex receptacles?

Fixed Multioutlet Assemblies

Per 220.14(H)(1) and (2), the load for fixed multioutlet assemblies is determined by how they are used. Where the load is unlikely to be used simultaneously, each 5′ or fraction of 5′ will be calculated at 180 VA. Where the load is likely to be used simultaneously, each 1′ will be calculated at 180 VA. **See Figure 12-7.**

For example, what is the receptacle load for 25′ of multioutlet assemblies that are used simultaneously?
Receptacle load = length of multioutlet assembly × VA/ft
Receptacle load = 25 × 180
Receptacle load = **4500 VA**

For example, what is the demand for 25′ of multioutlet assemblies that are not used simultaneously?

$$\text{Receptacle load} = \frac{\text{length of multioutlet assembly}}{5} \times \text{VA/ft}$$

$$\text{Receptacle load} = \frac{25}{5} \times 180$$

Receptacle load = **900 VA**

Practice Questions

Determining Receptacle Load for Multioutlet Assemblies—Nondwelling
1. What is the load for 35′ of multioutlet assemblies that are not used simultaneously?
2. What is the load for 35′ of multioutlet assemblies that are used simultaneously?

Heating and Air Conditioning

Commercial locations will have some type of heating and air conditioning (A/C) load. As with dwellings, 220.60 allows the smaller of the two loads to be left out of the calculation. Both the heating and A/C load will be calculated at 100% of the connected load. Many commercial-type buildings may have more than one unit, so the total connected load is calculated for all units.

For example, what is the load for a store that has four A/C units rated at 25,000 VA and four electric heating units rated at 15,000 VA?

First, the A/C load is calculated.
A/C load = rating × number of units
A/C load = 25,000 × 4
A/C load = 100,000 VA

Next, the heating load is calculated.
Heating load = rating × number of units
Heating load = 15,000 × 4
Heating load = 60,000 VA

Largest load = 100,000
Heating and A/C load = **100,000 VA**

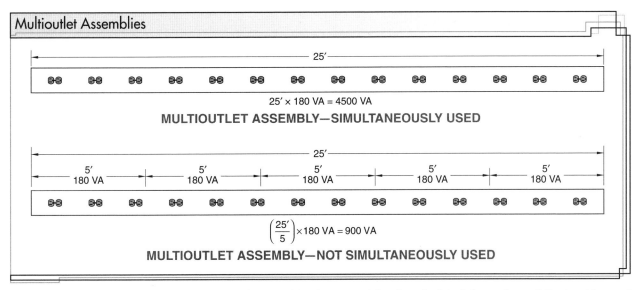

Figure 12-7. The way in which the load for fixed multioutlet assemblies is calculated depends on if the load is used simultaneously.

Commercial Cooking Equipment

Commercial cooking equipment regulations are found in 220.56. The demand rating for commercial cooking equipment is based on the number of appliances. The appliances that are considered commercial cooking equipment include: electric ranges, electric hot water heaters, dishwasher booster heaters, disposals, mixers, electric ovens, deep fryers, toasters, vent fans, and any other item in use that is thermostatically controlled or used intermittently. Commercial cooking equipment does not include electric heating or A/C equipment.

To determine the cooking equipment load, add the nameplate rating of each appliance and apply the demand factor for the number of appliances as found in Table 220.56. Next, add the two largest loads together. The cooking equipment demand load will be the largest amount between the total load, including the demand factor, and the two largest loads added together. **See Figure 12-8.**

For example, what is the load for a commercial kitchen that has the following appliances?
- two electric ranges at 14 kW
- two deep fryers at 8 kW
- one mixer at 1 kW
- one booster heater at 10 kW

14,000 + 14,000 + 8000 + 8000 + 1000 + 10,000 = 55,000 W

See Table 220.56.

Demand factor for six loads = 65%
55,000 × 65% = 35.75 kW
Add the largest two loads together.
14 kW + 14 kW = 28 kW

Largest of the two loads = 35.75 kW
Cooking equipment load = **35.75 kW**

Practice Question

Determining Commercial Cooking Equipment Load

1. What is the load for a commercial kitchen that has the following appliances?
   - one oven at 14 kW
   - two deep fat fryers at 10 kW
   - one mixer at 4 kW

TECH TIP

Where a kitchen is located in a classroom, such as a high school or culinary classroom, equipment such as electric ranges, electric ovens, and cooktops are calculated as cooking equipment for dwellings. In Table 220.55, Note 5 states that Table 220.55 applies to household cooking equipment used for educational purposes. If the appliances are not household-type cooking equipment, then they fall into Section 220.56 for commercial cooking equipment.

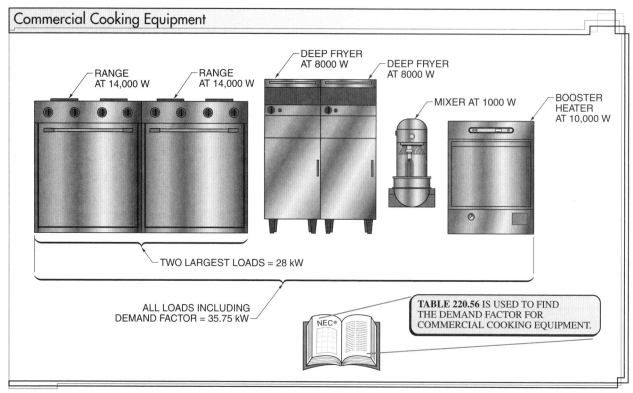

Figure 12-8. The demand factor for commercial cooking equipment is found in Table 220.56.

Fixed Appliances

An example of a fixed-appliance load for a nondwelling occupancy is an apartment building that has a common laundry area. Fixed appliances for nondwelling locations are rated at 100%.

Fixed appliances for nondwelling locations are calculated by adding the nameplate ratings to obtain a 100% connected load. Unlike dwelling units, there is no demand factor for four or more appliances because there is a greater chance of using multiple appliances at the same time in commercial buildings.

Largest Motor

The largest motor in a nondwelling location calculation shall be sized at 25%. If there are two or more motors of the same size, choose only one as the largest. The largest motor is not based on horsepower but on volt amps.

OPTIONAL SERVICE CALCULATION METHODS

Optional service calculation methods may be used for schools and new restaurants. The optional calculation method is usually easier to do but is only used on the exam when the question specifies it.

Optional School Calculation Method

Per 220.86, an optional calculation method can be used for schools. The demand factors in Table 220.86 can be applied to loads such as lighting, power, water, heating, cooking, and air-conditioning or space heating. The main difference in the optional school calculation method is that after determining the total connected load in volt amps, it is divided by the area of the school and demand factors are applied.

Heating and Air Conditioning. The heating and air conditioning are both sized at 100% of their respective loads. As with typical service calculations, the smaller of these two loads is omitted from the calculation. When loads are not likely to be used at the same time, Article 220.60 allows the smaller of the two loads to be dropped.

Lighting Loads. The lighting load is sized at 100% of the connected load. This includes both interior and exterior lighting loads. The interior lighting shall be sized at 3 VA per sq ft as found in Table 220.12. The outside load will be determined by the number of fixtures used for the exterior lighting.

Cooking Equipment. Per 220.56, cooking-equipment devices are items that are thermostatically controlled or used intermittently. **See Figure 12-9.** The cooking-equipment load is calculated by adding the nameplate rating of all the items. No demand factor is applied to the cooking-equipment load.

Figure 12-9. A convection oven is an example of cooking equipment that may be used in a school cafeteria.

Receptacles. Receptacles will be sized at 180 VA per strap or yoke. Per 220.44, the optional calculation method does not allow for the derating if the receptacle load is over 10,000 VA.

Multioutlet Assemblies. Per 220.14(H)(1) and (2), where the load is unlikely to be used simultaneously, each 5′ or fraction of 5′ will be calculated at 180 VA. Where the load is likely to be used simultaneously, each 1′ will be calculated at 180 VA.

Separate Circuits. Separate circuits are used for any special equipment that requires a separate circuit to operate. Loads that could be placed here are ventilation and exhaust systems, pumps for water systems, or pool filtration and circulation.

Demand Factor for Optional School Calculation. Once all of the loads are calculated and added together, the total connected load is divided by the square footage of the school to provide an average load in volt amps per square foot. Then, the demand factor is applied per Table 220.86. After the demand factor is applied, multiply by the square footage of the school to get the total load in volt amps, including the demand factor. To determine the service size, divide the total load by the supply voltage. To determine the service size for 3ϕ systems, divide the total load by the product of the supply voltage and the square root of 3 (1.732).

For example, what is the service size for a 20,000 sq ft school supplied by 208/120 V with the following loads?

Electric heat = 50,000 VA
A/C = 30,000 VA
Receptacles = 25,000 VA
Separate loads = 20,000 VA
Cooking equipment = 35,000 VA
Lighting load = 60,000 VA

Total load = 50,000 + 30,000 + 25,000 + 20,000 + 35,000 + 60,000
Total load = 220,000

$$\text{Average VA/sq ft} = \frac{\text{total load}}{\text{area/sq ft}}$$

$$\text{Average VA/sq ft} = \frac{220,000}{20,000}$$

Average VA/sq ft = 11 VA/sq ft

See Table 220.86.
First 3 VA is rated at 100%.
3 × 1.00 = 3
Over 3 VA to 20 VA is rated at 75%.
11 − 3 = 8
8 × 0.75 = 6
Average VA/sq ft with demand factor = 6 + 3
Average VA/sq ft with demand factor = 9

Total VA including demand factor = average VA/sq ft with demand factor × area/sq ft
Total VA including demand factor = 9 × 20,000
Total VA including demand factor = 180,000 VA

$$\text{Service size} = \frac{\text{total VA including demand factor}}{\text{supply voltage} \times 1.732}$$

$$\text{Service size} = \frac{180,000}{208 \times 1.732}$$

$$\text{Service size} = \frac{180,000}{360.26}$$

Service size = 499.64
Service size = **500 A**

> **Practice Question**
>
> *Determining the Load for a School Using the Optional School Calculation Method*
>
> 1. What is the load for a 35,000 sq ft school supplied by 208/120 V with the following loads?
>   - Electric heat = 50,000 VA
>   - A/C = 40,000 VA
>   - Receptacles = 30,000 VA
>   - Separate loads = 20,000 VA
>   - Cooking equipment = 35,000 VA
>   - Lighting load = 75,000 VA

Optional Restaurant Calculation Method

The optional restaurant calculation method requires all of the loads in the restaurant to be added, and then a demand factor is applied per Table 220.88. The demand will depend on whether the entire load is electric or whether only part of it is electric. As with the optional school calculation method, all of the loads are rated at 100%. One major difference in the optional restaurant calculation method is in the way the heating and A/C loads are calculated. According to the note under Table 220.88, both the heating and A/C loads are included in the total connected load.

Entourage
When using the optional restaurant calculation method, both the heating and A/C loads are included in the total connected load.

Demand Factors for Optional Restaurant Calculation Method. Table 220.88 gives the demand factors for the optional restaurant calculation method. The demand factors are determined by the total connected load and whether the restaurant is all electric.

Once all of the loads are calculated and added together, including both the heating and A/C loads, the total connected load is multiplied by the demand factors as found in Table 220.88. To determine the service size, divide the total load by the supply voltage. To determine the service size for 3ϕ systems, divide the total load by the product of the supply voltage and the square root of 3 (1.732).

For example, what is the load for a restaurant that has a connected load of 450 kVA and operates on a 480/277, 3ϕ system?

First, determine the service size for an all-electric restaurant.
See Table 220.88.
450 kVA is rated at 50% of the amount over 325 + 172.5
450 − 325 = 125
125 × 0.50 + 172.5 = 235 kVA or 235,000 VA

$$\text{Service size} = \frac{\text{total VA including demand factor}}{\text{supply voltage} \times 1.732}$$

$$\text{Service size} = \frac{235,000}{480 \times 1.732}$$

$$\text{Service size} = \frac{235,000}{831.36}$$

Service size = **282.67 A**

Next, determine the service size for a part-electric restaurant.
See Table 220.88.
450 kVA is rated at 45% of the amount over 325 + 262.5
450 − 325 = 125
125 × 0.45 + 262.5 = 318.75 kVA or 318,750 VA

$$\text{Service size} = \frac{\text{total VA including demand factor}}{\text{supply voltage} \times 1.732}$$

$$\text{Service size} = \frac{318,750}{480 \times 1.732}$$

$$\text{Service size} = \frac{318,750}{831.36}$$

Service size = **383.41 A**

> **Practice Question**
>
> *Determining the Load—Optional Restaurant Calculation*
>
> 1. What is the load for an all-electric restaurant that has a connected load of 500 kVA and operates on a 480/277, 3ϕ system?

Farm Loads

Total load calculations for farm loads have two parts. The first part includes calculating the loads for each individual building. The second part includes the total service load connection. Per Table 220.102, all loads that are expected to operate simultaneously, 125% of the full-load current of the largest motor, or the first 60 A of the load shall be sized at 100%. The next 60 A shall be 50% of its load, and the remainder of all other loads shall be at 25%.

After the building loads are determined, refer to Table 220.103. The largest load is sized at 100%, the second largest load is sized at 75%, the third largest is sized at 65%, and the rest of the load is sized at 50%. This total is added to any house load in the house calculation found in Unit 11.

For example, what is the service demand for a farm with the following loads?

Building 1 = 18,000 VA
Building 2 = 12,000 VA
Building 3 = 8,000 VA

Building 1 = 18,000 VA × 100% = 18,000 VA
Building 2 = 12,000 VA × 75% = 9000 VA
Building 3 = 8000 VA × 65% = 5200 VA
Total VA = 18,000 + 9000 + 5200 = **32,200 VA**

Practice Question

Determining Service Demand for Farm Loads

1. What is the service demand for a farm with the following loads?

   Building 1 = 22,500 VA
   Building 2 = 16,250 VA
   Building 3 = 6,150 VA

OTHER EQUIPMENT USED IN NONDWELLING OCCUPANCIES

Nondwelling occupancies may contain many different types of equipment. The NEC® is used to size conductors and overcurrent protective devices (OCPDs) for this equipment.

Arc Welders

The procedure for determining conductor size for arc welders is found in 630.11. To determine the current of the conductors, multiply the primary current rating listed on the nameplate of the welder by the duty cycle factor that is found in Table 630.11(A). **See Figure 12-10.** Once the current of the conductors is found in amps, use Table 310.15(B)(16) to size the conductors.

The procedure for determining overcurrent protection for arc welders is found in 630.12. Overcurrent protection for arc welders is determined by multiplying the primary current by 200%. If the OCPD does not correspond to a standard size, the next lower standard-sized OCPD should be used.

Figure 12-10. Section 630.11 specifies the procedure for determining the ampacity of conductors for arc welders.

Resistance Welders

The procedure for determining conductor size for resistance welders is found in 630.31. To determine the current of the conductors, multiply the primary current rating listed on the nameplate of the welder by the duty cycle factor that is found in Table 630.31(A)(2). Once the current of the conductors is found in amps, use Table 310.15(B)(16) to size the conductors.

The procedure for determining overcurrent protection for resistance welders is found in 630.32. Overcurrent protection for resistance welders is determined by multiplying the primary current by 300%. If the OCPD does not correspond to a standard size, the next higher standard-size OCPD should be used.

Single Non-Motor-Operated Appliances (Water Heaters)

Per 422.13, a water heater that holds 120 gal. or less is considered to be a continuous load when sizing branch circuits. To determine the current for the conductors, multiply the nameplate amp rating by 125%. If the rating is in watts, change it to amps and then increase it by 125%.

Per 422.11(E), overcurrent protection for non-motor-operated appliances shall comply with the following rules:
- The rating marked on appliance must not be exceeded.
- A maximum of 20 A is allowed if current is 13.3 A or less.
- For over 13.3 A, use 150% of the current. If the OCPD does not correspond to a standard size, then go up to the next standard-size OCPD as found in 240.6.

Per 422.11(F), electrical heating appliances that are rated at more than 48 A shall have the heating elements subdivided. Each load shall not exceed 48 A and the OCPD shall not exceed 60 A. Overcurrent protection for a subdivided load is determined by multiplying the current of the subdivided load by 125%. If the OCPD does not correspond to a standard size, then go up to the next standard-size OCPD as found in 240.6.

Capacitors

The procedure for determining conductor size for capacitors is found in 460.8(A). To determine the current of the conductors, multiply the rating of the capacitor by 135% or ⅓ of the motor circuit conductors, whichever is greater in size. Once the current of the conductors is found in amps, use Table 310.15(B)(16) to size the conductors. The rating or setting for the OCPD shall be as low as practicable.

Fixed Electric Space Heating Equipment

Per 424.3(B), fixed electric space heating equipment is considered to be a continuous load. To determine the current of the conductors, multiply the total current of the equipment by 125%. Once the current of the conductors is found in amps, use Table 310.15(B)(16) to size the conductors.

The procedure for determining overcurrent protection for fixed electric space heating equipment is also found in 424.3(B). Overcurrent protection for fixed electric space heating equipment is determined by multiplying the total current of the equipment by 125%. If the OCPD does not correspond to a standard size, then go up to the next standard-size OCPD as found in 240.6.

Refrigeration Equipment

The procedure for determining conductor size for refrigeration equipment is found in 440.32. To determine the current of the conductors, multiply the rating listed on the nameplate by 125%. Once the current of the conductors is found in amps, use Table 310.15(B)(16) to size the conductors.

The procedure for determining overcurrent protection for refrigeration equipment is found in 440.22. Overcurrent protection for refrigeration equipment is determined by multiplying the nameplate rating by 175%. If the OCPD does not correspond to a standard size, then go down to the next standard-size OCPD as found in 240.6. If this size of OCPD is not sufficient to start the refrigeration equipment, then the nameplate rating may be multiplied by 225%. If the OCPD does not correspond to a standard size, then go down to the next standard size.

The procedure for determining overload protection for refrigeration equipment is found in 440.52(A)(1). Overload protection for refrigeration equipment is determined by multiplying the nameplate rating by 140%.

Phase Converters

Per 455.6, the ampacity of the conductors for phase converters is determined by the type of load that is connected to it. The load can be variable or fixed.

A *variable load* is a load where the current fluctuates. When the loads are variable, the conductor ampacity shall not be less than 125% of the phase converter nameplate 1ϕ input full-load amps.

A *fixed load* is a load where the current does not fluctuate. For fixed loads, if the conductor ampacity is less than 125% of the phase converter nameplate 1ϕ input full-load amps, the conductors shall have an ampacity of not less than 250% of the sum of the full-load, 3ϕ current rating of the motors and other loads served where the input and output voltages of the phase converter are identical. Where the input and output voltages of the phase converter are different, the current (as determined by this section) shall be multiplied by the ratio of output to input voltage.

The procedure for determining overcurrent protection for phase converters is found in 455.7. The OCPD shall not have a rating of less than 125% of the phase convertor's 1ϕ input rating for variable and for fixed loads. If the required OCPD is not a standard rating, use the next higher rated OCPD per 240.6.

Generators

Per 445.13, the conductors from the generator windings to the first overcurrent protective device shall be sized at 115% of the nameplate rating of the generator. The conductors that are connected to the load side of the overcurrent device are sized at 100% of the overcurrent device.

CHAPTER 12
Nondwelling Occupancies
Review Questions

Name _____   Date _____

_____ 1. The ___ load for nondwelling occupancies is determined by multiplying the area of the building by the volt amps per square foot.
   A. fixed-appliance
   B. lighting
   C. dryer
   D. heating or A/C

_____ 2. Conductors and overcurrent protection devices used for continuous loads are sized at ___% of the connected load.
   A. 50
   B. 75
   C. 100
   D. 125

_____ 3. When used in commercial applications, track lighting is calculated at ___ VA per linear foot.
   A. 50
   B. 100
   C. 150
   D. 175

_____ 4. A continuous load is a load that is on for ___ hr or more at one time.
   A. 2
   B. 3
   C. 4
   D. 6

_____ 5. A 20,000 sq ft store with a 277/480 V service will have ___ 20 A circuits for the lighting load.
   A. fourteen
   B. twenty four
   C. fifty
   D. ninety-two

_____ 6. A multioutlet assembly that is not simultaneously used is calculated at 180 VA per ___'.
   A. 1
   B. 5
   C. 10
   D. 20

_____ 7. The procedure for determining conductor size for arc welding machines is found in Section ___.
   A. 422.13
   B. 455.6
   C. 460.8
   D. 630.11

_____ 8. A general lighting load that is on for 6 hr in a 4300 sq ft church is ___ VA.
   A. 4300
   B. 5375
   C. 10,750
   D. 16,125

_____ 9. The load for 50′ of multioutlet assemblies that are not used simultaneously is ___ VA.
   A. 1800
   B. 2000
   C. 9000
   D. 10,000

_____ 10. A 30,000 sq ft school supplied by 208/120 V with the following loads will have a service size of ___ A. Use the optional school calculation.
   - Electric heat = 50,000 VA
   - A/C = 20,000 VA
   - Receptacles = 25,000 VA
   - Separate loads = 25,000 VA
   - Cooking equipment = 40,000 VA
   - Lighting load = 65,000 VA

   A. 250
   B. 498.67
   C. 520.78
   D. 530.87

CHAPTER

13

Printreading

Digital Resources
ATPeResources.com/QuickLinks
Access Code: 637160

In addition to the knowledge and skills necessary to perform the various jobs in the field, the master electrician must be able to read and interpret job-specific documentation, including electrical engineers' written specifications, schedules, and drawings. While an electrician is mainly interested in electrical prints, it is also important to identify symbols and notes used by the other trades to reduce possible obstacles and problems. Many master electrician exams include questions that require printreading knowledge in order to answer them correctly.

OBJECTIVES

After completing this chapter, the learner will be able to do the following:

- Determine the type of plans that are more commonly used by electricians.
- Understand the three main subjects included in print specifications.
- Identify information typically displayed in a title block.
- List the type of items that are represented by architectural symbols.
- Describe the types of information that schedules provide.
- List several different types of schedules included in building plans.

PRINTS

A *print* is a detailed plan of a building drawn orthographically using a conventional representation of lines and includes abbreviations and symbols. **See Figure 13-1.** Prints are first drawn as working drawings, and copies are made to produce the prints. Each print of a set of working drawings represents a part of a building drawn as an orthographic projection. This method of drawing allows the individual parts of the building to be shown on a flat plane in their true shape. A substantial number of individual prints may be required to clearly show the shape and size of all components. The most essential plans for an electrician include plot plans, floor plans, elevations, and one-line drawings. Of these, the floor plan is the primary plan in that it gives the electrician the most information about the building.

Floor Plans

A *floor plan* is an orthographic drawing of a building as though cutting planes were made through it horizontally. **See Figure 13-2.** The cutting plane is generally taken 5′-0″ above the floor being shown. Objects below the 5′-0″ elevation are shown with object lines. Objects above the 5′-0″ elevation are shown with hidden lines.

A floor plan is essential to an electrician because it identifies the location of electrical equipment and devices both inside and outside the building. Electrical information on the floor plan may include receptacles, switches, kitchen appliances, fans, washing machines, dryers, and lighting.

Plot Plans

A *plot plan* is an orthographic drawing that shows the location and orientation of a structure on a lot and the size of the lot. **See Figure 13-3.** Information located on a plot plan of particular value to an electrical contractor includes the point of beginning, contour lines (if an underground distribution system is to be installed), locations of buildings on the property, new and existing roadways, electrical overhead transmission lines, overhead and underground electrical distribution lines, telephone lines, and any other utility lines or equipment.

TECH TIP

The point of beginning is a fixed point from which all measurements are made in order to show the building lot. The point of beginning, in turn, is related to a benchmark or city datum point. A datum point may be a concrete marker within sighting distance of the lot, a mark on a fire hydrant, the top of a sewer in the street, or any other point established by the city as a local point of reference. A datum point is not only a point from which the point of beginning is located by horizontal measurement, it is also the beginning point for vertical dimensions.

Elevation View

An *elevation view* is an orthographic drawing showing vertical planes of a building. Depending on building size, shape, and complexity, four or more elevations may be required to clearly show all exterior walls. Generally, four exterior views are sufficient. These are the front, rear, right side, and left side, or North, South, East, and West. Interior elevations are used to identify cabinets, door and window openings, and the type of materials used.

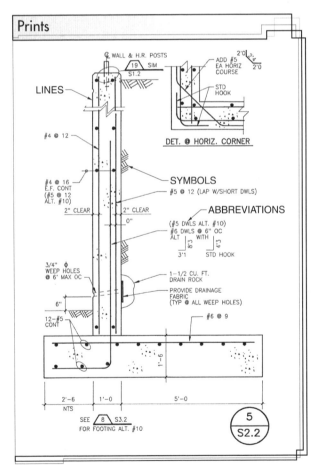

Figure 13-1. Lines, symbols, and abbreviations are elements that make up a detailed print of a building.

Floor Plans

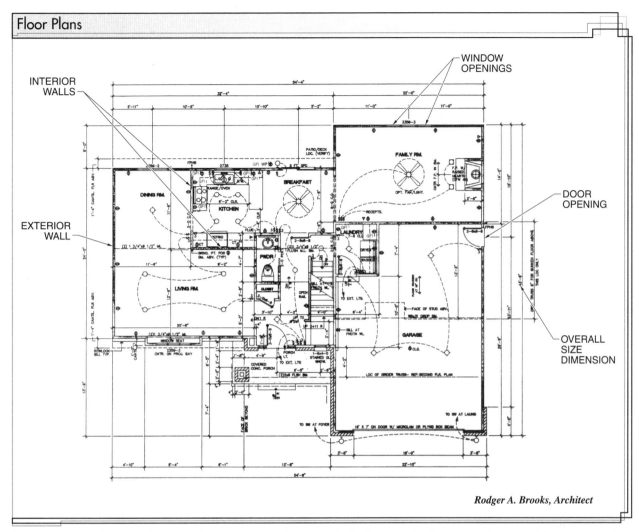

Figure 13-2. The location of lights, switches, and receptacles are just a few items that an electrician will look for on a floor plan.

An elevation view typically shows dimensions such as height and width. For example, an elevation view of a motor control center is necessary when wiring motor control circuits. A motor control center contains many cubicles that are similar. When wiring is installed, it must be installed in the correct cubicle. Such information is determined from the motor control center elevation view. **See Figure 13-4.**

TECH TIP

Elevations typically show the exterior of a house or building. Elevations are directly related to floor plans and are usually drawn at the same scale. In some cases, interior elevations may also be included in a set of plans to show interior wall details.

Motor control wire installation information may be listed on elevation views.

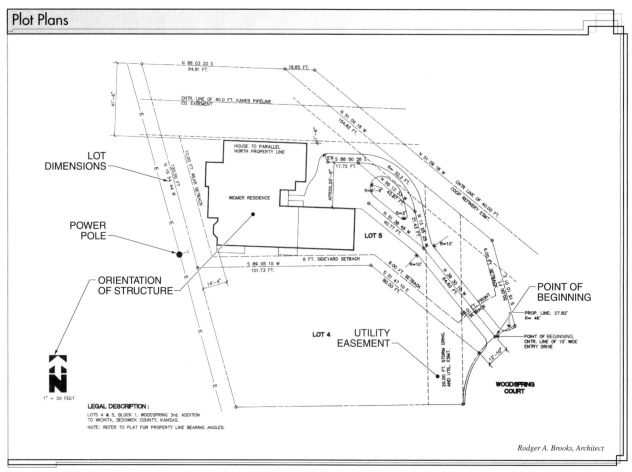

Figure 13-3. Plot plans show a bird's-eye view of the construction area.

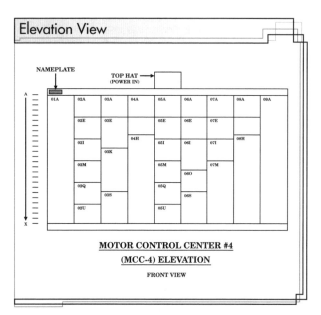

Figure 13-4. An elevation view typically shows dimensions such as height and width.

One-Line Riser Drawing

A *one-line riser drawing* is an electrical drawing that shows the arrangement of the electrical service and main distribution equipment in block form. **See Figure 13-5.** One-line riser drawings are often used to represent electrical distribution systems in commercial installations. One-line riser drawings not only show the flow path of electrical power into a facility, but also identify pertinent information regarding the specifics of the installation, including service conductor type, size, and whether underground or overhead; the size and ratings of the main distribution panel; the size and type of feeders to outlying major equipment and panelboards; and conduit size and type.

SPECIFICATIONS

Specifications are sheets of paper, gathered together into a pamphlet (or into several volumes for a large building) covering a number of subjects in detail. These subjects include information in paragraph form that describes the building to be constructed, the materials to be used, and the responsibilities of those involved.

The simplest type of construction job should have specifications. They are included with prints and a written contract to become the basis for agreement between the owner and the contractor. The specifications may be used to determine the type of equipment or materials that are to be used. Many times the specifications will list the exact brand and part number that is required.

The Construction Specifications Institute (CSI) has developed standard procedures that are helpful in writing specifications, particularly in writing for the CSI MasterFormat™. Specification writing is divided into 50 technical divisions, each with several broad (smaller subheads) section headings. **See Figure 13-6.** Section 26 is for electrical prints.

SUPPORT INFORMATION

Supporting information for electrical construction drawings is located on individual drawing sheets. Supporting information concerning a specific drawing sheet is located in the supplemental-information section and is often required for accurate interpretation of electrical construction drawings.

Title Blocks

A *title block* is an area on a print used to provide information about the print. The title block generally appears in the lower right corner and gives the subject, or name of the sheet. The name and company logo of the AE firm are usually displayed in the title block. Other information that may be displayed in the title block includes the project name and location, the drafter's initials, the architect's seal, the plan completion date, the number of sheets in a set of prints and the particular sheet number, the scale, and the architect's name and location. **See Figure 13-7.** An *architect's seal* is an embossed stamp or insignia affixed to the construction drawing that signifies that a licensed professional architect has reviewed and approved the project drawings.

In a large set of plans, initials may precede the sheet number to indicate a particular trade area. For example, E1 OF 3 denotes the first electrical sheet of three electrical sheets. Other letters commonly used to denote specific trade area prints are P (plumbing) and M (mechanical), which includes heating, ventilating, and air-conditioning.

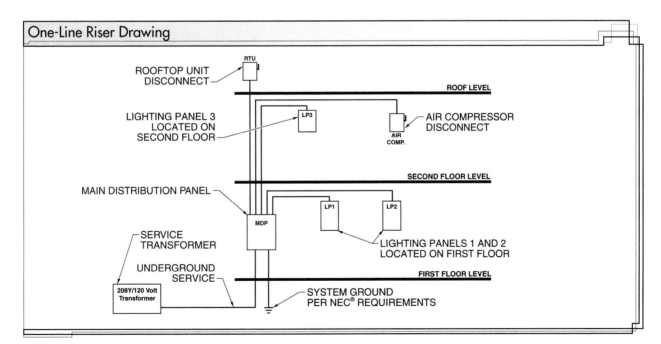

Figure 13-5. A one-line riser drawing is an electrical drawing that shows the arrangement of the electrical service and main distribution equipment in block form.

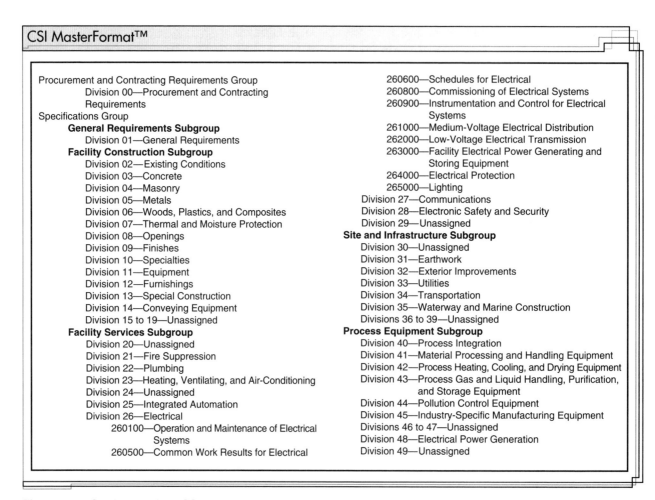

Figure 13-6. Section 26 of the CSI MasterFormat™ covers specification writing for electrical prints.

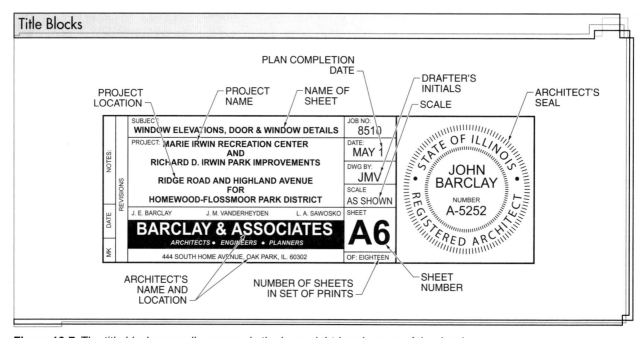

Figure 13-7. The title block generally appears in the lower right-hand corner of the drawing.

Drawing Scale

A *drawing scale* is a system of drawing representation in which drawing elements are proportional to actual elements. Dimensions of the drawing can be larger or smaller than the actual size of the object. For example, a typical scale as it appears on a drawing can be ¼″ = 1′, which indicates that one-quarter inch on the drawing represents one foot in the actual field. The scale provides the electrician a way to determine the distance between points on a print. Common scales sizes include; ¼″ = 1′, ⅛″ = 1′, ¹⁄₁₆″ = 1′, ¹⁄₃₂″ = 1′. The scale can be located anywhere on the print, but it is usually located toward the bottom of the page. Many electrical drawings, such as wiring diagrams and one-line diagrams, do not have a scale.

Electricians use an architect's scale to determine distances on a print. Common architect's scales are triangular in cross section and have six edges. Five of the edges have two scales each, and one edge is divided into inches and fractions of an inch.

Architect's scales are read from left to right or right to left depending on the scale. A ¼″ = 1′-0″ scale is read from right to left beginning at the 0 on the right end of the scale. The same set of markings is used for both the ¼″ = 1′-0″ scale and the ⅛″ = 1′-0″ scale. The correct line in relation to the scale used must be read. For example, 18′-0″ on the ¼″ = 1′-0″ scale is on the line representing 57′-0″ on the ⅛″ = 1′-0″ scale. **See Figure 13-8.**

Drawing Notes

A *drawing note* is a note on a drawing that contains information specific to the drawing on that particular page or sheet. Notes will tell the electrician the meaning of certain subscript or superscript letters or numbers. They may also give height requirements, the fixture type, and box depth for a location with a different type of wall covering. Drawing notes are numbered, with the number correlating to a number on a specific section of the drawing.

When performing commercial or industrial wiring, it is important to understand the information located in the drawing notes sections. For example, drawing notes on a wiring drawing could read, "Coordinate installation of junction box with mechanical contractor" or "All 15 A and 20 A branch-circuit conductors to be 12 AWG THHN." Failure of a field electrician or contractor to comply with drawing notes could result in increased time and costs due to rework. The notes on a page will typically cover items on that page unless otherwise noted.

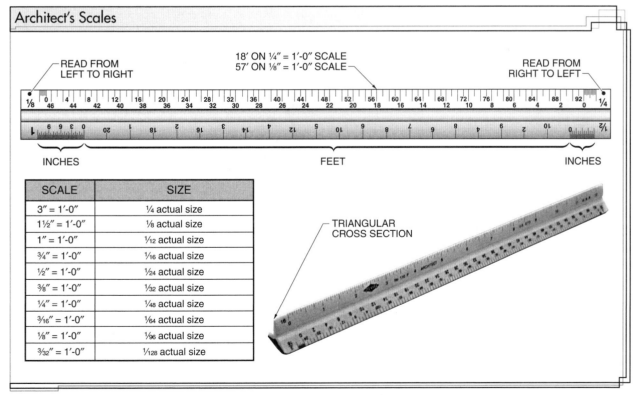

Figure 13-8. An architect's scale is used to determine distances on a print.

Prints and specifications are used by contractors and subcontractors to determine bids.

Symbols

A *symbol* is a graphic representation of an object. **See Figure 13-9.** Symbols provide a uniform representation of building materials, fixtures, and structural parts that are easily recognizable. Symbols are universal in that the symbol can be understood no matter what language a person speaks. Symbols are shown on plan views (plot, floor, framing, etc.), elevation views, sections, and details.

Architectural symbols show building materials such as wood, brick, concrete, glass, roofing, and structural steel. Electrical symbols show various types of outlets, receptacles, lights, wiring, signaling systems, etc.

SCHEDULES

A *schedule* is a detailed list on a print that provides information about building components such as doors, windows, and electrical and mechanical equipment. Schedules include tables and charts that supply supporting information for drawings. Numbers and/or letters on the print refer to the schedule. Schedules appear on separate sheets and are listed in the drawing index. Depending on the size, complexity, and type of project, separate schedules can exist for panelboards, lighting systems, and other mechanical equipment.

Panelboard Schedules

A *panelboard schedule* is a schedule that supplies detailed information about individual panelboards. Information located on a panelboard schedule includes panelboard designations, circuit-breaker sizes, loads supplied by individual breakers, panelboard ratings, the load per phase in kilovolt amps, and the total load. A panelboard schedule is used to determine future load capabilities, minimize phase unbalance as loads are added, and schedule lockout and tagout for routine maintenance.

Electricians can use panelboard schedules to order panels with the required breakers already installed, which will save time during the construction period. Panelboard schedules can also be used to verify connections to overcurrent protective devices.

All panels will have two rows of breakers, even 3φ panels. Panels are available for different current ratings. Odd numbered circuit breakers are listed on the left side and the even numbered circuit breakers are on the right side of the panel. The size of the load for each circuit shall be identified in the panel as required by the NEC® in case of an emergency.

Luminaire Schedule

A *luminaire schedule* is a schedule that supplies specific information regarding the lighting distribution systems. Typical information on a luminaire schedule includes the manufacturer's name, the model, number of luminaires, number of lamps per fixture, lamp types and ratings, and type of lamp lens.

There are many different types of luminaires that may be used in a structure. If the name, part number, and lamp type were written next to each fixture, the print would become filled with writing and be hard to read. Each luminaire has a designated letter or letter and number. Each letter and number corresponds to a schedule that is placed on the page to the side of the actual drawing. **See Figure 13-10.**

Mechanical Equipment Circuit Schedule

A *mechanical equipment circuit schedule* is a schedule that provides specific information about the loads supplied by the distribution system. Typical information on mechanical equipment schedules includes motor or equipment function, equipment identification number, the manufacturer name and model, horsepower or kilovolt-amp rating, motor RPM, and power source. The conductor number and size will not include the equipment grounding conductor. The electrician must know the required size for the equipment grounding conductor.

Chapter 13—Printreading **223**

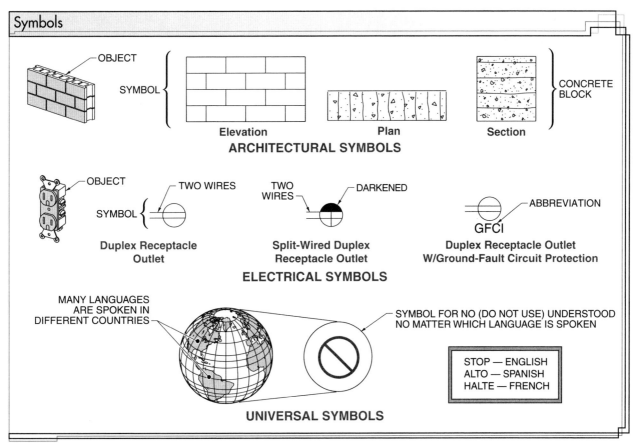

Figure 13-9. Symbols are universal in that the symbol can be understood no matter which language a person speaks.

Type	Lamps		Description	Volt	Mounting	Manufacturer Cat. Number
	Qty.	Cat. No.				
F1	3	F32T8 3500K	Recessed static troffer	120	Recess T-grid	METALUX 2GP-332A-120V-EB82
F2	3	F32T8 3500K	Recessed static troffer w/ self-contained emergency lighting	120	Recess T-grid	METALUX 2GP-332A-120V-EB82-EL4
F3			Not used			
F4	2	26W DTT	Compact fluorescent recessed downlight	120	Recess ceiling	HALO 8″ aperture H-7871-99870BA
F5	1 1	T5/8W T5/13W	Under cabinet task light	120	Under cabinet	METALUX 8F-CL-1-0813T5-EB-120
F6	1	100W A-21	Explosion proof incandescent	120	Surface ceiling	APPLETON AC1050
F7	1	100W A-21	Porc. keyless lampholder	120	Surface ceiling	PASS and SEYMOUR Series 44
F8	1	50W/Mogul MH	Surface wall pack	120	Wall 9′-6″	LUMARK MH-WL-50-120-FI-PE/MT
X1		LED	Polycarbonate exit; self powered; single face	120	Surface ceiling	SURELIGHT CCX-7-1-70-G-WH
X2		LED	Polycarbonate exit; self powered; double face	120	Surface ceiling	SURELIGHT CCX-7-2-70-G-WH

Figure 13-10. Luminaire schedules are tables and charts that supply drawing support information.

CHAPTER 13
Printreading
Review Questions

Name _____ Date _____

_____ 1. Blueprints use ___ instead of words to represent the electrical equipment installed in a building.
   A. symbols
   B. colors
   C. letters
   D. pictures

_____ 2. If an electrician measures 3.5″ for the length of a wall on a print, and the scale for the print is ⅛″ = 1′, then the total length of the wall is ___′.
   A. 12
   B. 14
   C. 28
   D. 35

_____ 3. The revision date of a print can be found on the ___ of the print.
   A. front page
   B. binding
   C. title block
   D. back

_____ 4. Plot prints show where the ___ is (are) located on a plot of land.
   A. grade level
   B. gas lines
   C. power lines
   D. all of the above

_____ 5. The Construction Specifications Institute refers to section ___ for writing specifications for electrical prints.
   A. 2
   B. 8
   C. 10
   D. 16

_____ 6. The ___ is the primary plan in that it gives the electrician the most information about the building.
   A. plot plan
   B. floor plan
   C. elevation drawing
   D. detail drawing

_____ 7. A ___ schedule is a schedule that supplies specific information regarding the lighting distribution systems.
   A. luminaire
   B. panelboard
   C. mechanical equipment
   D. fire alarm

_____ 8. The ⊖ symbol is used on prints to identify a ___ outlet.
   A. single receptacle
   B. duplex receptacle
   C. triplex receptacle
   D. split-wired duplex receptacle

_____ 9. A common scales size for a drawing is ___″ = 1′.
   A. ¼
   B. ⅛
   C. ¹⁄₁₆
   D. all of the above

_____ 10. A(n) ___ is an orthographic drawing showing vertical planes of a building.
   A. symbol
   B. elevation view
   C. floor plan
   D. plot plan

Digital Resources
ATPeResources.com/QuickLinks
Access Code: 637160

CHAPTER

14

Practice Exams

This chapter contains four sample master electrician's exams. Each exam has a 4-hour time limit. On average, the estimated time to answer each question is two minutes and twenty-four seconds. On the first pass through the exam, answer only the questions that can be answered with absolute certainty. If a question is skipped, be sure to skip that number on the answer sheet. Answers can be recorded on the answer sheet found in the appendix of this book. The answers to each question can be found in the Answer Key. Most tests require a score of 70% to pass the exam.

Practice Exam 14-1. 229

Practice Exam 14-2. 243

Practice Exam 14-3. 257

Practice Exam 14-4. 271

CHAPTER 14
Practice Exams

Practice Exam 14-1

Name _____   Date _____

_____ 1. The material that has the least amount of specific resistance (lowest resistance) is ___.
   A. gold
   B. silver
   C. copper
   D. aluminum

_____ 2. Secondary tie circuits shall have overcurrent protection at ___.
   A. one end only
   B. the load
   C. both ends
   D. the load and at both ends

_____ 3. All wiring installations shall be free from ___ other than as required or permitted in Article 250.
   A. ground faults
   B. short circuits
   C. resistance
   D. A and B

_____ 4. The minimum size for a single conductor used for power-limited fire alarm circuits shall be No. ___ AWG.
   A. 14
   B. 16
   C. 18
   D. 20

_____ 5. A wall space between two closet doors shall have a receptacle if the space is ___' or greater.
   A. 2
   B. 3
   C. 4
   D. 6

_____ 6. A legally required stand-by system shall have stand-by power ready within ___ sec after normal power failure.
   A. 10
   B. 60
   C. 90
   D. 120

229

7. Floor receptacles shall not be counted as part of the required number of receptacles unless located within ___" from the wall.
   A. 12
   B. 18
   C. 20
   D. 24

8. The first 3000 W is calculated at 100% of the general lighting load. The amount from 3001 W to 120,000 VA shall be calculated at ___%.
   A. 25
   B. 30
   C. 35
   D. 50

9. MI cable shall be permitted ___.
   A. in any hazardous classified locations where permitted by other code articles
   B. where exposed to moderate physical damage
   C. in underground circuits
   D. all of the above

10. Flat conductor cable (FCC) shall be permitted to be used in ___ locations.
    A. damp
    B. wet
    C. outdoor
    D. hazardous

11. The service head on a mast shall be ___ at the point of connection to the service drop conductors.
    A. weatherproof
    B. rainproof
    C. watertight
    D. listed for use in wet locations

12. The bending radius for the inner edge of AC cable shall not be less than ___ times the diameter of the cable.
    A. 5
    B. 8
    C. 10
    D. 12

13. The following can be used for protection of service entrance conductors except ___.
    A. IMC
    B. EMT
    C. FMC
    D. Schedule 80 PVC

_____ 14. A PVC box must be at least ___ cubic inches if containing the following:
two No. 14-2 Romex cables with grounding conductor
two No. 14-3 Romex cables with grounding conductor
two switches
two internal clamps
   A. 20
   B. 24
   C. 32
   D. 36

_____ 15. Service heads or goosenecks shall be placed ___ the point of the service drop attachment point.
   A. below
   B. above
   C. all of the above
   D. none of the above

_____ 16. Receptacles that are located within ___′ of the outer edge of a bathtub or shower stall shall be GFCI protected in a dwelling.
   A. 2
   B. 4
   C. 6
   D. 8

_____ 17. The ___ is used when sizing branch circuit conductors, overcurrent protection for short circuits, and ground faults for motors.
   A. nameplate
   B. full-load current
   C. voltage
   D. lock rotor current

_____ 18. Two power fuses over 1000 V shall be permitted to be placed in parallel on the same load as long as the two fuses ___.
   A. have the same rating
   B. are installed indoors
   C. are installed underground
   D. all of the above

_____ 19. When the length of the tap conductors does not exceed ___′, the ampacity of the conductors is not less than the combined calculated load of the circuits and not less than the rating of the overcurrent protective device supplied by the secondary conductors.
   A. 10
   B. 35
   C. 50
   D. 75

_____ 20. The effective ground fault current path of equipment, raceways, and other non-current-carrying parts shall ___.
   A. be an intentionally constructed, low impedance path
   B. be able to carry current underground fault conditions
   C. help facilitate the operation of overcurrent protective devices or ground fault detectors
   D. all of the above

21. Each 1000 sq ft apartment in a multifamily dwelling with 40 units contains an 8 kW range, a 1.2 kVA dishwasher, a 600 VA disposal, a 120 V motor that draws 18 A, and a 2.5 kW water heater. When using the optional calculation, the current required to size the service for a 1φ, 120/240 service is ___ A. *Note:* Do not include the laundry or any house loads.
    A. 551
    B. 955
    C. 1250
    D. 3410

22. Electrical systems of 50 V to 1000 V shall be grounded if ___.
    A. the supply system is 3φ, 4-wire delta connected in which the midpoint of one winding is used as a circuit conductor
    B. the system can be grounded so that the maximum voltage to ground on the ungrounded conductors does not exceed 150 V
    C. the system is 3φ, 4-wire, wye connected in which the neutral conductor is used as a circuit conductor
    D. all of the above

23. The grounding connection shall not be installed using ___.
    A. pressure connectors
    B. listed clamps
    C. solder only
    D. exothermic welding

24. A 200 A service to a dwelling has 2/0 THHN conductors used for the service entrance. The required size of the grounding electrode conductor to the concrete encased electrode is No. ___.
    A. 2 AWG copper or 1/0 AWG aluminum
    B. 2/0 AWG copper
    C. 4 AWG copper
    D. 6 AWG copper or 4 AWG aluminum

25. The capacitive reactance for a capacitor rated 28.5 µF and operating at 60 Hz is ___ Ω.
    A. 0.0107
    B. 93.1
    C. 107.5
    D. 10,739

26. The required size for a grounding electrode conductor for a service using 500 kcmil aluminum conductors is No. ___.
    A. 2 AWG copper or 1/0 aluminum
    B. 2/0 AWG copper or 4/0 aluminum
    C. 4 AWG copper or 2 AWG aluminum
    D. 6 AWG copper or 4 aluminum

27. A bending space of ___" is required for 1/0 THW conductors that do not leave the enclosure through the opposite wall from the terminals.
    A. 3.5
    B. 4
    C. 5
    D. 6

28. When used at a point on a circuit, a surge-protective device shall be connected to each ___.
    A. grounded conductor
    B. grounding conductor
    C. ungrounded conductor
    D. equipment grounding conductor

29. Attached and detached garages with electric power shall contain at least ___ 120 V, 20 A branch circuit(s) to supply receptacle outlets.
    A. one
    B. two
    C. three
    D. four

30. The minimum metric size designator for conduit, tubing, or other fittings is ___.
    A. 10
    B. 12
    C. 14
    D. 16

31. Insulated fittings shall be used for insulated conductors of size No. ___ AWG or larger when installed in a metal raceway.
    A. 1
    B. 2
    C. 4
    D. 8

32. The grounded conductor shall be routed with the phase conductors and the size shall not be smaller than specified from Table ___.
    A. 250.66
    B. 250.102(C)
    C. 250.122
    D. 310.15(B)16

33. The total capacitance of two 15 µF capacitors placed in parallel with each other is ___ µF.
    A. 3.75
    B. 7.5
    C. 15
    D. 30

34. The minimum burial depth for a 120 V residential circuit that is GFCI protected and rated at 20 A is ___".
    A. 6
    B. 12
    C. 16
    D. 18

35. Systems of less than 50 V shall be grounded where the supply is ___.
    A. less than 120 V
    B. supplied by a transformer if the transformer supply is ungrounded
    C. used on control circuits
    D. none of the above

36. If a transformer has 2000 W on the primary, with five equal secondaries, the wattage on each of the secondary windings will be ___ W.
    A. 400
    B. 500
    C. 1000
    D. 2000

37. In dwelling units, a receptacle shall be placed at least ___′ but not more than 20′ from a swimming pool.
    A. 5
    B. 6
    C. 10
    D. 15

38. Kitchen dishwasher branch circuits in a dwelling shall be ___.
    A. hardwired only to the dishwasher
    B. GFCI protected
    C. cord-and-plug-connected only to the dishwasher
    D. GFCI protected and cord-and-plug-connected only to the dishwasher

39. Conductors that can be used in a wet location include ___.
    A. moisture-impervious metal-sheathed
    B. types THW, RHW, TW, MTW, XHHW, ZW
    C. types listed for use in wet locations
    D. all of the above

40. The type of conductor that shall not be used in a wet location is ___.
    A. THHN
    B. XHHW
    C. THW
    D. TW

41. A box with a minimum size of ___ cu in. can be used if containing the following.
    two No. 12-2 Romex cables with grounding conductor
    two No. 14-2 Romex cables with grounding conductor
    one switch connected to the No. 12 conductors
    two internal clamps
    A. 18
    B. 26
    C. 32.25
    D. 39

42. The connected lighting load for a 4500 sq ft dwelling is ___ VA.
    A. 9900
    B. 10,234
    C. 13,500
    D. 18,000

43. When flat cable assemblies are installed up to 8′ above the floor, a(n) ___ is needed to protect them.
    A. PVC conduit
    B. EMT conduit
    C. ENT conduit
    D. cover identified for the use

44. When an enclosure is mounted on a wall in a wet location, the minimum distance that the enclosure has to be mounted off the wall is ___″.
    A. ¹⁄₁₆
    B. ⅛
    C. ¼
    D. ½

45. A disconnecting means shall be provided for all ___ conductors derived from an energy storage system (ESS).
    A. grounded
    B. ungrounded and equipment grounding
    C. ungrounded
    D. grounded and equipment grounding

46. The classification of Class I, Division 2 is based on the ___.
    A. volatility of the flammable gases or vapors
    B. adequacy of the ventilation
    C. record of the industry with the respect to explosions or fires
    D. all of the above

47. Metal clad cable shall be permitted to be ___ where identified for such use.
    A. direct buried
    B. in concrete
    C. in cinder fill
    D. none of the above

48. The minimum burial depth for a UF cable used for a landscape lighting system operating at 24 V is ___″.
    A. 6
    B. 12
    C. 18
    D. 24

49. Single conductor SE cable outdoors shall be supported by ___.
    A. staples
    B. cable ties listed and identified for securement and support
    C. straps
    D. all of the above

50. Locations where type UF cable can be used include ___.
    A. underground by direct burial
    B. wet locations
    C. nonmetallic-sheathed cable
    D. all of the above

51. The sizes of IMC conduit are ___″ to ___″.
    A. 3/8; 4
    B. 1/2; 4
    C. 1/2; 6
    D. 1; 4

52. The information found in the specification section in a set of prints includes the ___.
    A. type of building being built
    B. materials being used
    C. responsibilities of the contractor and owner
    D. all of the above

53. Multioutlet assemblies are not permitted ___.
    A. in dry locations
    B. where subject to severe physical damage
    C. in hoistways
    D. B and C

54. Underfloor raceways that are over ___ mm wide but not more than ___ mm wide shall have at least 25 mm between raceways and shall be covered by at least 25 mm of concrete.
    A. 25; 200
    B. 100; 150
    C. 100; 200
    D. 200; 250

55. The disconnecting means for an adjustable speed drive shall have a rating of not less than ___% of the rated input current of the conversion unit.
    A. 100
    B. 115
    C. 125
    D. 130

56. Nonflammable fluid-insulated transformers installed indoors shall ___.
    A. have a liquid containment area
    B. have a pressure relief vent
    C. be equipped with a means for absorbing any gases generated by arcing inside the tank
    D. all of the above

57. The disconnecting means for a phase convertor shall disconnect simultaneously ___.
    A. all three phase conductors to the load
    B. any two of the three conductors to the load
    C. all single phase ungrounded supply conductors to the phase convertor
    D. any grounded conductors to the phase convertor

58. Flexible metal conduit connected to an 18 cu in. device box must be supported within ___″ from the device box.
    A. 12
    B. 18
    C. 24
    D. 30

_____ 59. Locations such as cotton or textile mills and woodworking plants are considered ___ locations.
   A. Class 1
   B. Class II
   C. Class III, Division 1
   D. none of the above

_____ 60. A patient care space would include ___.
   A. lounges
   B. offices
   C. corridors
   D. none of the above

_____ 61. A courtroom would be considered a place of assembly if it held ___ people or more.
   A. 50
   B. 75
   C. 100
   D. 150

_____ 62. Temporary wiring installations can be used for ___.
   A. testing
   B. experiments
   C. development
   D. all of the above

_____ 63. Stand-alone fuel cell systems shall be permitted to operate at 120 V to 120/240 V and shall not have ___ loads.
   A. 120 V
   B. 240 V
   C. multiwire circuit
   D. B and C

_____ 64. Industrial radiographic and fluoroscopic equipment shall be effectively ___ or shall have interlocks that deenergize the equipment automatically to prevent access to energized parts.
   A. enclosed
   B. guarded
   C. protected
   D. shielded

_____ 65. For PV systems installed on roofs of buildings, photovoltaic source circuits shall be deenergized from all sources within ___ sec of when rapid shutdown is initiated or when the PV power source disconnecting means is opened.
   A. 10
   B. 30
   C. 60
   D. 45

_____ 66. The conductor that is identified by a white or gray color is the ___ conductor.
   A. ungrounded
   B. grounding
   C. isolated
   D. grounded

_____ 67. The equal-spaced dashed line on blueprints represents ___.
    A. underground
    B. hidden edges
    C. ceiling lines
    D. home runs

_____ 68. The NEC® does not cover ___.
    A. public buildings
    B. automotive vehicles
    C. mobile homes
    D. warehouses

_____ 69. The high-leg conductor shall be marked with a(n) ___ color on 4-wire, delta-connected systems where the midpoint of one phase winding is grounded.
    A. purple
    B. red
    C. white
    D. orange

_____ 70. When sizing a bonding jumper for the supply side of the service disconnect, the bonding jumper shall be sized at 12.5% of the ungrounded phase conductors if the conductors are greater than ___ kcmil copper and ___ kcmil aluminum.
    A. 750; 1500
    B. 1100; 1750
    C. 1250; 2000
    D. 1750; 1100

_____ 71. Recognized as suitable for the specific purpose, function, use, environment, application, and so forth is defined as ___.
    A. labeled
    B. listed
    C. identified (as applies to equipment)
    D. approved

_____ 72. The load requirement per linear foot for a show window is ___ VA.
    A. 100
    B. 150
    C. 200
    D. 250

_____ 73. The area around electrical equipment shall have sufficient access, and ___ space shall be provided and maintained around the equipment to permit ready and safe operation and maintenance of such equipment.
    A. working
    B. open
    C. clear
    D. enclosed

74. Conductors other than flexible cords shall be protected against overcurrent in accordance with the specified ___ in 310.15 unless otherwise permitted.
    A. insulation
    B. terminations
    C. length
    D. ampacities

75. Type TFE conductors come in sizes of No. 14 to No. ___ AWG.
    A. 2/0
    B. 3
    C. 4/0
    D. 250

76. When conductors are installed in a location subject to earth movement due to settlement or frost, ___ to keep the conductors from being pulled out of the enclosure.
    A. stake down the conductors
    B. use S loops in the conductors to allow for movement
    C. install in conduit
    D. leave extra wire in the box

77. An RV parking stall equipped with both 20 A and 30 A receptacles shall use ___ VA for calculating the load for the service demand.
    A. 2400
    B. 3600
    C. 4800
    D. 9600

78. Coaxial cable used for a community TV cable system shall be buried at least ___″ from the power conductors.
    A. 6
    B. 12
    C. 16
    D. 24

79. The minimum size conductor for an equal-potential grid around a swimming pool shall not be less than a No. ___ AWG.
    A. 4
    B. 6
    C. 8
    D. 10

80. The size of rigid metal conduit required for 10 No. 4 THHN copper conductors is ___″.
    A. 1¼
    B. 1½
    C. 2
    D. 2½

81. The A-phase is connected to the ___ of a panelboard containing a set of 3ϕ busbars.
    A. left side
    B. right side
    C. middle
    D. all of the above

_____ 82. Conductors on the outside of buildings within ___′ of the building shall be insulated.
  A. 4
  B. 6
  C. 8
  D. 10

_____ 83. Cartridge fuses and fuseholders shall be classified by their ___.
  A. voltage only
  B. wattage
  C. voltage and amperage
  D. amperage only

_____ 84. Receptacles sized at ___ A can be installed only on a 30 A branch circuit.
  A. 20
  B. 20 A and 30
  C. 30
  D. 30 A and 40

_____ 85. The current of a 480/277 wye, 150 kVA transformer is ___ A.
  A. 90
  B. 180
  C. 312
  D. 360

_____ 86. The service demand for the required sign circuit for a commercial building that has grade access to the public is ___ VA.
  A. 1000
  B. 1200
  C. 1500
  D. 2400

_____ 87. A noncontinuous lighting load of 15,700 VA on 120 V circuits is required to have ___ 15 A circuits.
  A. six
  B. seven
  C. eight
  D. nine

_____ 88. The required size for an aluminum grounding-electrode conductor connected to a water pipe for two sets of paralleled 350 kcmil conductors per phase is No. ___ AWG.
  A. 1/0
  B. 2/0
  C. 3/0
  D. 4/0

_____ 89. The allowable ampacity for 2/0 THW conductors with eight current-carrying conductors is ___ A.
  A. 87.5
  B. 122.5
  C. 150
  D. 175

90. Cable trays shall not be installed in ___.
    A. hoistways
    B. attics
    C. wet locations
    D. damp locations

91. A feeder conductor size of ___ A is required for three motors drawing 30 A, 20 A, and 10 A.
    A. 60
    B. 67.5
    C. 75
    D. 80

92. When using a 15 A AC general-use snap switch to control a motor, the maximum load that can be placed on the switch is ___ A.
    A. 10
    B. 12
    C. 15
    D. 20

93. An electric sign must be mounted ___′ over areas that are accessible to traffic.
    A. 10
    B. 12
    C. 14
    D. 18

94. A device that is considered to be an outlet is a ___.
    A. lighting fixture
    B. controller
    C. circuit breaker
    D. snap switch

95. The service demand for three 12.4 kW electric ranges in a triplex is ___ kW.
    A. 8
    B. 14
    C. 36
    D. 37.2

96. THW conductors sized at ___ AWG are required for a 3600 VA heating device that operates on 120 V.
    A. 4
    B. 6
    C. 8
    D. 10

97. When using batteries for emergency lighting systems, the required voltage after 1½ hr on a 120 V system is ___ V.
    A. 90
    B. 105
    C. 110
    D. 120

_____ 98. A cord and plug used on a room air conditioner can be used as the disconnect when ___.
- A. the controls are 7′ high from the floor
- B. it is a 3φ unit
- C. the voltage rating is 250 V or less
- D. an approved manually operated disconnect is installed out of sight from the unit

_____ 99. The maximum setting for ground fault protection shall be ___ A, and the maximum time delay shall be one second for ground fault protection equal to or greater than ___ A.
- A. 1000; 3000
- B. 1200; 2400
- C. 1200; 3000
- D. 3000; 1200

_____ 100. When No. 14 THWN motor control circuit conductors are tapped to the load side of the overcurrent protective device, the maximum rating for the overcurrent protective device is ___ A if the control conductors do not leave the enclosure.
- A. 15
- B. 45
- C. 50
- D. 100

CHAPTER 14

Practice Exams

Practice Exam 14-2

Name _____ Date _____

_____ 1. A 3φ, 480 V switchboard that is rated at 1000 A is required to have ___ exit(s) from the working space.
- A. one
- B. two
- C. three
- D. four

_____ 2. A conductor carrying 30 A and 0.5 Ω will use ___ W of power.
- A. 7.5
- B. 15
- C. 200
- D. 450

_____ 3. The color white or gray shall be used for the ___ conductor.
- A. ungrounded
- B. grounded
- C. grounding
- D. phase

_____ 4. A(n) ___ bonding jumper is used to bond around an expansion fitting to ensure continuity.
- A. equipment
- B. main
- C. system
- D. none of the above

_____ 5. The overcurrent protective device shall be sized based on ___% of the noncontinuous loads and ___% of the continuous loads.
- A. 100; 100
- B. 100; 125
- C. 125; 100
- D. 125; 125

_____ 6. PVC conduit will expand or contract ___″ for a 10°F change in temperature for every 100′ of raceway.
- A. 0.21
- B. 0.3
- C. 0.41
- D. 0.61

7. Light switches shall be at the top and bottom of stairways if there are ___ or more steps between each floor.
   A. four
   B. five
   C. six
   D. seven

8. A service amperage of ___ A is required for an 1100 sq ft single-family dwelling containing the following loads:
   one 10 kW range
   one 6 kW water heater
   one electric dryer
   Note: There is no AC and the dwelling uses gas heat.
   A. 96
   B. 99
   C. 110
   D. 125

9. The maximum voltage for a Class 1 remote control and signaling circuit is ___ V.
   A. 30
   B. 120
   C. 250
   D. 600

10. The minimum size for an aluminum conductor shorter than 50′ in length used for 1000 V or less for outdoor circuits without a messenger wire is No. ___ AWG.
    A. 2
    B. 6
    C. 8
    D. 10

11. Service-entrance cable shall be supported within ___″ of the weatherhead.
    A. 6
    B. 8
    C. 12
    D. 18

12. A tap that is less than 25′ from a 400 A feeder using No. 500 THHN copper conductors is required to have at least No. ___ AWG THW copper conductors.
    A. 1
    B. 2/0
    C. 1/0
    D. 500

13. When a metal water pipe is used as a grounding electrode conductor, interconnection is only allowed within ___′ of the building entrance.
    A. 5
    B. 10
    C. 15
    D. none of the above

14. How many seal-off fittings are required in the system depicted below?
    A. 6
    B. 8
    C. 10
    D. 12

15. The minimum length of a ground rod is ___'.
    A. 6
    B. 8
    C. 10
    D. 12

16. A stationary motor shall be grounded except when ___.
    A. supplied by metal enclosed wiring
    B. in a hazardous location
    C. the motor operates at over 150 V to ground
    D. in a wet location and isolated or guarded

17. The requirements for using ARC fault protection shall be applied to ___.
    A. receptacle outlets
    B. lighting outlets
    C. the entire branch circuit
    D. conductors only in raceways

18. A 480 V, 3φ service that uses 500 kcmil conductors is required to have a No. ___ AWG grounding electrode conductor.
    A. 1
    B. 1/0
    C. 2/0
    D. 3/0

19. A 30 A, 240 V circuit that is 100' long and has a voltage drop of 5 V is required to have a conductor size of No. ___ AWG.
    A. 4
    B. 6
    C. 8
    D. 10

246 MASTER ELECTRICIAN'S EXAM WORKBOOK

_____ **20.** A heating element that is protected by a 40 A inverse time breaker is required to have a No. ___ AWG equipment grounding conductor.
  A. 6
  B. 8
  C. 10
  D. 12

_____ **21.** The metric designator for 1½″ is ___.
  A. 16
  B. 27
  C. 41
  D. 61

_____ **22.** Equipment that is to be installed in an area with deteriorating agents shall be ___.
  A. labeled for the use
  B. approved for the use
  C. identified for the use
  D. placed in another area

_____ **23.** If a raceway has four 3φ motor conductors installed in it, plus ground wires for each motor, the conductors shall be derated by ___%.
  A. 45
  B. 50
  C. 70
  D. 80

_____ **24.** A receptacle shall be placed for every ___′ or major fraction thereof of a show window.
  A. 6
  B. 8
  C. 10
  D. 12

_____ **25.** The maximum service size for a dwelling that is using 2/0 AL THW conductors is ___ A.
  A. 100
  B. 110
  C. 125
  D. 150

_____ **26.** A box with a minimum size of ___ cu in. is required be used if containing the following:
  two duplex receptacles
  five equipment-grounding conductors
  four locknuts
  six No. 12 AWG conductors entering into the box
  A. 20
  B. 22.5
  C. 24.75
  D. 30

27. A troffer luminaire shall be connected to a suspended ceiling grid by all of the following except ___.
    A. clips
    B. rivets
    C. screws
    D. fixture wires

28. Type NM cables shall not be used in ___.
    A. exposed locations
    B. concealed locations
    C. attics
    D. wet or damp locations

29. A 2″ horizontal rigid-metal conduit with screw-on couplings shall be permitted to be supported every ___′.
    A. 10
    B. 14
    C. 16
    D. 20

30. Fixed luminaires shall be placed at least ___′ above vehicle lanes in a garage used to service cars or trucks.
    A. 8
    B. 10
    C. 12
    D. 14

31. Flexible metal conduit shall be supported every ___′.
    A. 1
    B. 2
    C. 4.5
    D. 6

32. When installing cable trays, a fitting is used for ___.
    A. stopping the tray run
    B. changing direction or elevation
    C. starting a run
    D. reducing the width of the tray

33. The temperature of combustible materials around a recessed luminaire shall not be greater than ___°C.
    A. 45
    B. 60
    C. 90
    D. 105

34. A circuit breaker may be used as a disconnect for an appliance rated over ___ VA if the circuit breaker is in sight of the appliance.
    A. 200
    B. 300
    C. 450
    D. 600

35. The required size circuit breaker for a 3 kW baseboard heater and a 1.5 kW baseboard heater both operating at 240 V is ___ A.
   A. 15
   B. 20
   C. 25
   D. 30

36. The maximum rating for an instantaneous breaker for a 7½ HP, 480 V, 3ϕ motor is ___ A.
   A. 90
   B. 100
   C. 125
   D. 150

37. If a conductor is run through a hole in a stud that is 1″ from the nearest edge of the stud, an unlisted steel plate of at least ___″ in thickness is required to protect the conductor.
   A. 1/16
   B. 1/8
   C. 3/16
   D. 1/2

38. A hermetic refrigeration motor that has a nameplate rating of 20 A and a branch selection current of 26 A is required to have a disconnect sized for ___ A.
   A. 23
   B. 30
   C. 35
   D. 40

39. A current of ___ A is used to size the feeder conductors feeding motors of 28 A, 16 A, and 12 A.
   A. 59.25
   B. 63
   C. 70
   D. 80

40. A 3ϕ, 250 V room air conditioner shall be ___ connected.
   A. directly
   B. cord and plug
   C. pin and sleeve
   D. all of the above

41. If a motor starter has 1.5 A of current on the primary side of a control transformer, the maximum overcurrent protection on the primary side is ___ A.
   A. 2
   B. 4
   C. 6
   D. 7.5

42. A raceway that has multiple 15 A circuits is required to have at least one No. ___ AWG equipment grounding conductors.
    A. 10
    B. 12
    C. 14
    D. 16

43. The walls, floors, and roof of a transformer vault shall have a fire rating of ___ hr.
    A. 1
    B. 1.5
    C. 3
    D. 4

44. A 40 A inverse time breaker used to protect a motor is required to have a No. ___ AWG equipment grounding conductor.
    A. 6
    B. 8
    C. 10
    D. 12

45. Dampers used for air in transformer vaults shall have a fire rating of ___ hr.
    A. 1
    B. 1.5
    C. 2
    D. 3

46. A kilowatt (kW) is a unit of ___.
    A. voltage
    B. power
    C. current
    D. resistance

47. The overcurrent protection of a phase converter that does not have a fixed load shall be at least ___% of the single-phase input amperage of the converter.
    A. 100
    B. 125
    C. 150
    D. 300

48. A Class II location is used for ___.
    A. combustible dust
    B. gases and vapors
    C. flyings
    D. grease

49. A load that can be connected to the critical branch of the essential electrical system of a hospital is the ___ system.
    A. exit light
    B. nurse call
    C. office light
    D. emergency light

50. The circuit size required for a sign on a commercial location with ground access is ___ A.
    A. 15
    B. 20
    C. 25
    D. 30

51. When the wireway of an audio reproduction location does not contain power supply conductors, the minimum size for the equipment grounding conductor shall be No. ___ AWG.
    A. 10
    B. 12
    C. 14
    D. 16

52. The maximum weight a luminaire can put on a ceiling outlet box shall be ___ lb.
    A. 25
    B. 35
    C. 50
    D. 70

53. The minimum height for a coaxial cable over a permanent swimming pool is ___'.
    A. 10
    B. 12
    C. 15
    D. 18

54. The maximum number of disconnects that can be used to control a photovoltaic system is ___.
    A. two
    B. four
    C. six
    D. eight

55. The type of system that could possibly create a hazard to firefighters if it stopped working is the ___ system.
    A. legally required standby
    B. emergency
    C. optional standby
    D. fuel cell

56. The minimum size conductor used for a ground ring shall be a No. ___ AWG.
    A. 1
    B. 2
    C. 3
    D. 4

57. An electrode permitted for grounding includes one or more metal in-ground support structures in direct contact with the earth for ___' or more with or without concrete encasement.
    A. 8
    B. 10
    C. 15
    D. 20

58. The number of wires required for a remote start/stop station to be connected to the motor starter is ___ wires.
    A. two
    B. three
    C. four
    D. six

59. The minimum size of a Class 1 single-circuit conductor is No. ___ AWG.
    A. 16
    B. 18
    C. 20
    D. 22

60. Communication conductors shall be placed at least ___' apart from lightning conductors.
    A. 2
    B. 3
    C. 6
    D. 8

61. Identify the overload contact in the motor control circuit below.
    A. A
    B. B
    C. C
    D. D

62. A circuit that draws 20 A at 120 V and operates at a 90% power factor will have a true power of ___ W.
    A. 2160
    B. 2205
    C. 2400
    D. 2667

63. Industrial control panels shall not be installed where the available ___ exceeds the short-circuit current rating of the control panel.
    A. short-circuit current
    B. voltage rating
    C. overcurrent rating
    D. resistance rating

64. If the phase conductor has 0.02 Ω, the equipment grounding conductor has 0.156 Ω, and the circuit operates on 120 V, then the ground-fault current will be ___ A.
    A. 21
    B. 682
    C. 769
    D. 6000

65. A ½″ ENT conduit can hold up to ___ No. 12 THWN conductors.
    A. 8
    B. 9
    C. 12
    D. 16

66. Which of the following circuits represents a push-to-test motor control circuit?
    A. A
    B. B
    C. C
    D. D

67. A 480 V feeder and branch circuit shall be permitted to have a maximum voltage drop of ___ V.
    A. 12
    B. 14.4
    C. 24
    D. 28.8

68. The connection for a low-voltage wye-connected motor is ___.
    A. 4, 5, and 6 tied together, T1 to 1 and 7, T2 to 2 and 8, T3 to 3 and 9
    B. 1 to 1, 2 to 2, 3 to 3, 4 to 7, 5 to 8, 6 to 9
    C. 1 to 4, 2 to 5, 3 to 6, 4 to 7, 8 to 9
    D. 1 to 1, 6 and 7, 2 to 2, 4 and 8, 3 to 3, 5 and 9

69. Each small appliance circuit in a dwelling shall be calculated at ___ VA.
    A. 1500
    B. 1800
    C. 3000
    D. 4500

70. A 60' × 40' dwelling is required to have ___ 15 A circuits.
    A. four
    B. six
    C. eight
    D. nine

71. When conductors are derated, Romex cable conductors shall use a temperature rating of ___°C.
    A. 40
    B. 60
    C. 75
    D. 90

72. Fixed resistance and electrode industrial process heating equipment where there is more than one source, feeder, or branch circuit shall have their disconnecting means ___.
    A. grouped
    B. isolated
    C. identified as having multiple disconnecting means
    D. both A and B

73. If a panelboard is mounted onto a grounded metal rack, then the panelboard ___.
    A. shall be considered to be effectively grounded
    B. requires a bonding jumper
    C. is not bonded
    D. could be an effective ground-fault path

74. A dwelling service that has a 100 A panel using No. 2 AWG THHN aluminum conductors is required to have a No. ___ AWG aluminum supply-side bonding jumper.
    A. 4
    B. 6
    C. 8
    D. 10

75. ___ cannot be used to protect service entrance conductors where they are subject to physical damage.
    A. EMT
    B. RMC
    C. Schedule 80 PVC
    D. Liquid-tight flexible metal conduit

76. A 240 V continuous-duty motor rated at 22 A with a service factor of 1.1 and a temperature rise of 42°C will have a maximum-size running overload of ___ A.
    A. 25.3
    B. 27.5
    C. 28.6
    D. 30.8

77. Raceways containing service conductors embedded in concrete shall be arranged to ___ water.
    A. drain
    B. hold
    C. avoid
    D. none of the above

78. The AHJ can waive rules by special permission only if ___.
    A. allowed by the engineer
    B. assured that equivalent objectives are met and can maintain safety
    C. the electrical system is under 250 V
    D. the electrical system is exposed

79. ___ cannot be placed on the temporary receptacle circuit on a construction site.
    A. Lighting outlets
    B. Refrigerator receptacles
    C. Power tool receptacles
    D. Office receptacles

80. An arc flash warning shall be field or factory marked on switchboards and other similar equipment and shall be ___.
    A. placed inside the equipment
    B. placed on the bottom of the equipment
    C. clearly visible
    D. in capital letters

81. An assured equipment grounding conductor program shall have the equipment grounding conductors tested at intervals not exceeding ___.
    A. three weeks
    B. two months
    C. three months
    D. six months

82. A(n) ___ can be connected before the service disconnecting means.
    A. instrument transformer
    B. surge protective device
    C. tap conductor used for load management
    D. all of the above

83. The size of EMT conduit required for three THHN No. 4/0 conductors and one No. 4 THHN conductor is ___″.
    A. 1
    B. 1¼
    C. 1½
    D. 2

84. A locknut can also be called a ___.
    A. bushing
    B. fitting
    C. device
    D. connector

85. A(n) ___ cannot be used as a grounding electrode.
    A. metal water pipe
    B. aluminum rod
    C. steel building in contact with the earth
    D. concrete-encased electrode

86. Branch circuit raceways on the exterior wall of a building shall be ___.
    A. suitable for use in wet locations and arranged to drain
    B. weatherproof
    C. sealed
    D. in PVC conduit

87. To mix Class 2 wiring circuits with Class 3 circuits, ___.
    A. a barrier shall be install between them
    B. the insulation of the Class 3 circuits shall match the insulation of the Class 2 circuits
    C. the insulation of the Class 2 circuits shall match the insulation of the Class 3 circuits
    D. none of the above

88. The required working-space width from the front of an electrical panel is ___″.
    A. 24
    B. 30
    C. 36
    D. 48

89. A neutral that has 580 A of unbalanced load on a 120/240 V, 1φ system shall be permitted to be sized for ___ A.
    A. 380
    B. 406
    C. 466
    D. 580

90. A 1 HP induction motor has ___ W of power.
    A. 500
    B. 746
    C. 1000
    D. 1500

91. Not counting the equipment grounding conductor of the same size, an electrician can install ___ No. 14 THHN wires in a ⅜″ metal flex conduit with outside flex connectors.
    A. 2
    B. 3
    C. 4
    D. 5

92. If there are 50 Ω to ground after a ground rod is installed, the electrician should ___.
    A. add rods until the resistance is less than 25 Ω to ground
    B. add only one more electrode even if the resistance is more than 25 Ω
    C. add two ground rods
    D. not add any more ground rods

93. In a return air duct, nonmetallic-sheathed Romex cable should be installed ___.
    A. parallel to the stud space or joist
    B. in plastic conduit
    C. at a 90° angle to the long dimension of the stud space or joist
    D. in ENT conduit

94. The service demand for three 3000 W cooktops and three 6000 W ovens is ___ W.
    A. 6300
    B. 9900
    C. 16,200
    D. 21,000

95. For each yoke or strap containing one or more devices or equipment, a ___ volume allowance in accordance with Table 314.16(B) shall be made based on the largest conductor connected to the device or yoke.
    A. single
    B. double
    C. triple
    D. quadruple

96. The maximum overcurrent protection that is allowed on a No. 16 THHN motor control circuit conductor is ___ A unless tapped from the load side of the motor overcurrent protection device.
    A. 7
    B. 10
    C. 15
    D. 20

97. The maximum load on 15 A receptacles that are connected to a 20 A overcurrent protection device is ___ A.
    A. 12
    B. 15
    C. 16
    D. 20

98. The allowable ampacity of No. 6 AWG SO cord with two current-carrying conductors is ___ A.
    A. 45
    B. 50
    C. 55
    D. 60

99. If the circuit conductors are No. 10 AWG, a 240 V power source that is 250′ from a load and draws 10 A will have a power loss of ___ W.
    A. 6.21
    B. 31.5
    C. 62.10
    D. 124.2

100. An area around an outdoor fuel dispenser up to 18″ high and 20′ around is classified as Class ___.
    A. I, Division 1
    B. I, Division 2
    C. II, Division 1
    D. II, Division 2

CHAPTER 14
Practice Exams

Practice Exam 14-3

Name _____   Date _____

_____ 1. A 15,000 W, 3φ, 480 V, electric induction motor has a horsepower of ___ HP.
   A. 18
   B. 20
   C. 25
   D. 30

_____ 2. Cord-and-plug-connected motors can use the plug as the disconnecting means as long as the motor is rated ___ HP or less.
   A. ⅛
   B. ¼
   C. ⅓
   D. ½

_____ 3. The circuit depicted below has a current of ___ A.
   A. 2.55
   B. 5
   C. 10
   D. 20

_____ 4. Air conditioning equipment, which has multimotor or combination loads, installed on a roof shall have an equipment grounding conductor of the ___ type installed on the outdoor portion of the metallic raceway system that uses threadless connectors.
   A. bus-bar
   B. wire
   C. copper-wire
   D. aluminum-wire

_____ 5. The depth of the working-space distance from in front of a panel, rated 750 V to ground, to an ungrounded wall is ___ ′.
   A. 3
   B. 3.5
   C. 4
   D. 4.5

257

6. A service using 500 kcmil conductors on a 400 A service will have a No. ___ AWG supply-side bonding jumper.
   A. 1
   B. 1/0
   C. 2/0
   D. 4/0

7. The required working space for a 3ϕ, 4,160/2400 V switchgear with another 4,160 V switchgear directly across from it shall be at least ___′.
   A. 4
   B. 5
   C. 6
   D. 8

8. Luminaries mounted on the side of a commercial building shall have a maximum voltage of ___ V.
   A. 120
   B. 240
   C. 277
   D. 480

9. ___ cable is permitted to be installed as direct burial without any other protection.
   A. NM
   B. UF
   C. SE
   D. THHN

10. A 50 A-range circuit shall be permitted to be tapped with conductors of at least ___ A or more.
    A. 15
    B. 20
    C. 25
    D. 30

11. When multiple driving machines are connected to a single elevator, there shall be ___ disconnecting means allowed.
    A. one
    B. two
    C. three
    D. four

12. The maximum voltage drop on a 120 V branch circuit is ___ V.
    A. 3.6
    B. 5
    C. 10
    D. 12

13. A(n) ___ can be connected to the critical branch of an essential electrical system of a hospital.
    A. nurse call system
    B. automatic door
    C. exit sign
    D. communication system

_____ 14. Conductors emerging from grade shall be protected for at least ___′ above the final grade.
  A. 6
  B. 8
  C. 10
  D. 12

_____ 15. Conductors and overcurrent protective devices connected to continuous loads shall be sized at ___% of the load.
  A. 100
  B. 115
  C. 125
  D. 135

_____ 16. The current rating of a disconnecting means for a 22 A motor circuit shall be ___.
  A. 15
  B. 30
  C. 60
  D. 100

_____ 17. Listed boxes or handhole enclosures shall be permitted to be covered with all of the following except ___.
  A. clay
  B. gravel
  C. light aggregate
  D. noncohesive granulated soil

_____ 18. A 6′ × 3′ island in the middle of a kitchen requires ___ receptacle(s).
  A. one
  B. two
  C. three
  D. four

_____ 19. The current rating for a 120 V, ½ HP induction motor is ___ A.
  A. 4.9
  B. 9.8
  C. 13.8
  D. 16

_____ 20. Receptacles in a dwelling shall be GFCI-protected for ___.
  A. kitchen countertops
  B. bathrooms
  C. laundry areas
  D. all of the above

_____ 21. Exposed multiconductor, non-power-limited fire alarm cables located within 7′ above the floor shall be supported every ___″.
  A. 12
  B. 18
  C. 24
  D. 36

22. Coaxial cables shall have a vertical clearance of ___′ above a roof.
    A. 3
    B. 6
    C. 8
    D. 10

23. For electric vehicle supply equipment rated more than 60 A or more than 150 V to ground, the disconnecting means shall be provided and installed in a(n) ___ location.
    A. accessible
    B. readily accessible
    C. isolated
    D. protected

24. In dwellings where branch-circuit wiring is modified, replaced, or extended, the branch circuit shall be protected by ___.
    A. a listed combination-type AFCI breaker located at the origin of the branch circuit
    B. a listed branch/feeder type AFCI breaker with a listed outlet branch-circuit type AFCI device located at the first receptacle outlet of the existing branch circuit
    C. either A or B
    D. none of the above

25. What is the neutral ampacity for a 16-unit apartment if each unit is 832 sq ft and contains the following loads:
    one 8 kW range
    one 6 kW electric water heater
    one 8000 W electric heater
    Note: There are no laundry facilities in the building and there are no house loads on the 120/240 V service.
    A. 180 A
    B. 200 A
    C. 217 A
    D. 229 A

26. When a disconnecting means is required to be lockable open, it shall be capable of being locked in the ___ position.
    A. closed
    B. open
    C. loaded
    D. sealed

27. A ___ component is a component having contacts for making or breaking an incendive circuit and a contacting mechanism constructed so that the component is incapable of igniting the specific flammable gas-air or vapor-air mixture.
    A. nonincendive
    B. safety
    C. sealed
    D. none of the above

28. Communication conductors are installed below power conductors on a pole. If the power conductors are 480 V then the required distance between the two sets of conductors is ___".
    A. 24
    B. 30
    C. 42
    D. 48

29. A sealing fitting in a Class I, Division 1 location shall be installed within ___" of an enclosure containing arcing or sparking device.
    A. 12
    B. 18
    C. 30
    D. 36

30. The service disconnecting means shall be installed ___ and either inside the nearest point of entrance or outside the building.
    A. in a readily accessible location
    B. as a surge-protection device
    C. locked
    D. all of the above

31. The separate overload protection shall be calculated based on the ___ of the motor.
    A. dimensions
    B. nameplate ratings
    C. RPM
    D. voltage

32. ___ conductors shall be connected in series with the overcurrent protective devices.
    A. Grounded
    B. Grounding
    C. Neutral
    D. Ungrounded

33. A surface-mounted fluorescent luminaire shall be placed at least ___" away from the nearest storage space.
    A. 4
    B. 6
    C. 8
    D. 12

34. All of the following statements concerning the installation of a 23' feeder tap conductor are true except that ___.
    A. tap conductors shall be at least ⅓ the size of the OCPD
    B. tap conductors shall be permitted to terminate into more than one single set of fuses of circuit breakers
    C. tap conductors shall be protected from physical damage
    D. tap conductors shall be permitted to serve any number of OCPDs on the load side

_____ 35. An electrician may install up to ___ No. 12 conductors in a 4″ × 1½″ octagonal box.
　　　　A. three
　　　　B. four
　　　　C. five
　　　　D. six

_____ 36. Overcurrent protective devices shall not be located in a ___.
　　　　A. bedroom
　　　　B. hallway
　　　　C. bathroom
　　　　D. kitchen

_____ 37. A general care patient (Category 2) location in a hospital is required to have a minimum of ___ receptacles.
　　　　A. two
　　　　B. four
　　　　C. six
　　　　D. eight

_____ 38. An outdoor location where windblown dust is common requires a NEMA ___ enclosure.
　　　　A. 1
　　　　B. 2
　　　　C. 3
　　　　D. 3R

_____ 39. Where nonmetallic or metal-enclosed equipment over 1000 V is accessible to the general public and the bottom of the enclosure is less than ___′ above floor or grade level, the door or hinged cover shall be kept locked.
　　　　A. 4
　　　　B. 6
　　　　C. 8
　　　　D. 10

_____ 40. A common neutral may be utilized by ___ sets of 4-wire feeders.
　　　　A. 1
　　　　B. 2
　　　　C. 3
　　　　D. 4

_____ 41. All of the following can be used to mount equipment to a concrete block wall except ___.
　　　　A. toggle bolts
　　　　B. lead anchors
　　　　C. concrete screws
　　　　D. wooden plugs

_____ 42. If two sets of No. 500 kcmil aluminum conductors are installed in parallel, then a No. ___ AWG aluminum grounding electrode conductor is required.
　　　　A. 2/0
　　　　B. 4/0
　　　　C. 250 kcmil
　　　　D. 500 kcmil

43. The type of communication cable that does not have a permitted substitution is ___ cable.
    A. CMP
    B. CM
    C. CMG
    D. CMX

44. Meter disconnect switches nominally rated not in excess of 1000 V shall have a short-circuit rating equal to or greater than the ___.
    A. amperage rating
    B. available short-circuit current
    C. voltage
    D. wattage

45. Fuel-dispensing systems shall be provided with one or more clearly identified emergency shutoff devices or electrical disconnects. Such devices shall be placed at not less than ___′ but not more than ___′ from the devices that they serve.
    A. 10; 75
    B. 20; 50
    C. 20; 100
    D. 100; 200

46. A 400 A circuit using 600 kcmil copper conductors is required to have a No. ___ AWG equipment grounding conductor.
    A. 1/0
    B. 3
    C. 4
    D. 6

47. The maximum voltage on coaxial cable shall not exceed ___ V.
    A. 30
    B. 45
    C. 60
    D. 100

48. Conductors sized at No. ___ AWG and larger shall be protected by insulating bushings when installed in a raceway.
    A. 1
    B. 4
    C. 6
    D. 8

49. NM cables supplying power to a luminaire or other electrical equipment in a suspended ceiling shall be permitted to be unsupported for up to ___′.
    A. 4
    B. 4.5
    C. 6
    D. 8

50. Single conductors connected in parallel that are installed in metal wireways on AC systems shall be installed in groups consisting of ___ conductor(s) for each phase, neutral, or grounding conductor.
    A. one
    B. two
    C. three
    D. four

51. PVC conduit sized at ½″ shall be supported every ___′.
    A. 3
    B. 6
    C. 8
    D. 10

52. An ___ cord is suitable for a vacuum cleaner.
    A. SVO
    B. SO
    C. SPT-1
    D. SPE

53. The insulation rating for Class 1 circuits shall be a minimum of ___ V.
    A. 120
    B. 250
    C. 300
    D. 600

54. The minimum distance vertically from the rim of a bathtub to a hanging ceiling luminaire shall be ___′.
    A. 4
    B. 6
    C. 8
    D. 10

55. Hermetic motors shall be marked with all of the following except the ___.
    A. manufacturer's name
    B. voltage
    C. phase
    D. fuse size to be used

56. The width of working space required in front of electrical equipment that requires adjustment or maintenance shall be at least ___″.
    A. 24
    B. 30
    C. 36
    D. 48

57. The maximum setting of the ground fault protection shall be ___ A.
    A. 600
    B. 800
    C. 1000
    D. 1200

58. A 1ϕ, 120/240 V multiwire branch circuit that has 1000 W on line 1 and 500 W on line 2 will have ___ A on the neutral.
    A. 2.08
    B. 4.167
    C. 5
    D. 8.33

59. In high bay manufacturing buildings, the maximum length of a tap conductor shall not exceed more than ___'.
    A. 25
    B. 50
    C. 75
    D. 100

60. Receptacles installed in a dwelling crawl space below grade shall be protected by a(n) ___.
    A. GFCI
    B. AFCI
    C. GFPE
    D. standard overcurrent protective device

61. A 60' × 40' mobile home is required to have ___ 15 A circuits for lighting and receptacles.
    A. three
    B. four
    C. six
    D. eight

62. If the metal water piping is isolated from each occupancy in a multiple-occupancy building, it shall be bonded to the equipment grounding terminal at ___.
    A. the main service disconnecting means
    B. the utility transformer
    C. each panelboard or switchboard supplying the occupancy
    D. none of the above

63. No luminaire shall be installed less than ___" below the normal water level of a swimming pool.
    A. 4
    B. 6
    C. 12
    D. 18

64. Use the drawing below to determine the number of outlets that require GFCI protection. *Note:* Disregard any exceptions.
    A. 1
    B. 2
    C. 3
    D. 4

65. The minimum height of a wooden fence used to protect a substation that is over 600 V shall be ___′.
  A. 5
  B. 6
  C. 7
  D. 8

66. The maximum number of disconnects for a solar photovoltaic system shall be ___.
  A. one
  B. two
  C. six
  D. eight

67. ___ are not allowed to make connections on a grounding conductor.
  A. Sheet metal screws
  B. Terminal bars
  C. Pressure connectors
  D. Machine screws and nuts

68. The minimum size of copper grounding conductor for a TV antenna is No. ___ AWG.
  A. 8
  B. 10
  C. 12
  D. 14

69. A 12 A, 120 V circuit that has a 1000 W load will have a power factor of ___%.
  A. 1.20
  B. 69.4
  C. 83
  D. 90

70. A 1″ EMT conduit may have up to ___ No. 10 THWN conductors.
  A. 10
  B. 11
  C. 14
  D. 16

71. What is the total voltage of the circuit below?
  A. 55 V
  B. 64 V
  C. 110 V
  D. 120 V

72. A store has 40 fluorescent light fixtures with two ballasts per fixture. Each ballast is rated at 0.8 A. If each circuit is connected to a 20 A, 120 V breaker not rated for 100% loading, then ___ branch circuits are required for the store.
    A. three
    B. four
    C. five
    D. six

73. A 3φ, 208 V circuit connected to a 75,000 W load will have a current of ___ A.
    A. 120
    B. 208
    C. 288
    D. 360

74. Disconnects for a pool or spa pump shall be ___.
    A. 5′ above the deck
    B. readily accessible and within sight of the pool
    C. located at the panel
    D. installed at the motor

75. Lighting outlets that are installed in crawl spaces shall be ___.
    A. AFCI-protected
    B. GFCI-protected
    C. guarded
    D. connected to not more than a 15 A circuit

76. If two duplex receptacles are installed in an industrial location, the receptacle load will be ___ VA.
    A. 180
    B. 240
    C. 250
    D. 360

77. Using the optional calculation, the service demand for a 4000 sq ft dwelling with a 5000 W electric dryer and a 5000 W water heater is ___ VA.
    A. 10,000
    B. 16,600
    C. 26,500
    D. 32,340

78. A 3 Ω resistor and a 6 Ω resistor connected in parallel will have a total resistance of ___ Ω.
    A. 2
    B. 3
    C. 4.5
    D. 9

79. An extra service is allowed for a building operating on 480 V when the connected load is over ___ A.
    A. 1000
    B. 1500
    C. 2000
    D. 2500

80. A motor that draws an FLC of 45 A and is 20′ from the feeder will have No. ___ AWG copper tap conductor(s). *Note:* The feeder conductors are 500 kcmil THHN copper wire connected to a 400 A OCPD.
    A. one
    B. two
    C. three
    D. four

81. A grounded service that operates at 1000 V or less must have a grounded conductor connected to the ___.
    A. grounding electrode conductor
    B. service disconnect
    C. phase monitor
    D. grounding bar

82. Interior metal water piping located more than ___′ from the point of entrance to a building shall not be used as a part of the grounding electrode system.
    A. 5
    B. 6
    C. 8
    D. 10

83. If a second ground rod is required for a service it should be placed a minimum distance of ___′ from the other ground rod.
    A. 4
    B. 6
    C. 8
    D. 10

84. Where individual conductors pass through metal with magnetic properties the ___ effect shall be minimized by cutting slots in the metal between the holes.
    A. inductive
    B. capacitive
    C. magnetic
    D. resistive

85. To bond three parallel conduits for a service, a No. ___ AWG bonding jumper is required. *Note:* Each conduit contains 500 kcmil copper conductors and has a bonding jumper connected to it.
    A. 1/0
    B. 2
    C. 3/0
    D. 4

86. The efficiency of a 4 HP motor with an input of 4 kW is ___%. *Note:* 1 HP = 746 W.
    A. 50
    B. 75
    C. 90
    D. 134

87. If an emergency system relies on a single alternate source of power, which will be disabled for maintenance or repair, the emergency system shall include ___ to connect a portable or temporary alternative source of power.
    A. temporary switching means
    B. receptacles
    C. permanent switching means
    D. inlets

88. The allowable ampacity of a No. 14 THW conductor in a 49°C location with a total of eight other current-carrying conductors in the conduit is ___ A.
    A. 8.12
    B. 11.6
    C. 14
    D. 20

89. Three capacitors deliver the most capacitance when ___.
    A. they are all in series
    B. two are in series and one is in parallel
    C. one is in series and two are in parallel
    D. they are all in parallel

90. A minimum box size of ___ cu in. is required when containing the following items:
    three No. 12 THHN conductors
    three No. 10 THW conductors
    one switch connected to the No. 12 conductors
    Note: No clamps or ground wires are installed in the box.
    A. 12
    B. 16.5
    C. 18.75
    D. 21.25

91. Transformers sized at ___ kVA or less and under 1000 V can be installed above a suspended ceiling.
    A. 25
    B. 50
    C. 75
    D. 100

92. What size auxiliary gutter is required for the following conductors:
    six No. 1/0 THHN conductors
    four No. 2 THHN conductors
    six No. 3/0 THWN conductors
    A. 4″ × 4″
    B. 6″ × 6″
    C. 8″ × 8″
    D. 10″ × 10″

93. ___ is not allowed in a cable tray.
    A. Nonmetallic sheathed cable
    B. MI cable
    C. Multiconductor tray MC cable
    D. Single-conductor cable smaller than No. 1/0 AWG

_____ 94. ___ lamps give off the highest lumens per W.
    A. Incandescent
    B. Fluorescent
    C. Mercury-vapor
    D. High-pressure sodium

_____ 95. The neutral conductor of all three-wire, DC systems supplying premises wiring shall be ___.
    A. bonded
    B. isolated
    C. grounded
    D. removed

_____ 96. A 240 V motor rated at 12 A is required to have an AC snap switch rated at ___ A.
    A. 15
    B. 20
    C. 25
    D. 30

_____ 97. ___ conductors can be used within 3″ of a ballast.
    A. RHW
    B. THW and THHN
    C. TW
    D. UF

_____ 98. A service load of ___ VA is required for a set of 240 V electric heaters with the following loads:
    two heaters at 1000 W
    four heaters at 1500 W
    two heaters at 2000 W
    two heaters at 2500 W
    A. 12,000
    B. 17,000
    C. 20,000
    D. 21,750

_____ 99. A 20 A circuit containing 0.5 Ω that is 200′ long will have a wire resistance of ___ Ω for 100′ of conductor.
    A. 1.5
    B. 2
    C. 2.5 Electrical Theory
    D. 3

_____ 100. A current rating of ___ A is required for conductors feeding a set of three 440 V, 3ϕ motors with the following loads:
    one motor at 20 HP
    one motor at 30 HP
    one motor at 40 HP
    A. 100
    B. 119
    C. 132
    D. 149

CHAPTER 14

Practice Exams

Practice Exam 14-4

Name _____ Date _____

_____ 1. Fixture wires shall not be smaller than No. ___ AWG.
   A. 10
   B. 12
   C. 16
   D. 18

_____ 2. Nonconductive optical fiber may occupy the same raceway with conductors supplying light and power provided the operating voltage does not exceed ___ V.
   A. 250
   B. 480
   C. 600
   D. 1000

_____ 3. The maximum size of electrical metallic tubing that can be used to enclose service-entrance conductors is ___".
   A. 3
   B. 3.5
   C. 4
   D. 5

_____ 4. If the number of outlets in a bank is unknown, the general lighting load and receptacle load is calculated at ___ VA per sq ft.
   A. 3
   B. 3½
   C. 4½
   D. 5

_____ 5. Cables containing No.12 AWG conductors installed in a manufactured building using closed construction must be supported ___.
   A. every 3′
   B. every 4½′
   C. every 6′
   D. only at termination points

6. A branch circuit supplying more than one electric baseboard heater in a residential occupancy shall have a maximum rating of ___ A.
   A. 15
   B. 20
   C. 30
   D. 50

7. The supply-side bonding jumper on the supply side of a service is sized based on the ___.
   A. load
   B. overcurrent device
   C. service entrance conductors
   D. service drop

8. ___ wire insulation is not suitable for connecting a reactor.
   A. SA
   B. RHH
   C. THHN
   D. THWN

9. A device connected on the load side of a branch-circuit overcurrent protective device that transfers emergency lighting loads from the normal supply to an emergency supply is called a(n) ___.
   A. transfer switch
   B. branch-circuit emergency-lighting transfer switch
   C. unit device
   D. emergency light

10. EMT conduit shall be supported every ___′.
    A. 6
    B. 10
    C. 15
    D. 20

11. If two different size motors are supplied from a common feeder and are interlocked so that only one can run at a time, the ampacity calculation for the feeder is based on ___.
    A. the total FLC of the two motors
    B. 125% of the FLC of the larger motor
    C. the sum of the FLC of the two motors plus 25% of the FLC of the larger motor
    D. 125% of the total FLC of the two motors

12. Where threaded couplings are used to install IMC vertically from industrial machinery, the distance of the supports shall not exceed ___′.
    A. 3
    B. 5
    C. 10
    D. 20

13. An arc welding machine with a rated primary current of 60 A and a duty cycle of 30% will have a minimum size of No. ___ TW copper conductors.
    A. 4
    B. 6
    C. 8
    D. 10

14. A junction box is required in a single-family dwelling to connect two wall ovens and one cooktop. Each cooking unit has a No. 10-3 NM cable. The branch circuit from the service panel is a No. 6-3 NM cable with a No. 8 equipment grounding conductor. The minimum size required is a ___ square box.
    A. 4″ × 1¼″
    B. 4″ × 2⅛″
    C. 4 11⁄16″ × 2⅛″
    D. 5″ × 5″

15. The smallest overhead service conductors permissible for a limited load, single branch circuit is No. ___ AWG copper.
    A. 6
    B. 8
    C. 10
    D. 12

16. Conductors supplying branch circuits to receptacles installed under raised floors in an information technology equipment room are permitted to be installed in ___.
    A. flexible metal conduit
    B. metal wireways
    C. Type AC cable
    D. all of the above

17. A 1ϕ, 175 kVA transformer with a secondary voltage of 120/240 V has a current of ___ A on the secondary.
    A. 329
    B. 421
    C. 625
    D. 729

18. The minimum-size auxiliary gutter required for four No. 4/0 THHN kcmil conductors entering in through one end and leaving the top of the gutter is ___″. *Note:* The conductors are not spliced inside the gutter.
    A. 4
    B. 6
    C. 8
    D. 10

19. Live parts of motors must be guarded if they operate at over ___ V.
    A. 50
    B. 100
    C. 125
    D. 150

20. Conductors used to supply power to 1ϕ, 230 V, fixed electric space-heating equipment that includes a 10 HP motor and 1000 W of resistive elements will have a minimum size of No. ___ THW. *Note:* The supply voltage is 1ϕ, 240 V.
    A. 2
    B. 2/0
    C. 4
    D. 6

21. A lampholder operating at less than 50 V shall be rated for ___ W.
    A. 50
    B. 220
    C. 330
    D. 660

22. Switching devices must be located at least ___' from the inside walls of a pool.
    A. 5
    B. 10
    C. 15
    D. 20

23. The lighting load for a 12,000 sq ft apartment building is ___ VA.
    A. 12,000
    B. 14,550
    C. 16,250
    D. 36,000

24. The power for unit equipment emergency illumination must be supplied by ___.
    A. a local receptacle circuit
    B. any convenient branch circuit without a load
    C. the emergency panel
    D. the normal lighting circuit in the area

25. What size rigid conduit is required if it contains the following conductors:
    three No. 6 THW
    four No. 10 THW
    three No. 12 THW
    A. 1″
    B. 1¼″
    C. 1½″
    D. 2″

26. A circuit has an actual load of 87 A and is connected to a 90 A circuit breaker with a terminal rating of 60/75° C. If the load terminals are rated at 60° C, the required size of THHN copper conductors is ___.
    A. 1
    B. 2
    C. 4
    D. 6

27. If 240 V baseboard heaters are rated at 250 W per ft, a 240 V, 20 A circuit may supply ___ linear ft of baseboard heat.
    A. 12
    B. 15
    C. 16
    D. 18

_____ 28. When two or more conduits support a threaded enclosure containing devices, it shall be supported within ___″ of the enclosure.
   A. 6
   B. 12
   C. 18
   D. 24

_____ 29. A three-conductor NM cable attached to the bottom of a joist must have a minimum size of No. ___ AWG.
   A. 2
   B. 4
   C. 6
   D. 8

_____ 30. When more than three current-carrying conductors occupy a multiconductor cable in a cable tray, the current load must be reduced. The derating factor applies to the number of ___.
   A. current-carrying conductors in the cable
   B. conductors when the fill of the cable tray exceeds 40% of the cross-sectional area of the cable tray
   C. TC cable in the cable tray
   D. grounding conductors

_____ 31. A 1ϕ, 208 V, 3 HP motor has a full-load current of ___ A.
   A. 15.2
   B. 17
   C. 18.7
   D. 20.3

_____ 32. Transformers rated over 112.5 kVA shall be installed ___.
   A. in a fire-resistant room
   B. on the floor
   C. on the wall
   D. none of the above

_____ 33. An 8.8 A, 1 HP motor with a service factor of 1.15 is required to have a standard size overload of ___ A.
   A. 10.1
   B. 11
   C. 11.8
   D. 12.3

_____ 34. Nonheating leads of snow-melting equipment embedded in masonry or asphalt are ___.
   A. permitted if the cable has a grounding sheath or braid
   B. not permitted
   C. permitted if the equipment has overtemperature control
   D. installed in RMC or EMT

35. A 3ϕ, 220 V, 30 HP synchronous motor with a power factor of 90% will have a full-load current of ___ A.
    A. 63
    B. 69.3
    C. 76.23
    D. 86.62

36. ___ is permitted in a four-story building.
    A. MC cable
    B. EMT
    C. NM
    D. all of the above

37. When the premises wiring system has branch circuits supplied from more than one nominal voltage system, each ungrounded conductor of a branch circuit shall be identified by ___ at all termination, connection, and splice points.
    A. length
    B. phase or line and system
    C. size
    D. voltage

38. No. 10 THHN conductors installed in a location with an ambient temperature of 65°C will have an allowable ampacity of ___ A.
    A. 26
    B. 30
    C. 35
    D. 40

39. The receptacle for a built-in dishwasher shall be located in the space ___ the dishwasher.
    A. behind
    B. under
    C. above
    D. adjacent to

40. The minimum size SE copper cable that may be used for a 150 A residential feeder is ___ AWG.
    A. 1
    B. 1/0
    C. 2
    D. 3

41. No. ___ AWG is the minimum-size conductor permitted in a cable bus.
    A. 1
    B. 1/0
    C. 2
    D. 2/0

42. ___ is permitted to be used as the equipment grounding conductor at boatyard locations.
    A. Corrosion-resistant rigid metal conduit
    B. Electrical nonmetallic tubing with an insulated copper conductor based on 250.122
    C. An insulated conductor based on 250.122
    D. The outer sheath of MC cable

43. Lighting fixtures and lighting outlets located over a spa or hot tub shall not require ground-fault circuit-interrupter protection if located ___' above the maximum water level.
    A. 5
    B. 7.5
    C. 10
    D. 12

44. The maximum distance between supports and insulators for open branch-circuit conductors of No. 8 AWG and smaller is ___'.
    A. 4.5
    B. 6
    C. 8
    D. 10

45. A single receptacle installed on an individual branch circuit shall ___.
    A. be allowed only in commercial or industrial locations
    B. be of the locking type
    C. have an ampere rating at least equal to that of the load being served
    D. have an ampere rating not less than that of the branch circuit

46. When using a one-shot bender to bend a 2" rigid metal conduit, the minimum bend radius is ___".
    A. 7.5
    B. 9.5
    C. 12
    D. 15

47. ___ is the fixed wiring method used in a place of assembly that is fire rated and will seat more than 100 people.
    A. ENT
    B. Metal raceway
    C. NM cable
    D. Rigid nonmetallic conduit secured to the walls

48. An 80' × 120' store that has a 60' show window is required to have a minimum number of ___, 120 V, 20 A interior lighting circuits.
    A. 17
    B. 18
    C. 22
    D. 29

49. A recreational-vehicle park must have a 125 V, 30 A receptacle located at ___% of the parking spaces.
    A. 60
    B. 70
    C. 90
    D. 100

50. Lighting systems operating at less than 30 V that have bare conductors installed indoors shall not be less than ___′ above the finished floor, unless listed for a lower installation height.
    A. 6
    B. 7
    C. 8
    D. none of the above

51. A 4″ × 1½″ octagonal box containing two outside cable clamps may contain up to ___ No. 14 AWG conductors.
    A. four
    B. five
    C. six
    D. seven

52. Armored and nonmetallic-sheathed cable shall be supported every ___″.
    A. 36
    B. 42
    C. 48
    D. 54

53. A maximum number of ___ current-carrying conductors are allowed in a metallic wireway before derating is applied.
    A. 3
    B. 30
    C. 100
    D. 300

54. The minimum fixed bend radius for ½″ flexible metallic tubing is ___″.
    A. 3.5
    B. 4
    C. 5
    D. 12.5

55. Any motor of ___ HP or less that is started automatically shall be protected against overload.
    A. 1
    B. 2
    C. 5
    D. 10

56. Type MI cable must be supported at intervals not exceeding ___′.
    A. 3
    B. 6
    C. 8
    D. 12

57. An exterior pad-mounted transformer is supplying power to service equipment within a building. Two sets of four, 500 kcmil AL service-entrance conductors are pulled through a rigid nonmetallic raceway to a 600 A switch fuse. The grounded conductor between the switch and transformer shall be No. ___ AWG AL.
    A. 1/0
    B. 2/0
    C. 250 kcmil
    D. 500 kcmil

58. The maximum-size overcurrent device used to protect a water heater that is 240 V and rated at 4500 W is ___ A.
    A. 20
    B. 25
    C. 30
    D. 35

59. Article ___ covers building control circuits (e.g., elevator capture, fan shutdown) that are associated with a fire alarm system.
    A. 645
    B. 725
    C. 760
    D. 800

60. The nominal battery voltage per cell for lead-acid type batteries are computed on the basis of ___ V.
    A. 1.2
    B. 1.5
    C. 2
    D. 2.5

61. A motor controller is installed at the location of a motor. A remote start/stop station for the motor has been installed in a different location. The motor control circuit that runs between the controller and the start/stop station has been tapped from the line side of the controller with No. 12 AWG copper conductors. The maximum rating of the overcurrent device located in the motor disconnect is ___ A. *Note:* There is no separate overcurrent device within the motor controller for the motor control circuit conductors.
    A. 20
    B. 40
    C. 60
    D. 90

62. Underfloor flat-top raceways over 4″ wide but less than 8″ wide and spaced less than 1″ apart shall be covered with concrete to a depth of not less than ___″.
    A. ¾
    B. 1
    C. 1½
    D. 2

63. An overcurrent protective device rated for ___ A is required for a 480 V to 120 V, 1ϕ transformer that has primary-only protection with a primary current of 40 A.
    A. 40
    B. 45
    C. 50
    D. 60

64. A 1″ diameter, smooth-sheathed MC cable has a minimum bend radius of ___″.
    A. 7
    B. 10
    C. 12
    D. 15

65. An unventilated pit in the floor of a commercial major repair garage is classified as Class ___, Division ___.
    A. I; 1
    B. I; 2
    C. II; 1
    D. II; 2

66. Three metal raceways containing No. 1/0 AWG THW in parallel feeder conductors supplied by a 400 A, circuit breaker, will have a No. ___ AWG equipment grounding conductor in each raceway.
    A. 2 copper or No. 1/0 aluminum
    B. 2/0 copper or No. 4/0 aluminum
    C. 3 copper or No. 1 aluminum
    D. 3/0 copper or No. 250 aluminum

67. A box that is provided with one or more securely installed barriers made of nonmetallic materials not marked in cubic inches, shall take up ___ cubic inch(es) in the box.
    A. 0.50
    B. 1
    C. 1.25
    D. 2

68. A patient's bed located in a critical-care (Category 1) area of a health-care facility is required to have a minimum of ___ receptacle(s).
    A. four single or two double
    B. six duplex or four single
    C. two single or one duplex
    D. fourteen single or seven duplex

69. A 75 kVA transformer with a 3ϕ, 480 V primary and a 3ϕ, 120/208 V secondary will have a secondary FLC of ___ A.
    A. 156
    B. 208
    C. 360
    D. 362

70. If a 120 V, 13.2 A circuit is 460′ long and a minimum voltage drop of 3% must be maintained, the minimum conductor size that can be used is No. ___ AWG. *Note:* (K = 12.9)
    A. 3
    B. 3/0
    C. 4
    D. 6

71. Low-voltage gas-fired luminaires, or decorative fireplaces, shall be permitted to be within ___′ of the edge of a swimming pool.
    A. 5
    B. 10
    C. 12
    D. 20

72. The maximum voltage allowed for open wiring on insulators in an agricultural building is ___ V.
    A. 240
    B. 480
    C. 600
    D. 1000

73. What is the allowable demand factor for six clothes dryers in a multifamily dwelling with the following loads:
    two dryers at 4.5 kW
    two dryers at 5 kW
    two dryers at 5.5 kW
    A. 60%
    B. 75%
    C. 100%
    D. 115%

74. Receptacles located in pediatric locations (Category 2) shall be of the ___ type.
    A. GFCI
    B. tamper-resistant
    C. AFCI
    D. weather-resistant

75. When installing a header for concrete cellular floor raceways, the header shall be installed in a straight line at a ___° angle to the cell.
    A. 30
    B. 45
    C. 60
    D. 90

76. When a motor has limited flexibility in a Class I, Division 2 location, ___ may be permitted to be used.
    A. flexible metal fittings
    B. flexible metal conduit with listed fittings
    C. liquidtight flexible metal conduit with listed fittings
    D. all of the above

77. What is the minimum service for a 2,700 sq ft single-family dwelling with the following loads:
    one range at 27 kW
    one dryer at 6.5 kW
    A. 100 A
    B. 125 A
    C. 150 A
    D. 200 A

78. The largest service entrance conductor that can use a No. 8 AWG copper grounding electrode conductor connection to a metal underground water pipe is No. ___ AWG.
    A. 1/0
    B. 2
    C. 2/0
    D. 1000 kcmil

79. The minimum-size rigid nonmetallic conduit used for a 3ϕ, four wire, 200 A commercial feeder using THW aluminum conductors, a full-size neutral, and the required-size equipment grounding conductor is ___″.
    A. 1½
    B. 2
    C. 2½
    D. 3

80. Branch-circuit conductors connecting one or more units of informational technology equipment to a supply source shall have an ampacity of not less than ___% of the total connect load.
    A. 80
    B. 115
    C. 125
    D. 300

81. A bonding jumper located on the outside of a service conduit has a maximum length of ___′.
    A. 4
    B. 5
    C. 6
    D. 8

82. Type S fuses come in all classifications except ___.
    A. 0–15
    B. 16–20
    C. 21–30
    D. 31 and above

83. The output HP of a 1ϕ, 230 V, 15 kW motor is ___ HP.
    A. 15
    B. 20
    C. 30
    D. 230

84. If a 50 A branch circuit is feeding a counter-mounted cooking unit and a wall-mounted oven, the minimum ampacity of the tap to the wall-mounted oven is ___ A. *Note:* Disregard the nameplate rating of the wall-mounted oven.
    A. 1
    B. 20
    C. 30
    D. 40

85. Disconnecting means for rides at fairs or carnivals that are accessible to unqualified person shall be ___.
    A. isolated
    B. guarded
    C. lockable
    D. none of the above

86. A wood building is built on the earth without a footing or water system. If the 3000 A, 460 V service is grounded by ground rod electrodes, the minimum-size copper GEC is No. ___ AWG.
    A. 3/0
    B. 4
    C. 6
    D. 400 kcmil

87. A 36″ conduit with two cables installed has a maximum of ___% conduit fill.
    A. 31
    B. 40
    C. 53
    D. 60

88. The branch circuit for a hermetic compressor rated at 28 A uses a No. ___ AWG, THW copper conductor.
    A. 8
    B. 10
    C. 12
    D. 14

89. A Class ___ circuit has a higher level of voltage and current and shall have additional safe guards from electric shock.
    A. 1
    B. 2
    C. 3
    D. 4

90. The conductors for a 40 A nonmotor generator arc welding machine with a duty cycle of 90% are sized for ___ A.
    A. 30
    B. 36
    C. 38
    D. 40

91. What size auxiliary gutter is required if it contains the following conductors:
    six No. 1/0 AWG THHN
    four No. 2 AWG THHN
    six No. 3/0 AWG THWN
    A. 4″ × 4″
    B. 6″ × 6″
    C. 8″ × 8″
    D. 10″ × 10″

92. A conduit that has a 100 A, 1ϕ circuit and a 300 A, 3ϕ circuit shall have at least one No. ___ AWG equipment grounding conductor.
    A. 2
    B. 4
    C. 6
    D. 8

93. PV circuit or source conductors rated for 75°C and installed in a location with an ambient temperature of 56°C shall have a correction factor of ___% applied.
    A. 45
    B. 58
    C. 71
    D. 111

94. A dwelling service calculated at 168 A requires No. ___ AWG service entrance conductors.
    A. 1 AWG
    B. 1/0 AWG
    C. 3/0 AWG AL
    D. either B or C

95. EMT raceway shall be permitted to be installed ___.
    A. exposed
    B. where subject to severe physical damage
    C. in direct contact with the earth where protected from corrosive influences
    D. both A and C

96. What is the total current on the neutral for a 120/240 V, 1ϕ service in a 16-unit apartment building when each apartment contains the following loads:
    one 240 V, 1.5 HP air conditioner
    one 240 V, 6 kW baseboard heater
    one 240 V, 8 kW range
    one 240 V, 2.5 kW water heater
    Note: Each apartment is 735 sq ft and there are no house or laundry loads.
    A. 200 A
    B. 214 A
    C. 220 A
    D. 241 A

97. A voltage value assigned to a circuit for the purpose of conveniently designating its voltage class is ___ voltage.
    A. peak
    B. average
    C. nominal
    D. effective

98. When oversized, concentric, or eccentric knockouts are not encountered in a raceway system containing over 250 V circuits, ___ can be used for bonding.
    A. double locknuts
    B. threadless EMT connectors
    C. bonding bushings
    D. all of the above

99. A 60 gal. storage water heater shall be sized at ___% of its nameplate rating.
    A. 80
    B. 100
    C. 125
    D. 150

100. If the output of a load is 8000 W, and 7.2 Ω at 240 V, the output at 208 V is ___ W.
    A. 1111
    B. 1728
    C. 6009
    D. 10,783

Practice Question Answers

Chapter 2—Direct Current Principles

Atoms 8
1. Electrons
2. Opposite charged
3. Valance Shell
4. 3
5. 7

Current 9
1. current
2. negative; positive

Resistance 10
1. resistance
2. Silver

Voltage 10
1. Voltage
2. Piezoelectric
3. 2 or both

Batteries 10
1. 1.5
2. 2.0

Determining Voltage Using Ohm's Law 11
1. 120 V
2. 3.2 V

Determining Current Using Ohm's Law 12
1. 0.74 A
   120 ÷ 162 = 0.74 A
2. 3 A
   60 ÷ 20 = 3 A

Determining Resistance Using Ohm's Law 12
1. 1.67 Ω
   25 ÷ 15 = 1.67 Ω
2. 192 Ω
   120 ÷ 0.625 = 192 Ω

Determining Power Using the Power Formula 13
1. 135 W
   30 × 4.5 = 135 W
2. 100 W
   120 × 0.833 = 100 W
3. 1150 W
   115 × 10 = 1150 W

Determining Voltage Using the Power Formula 14
1. 115 V
   150 ÷ 1.3 = 115 V
2. 239 V
   550 ÷ 2.3 = 239 V

Determining Current Using the Power Formula 15
1. 20.8 A
   5000 ÷ 240 = 20.8 A
2. 65.2 A
   7500 ÷ 115 = 65.2 A

Using the Formula Wheel 16
1. 28.8 Ω
   $$\frac{120 \times 120}{500} = 28.8\ \Omega$$
2. 3312 W
   12 × 12 × 23 = 3312 W
3. 5760 W
   $$\frac{240 \times 240}{10} = 5760\ W$$

Meters 18
1. parallel
2. series
3. wattmeter
4. 90°

Chapter 3—Alternating Current Principles

Determining the Frequency of a Sine Wave 25
1. 5 Hz

Determining the Period of a Sine Wave 25
1. 0.005 sec

Determining the Frequency of a Sine Wave When the Period Is Known 25
1. 100 Hz

Determining Inductive Reactance 27
1. 15,700 Ω

Determining Capacitive Reactance 29
1. 6.37 Ω

Determining Impedance 29
1. 14.14 Ω

Power Factor 30
1. 83.3%
2. 6485 W
3. 3820 VA

Reactive Power 32
1. 7701 VA

Chapter 4—Circuits

Determining Total Current in a Series Circuit 37
1. 0.25 A

Determining Total Voltage in a Series Circuit 37
1. 67 V

Determining Total Resistance in a Series Circuit 37
1. 70 Ω

Determining Total Power in a Series Circuit 38
1. 12 W

Determining Total Current in a Parallel Circuit 39
1. 16.84 A

Determining Total Voltage in a Parallel Circuit 39
1. 200 V

Determining Total Resistance in a Parallel Circuit 40
1. 11.1 Ω
2. 54.5 Ω

Determining Total Power in a Parallel Circuit 40
1. 245 W

Determining Total Resistance in a Combination Circuit 43
1. 66 Ω

Multiwire Circuits 44
1. unbalanced
2. Smallest wattage

Chapter 5—Transformers

Determining Transformer Turns Ratio 48
1. 3:1 step down
2. 1:5 step up

Determining Transformer Voltage Ratio 49
1. 300 turns
$$\frac{120 \times 60}{24} = 300$$
2. 360 V
$$\frac{255 \times 120}{85} = 360 \text{ V}$$

Determining Current Ratios 50
1. 2 A
$$\frac{240 \times 0.25}{30} = 2 \text{ A}$$
2. 8.1 A
$$\frac{9 \times 1.8}{2} = 8.1 \text{ A}$$
3. 2 A
20 VA ÷ 10 V = 2 A

Determining Transformer Volt-Amps 51
1. 72,000 VA
480 × 150 = 72,000 VA
2. 124,704 VA
480 × 150 × 1.732 = 124,704 VA

Determining Transformer Kilovolt-Amp Rating 51
1. 332.544 kVA
480 × 400 × 1.732 = 332,544 ÷ 1000 = 332.544 kVA
2. 84 kVA
2400 × 35 = 84,000 ÷ 1000 = 84 kVA

Determining Transformer Efficiency 52
1. 475 VA
500 × 95% = 475 VA
2. 3205 VA
2500 ÷ 78% = 3205 VA
3. 1052 VA
1000 ÷ 95% = 1052 VA
4. 91%
2500 kVA ÷ 2750 kVA = 91%

Determining Transformer Size—Single-Phase (1φ) 52
1. 40 kVA
20 kVA + 20 kVA = 40 kVA

Determining Transformer Size—Wye 53
1. 43 kVA or 45 kVA
20 + 23 = 43 kVA or a standard size of 45 kVA
2. 14.3 or 15 kVA
20 + 23 = 43 kVA ÷ 3 = 14.3 or a standard size of 15 kVA

Determining Transformer Size—Three-Phase (3φ) Closed Delta 55
1. 15.84 kVA = Power
18 kVA × 33% = 5.94
30 kVA × 33% = 9.9
5.94 + 9.9 = 15.84 kVA
21.96 kVA = Lighting
18 kVA × 67% = 12.06
30 kVA × 33% = 9.9
12.06 + 9.9 = 21.96 kVA

Determining Transformer Size—Three-Phase (3φ) Open Delta 55
1. 11.6 kVA = Power
20 kVA × 58% = 11.6 kVA
26.6 kVA = Lighting
15 kVA × 100% = 15
20 kVA × 58% = 11.6
15 + 11.6 = 26.6 kVA

Determining Transformer Protective Device Size— Over 1000 Volts 58
1. 125 on primary, 175 on secondary
40 × 300% = 120 up to 125 A
69 × 225% = 155.25 up to 175 A

Determining Transformer Protective Device Size— Under 1000 Volts 58
1. 125 on primary, 125 on secondary
50 × 250% = 125, 100 × 125% = 125 A

Delta-Wye Conversions 59
1. 200 A

Chapter 6—Grounding, Bonding, and Neutrals

Determining Grounded Conductor Size 67
1. 1/0 AWG Copper

Determining Main Bonding Jumper Size 68
1. 4/0 AWG
500 kcmil × 3 = 1500 kcmil or 1,500,000 cmil
1,500,000 × 12.5% = 187,500 = 4/0 AWG

Determining Grounding Electrode Conductor Size 69
1. 2/0 AWG 250.66

Determining Equipment Grounding Conductor Size 71
1. No. 6 AWG 250.122

Determining Supply-Side Bonding Jumper Size 73
1. 1/0 AWG

Determining Equipment Bonding Jumper Size— Load Side 73
1. No. 4 AWG

Determining System Bonding Jumper Size 74
1. 1/0 AWG

Determining System Grounding Electrode Conductor Size 75
1. 3/0 AWG
500 kcmil × 3 = 1500 kcmil
250.66

Determining Grounded Conductor Size 76
1. No. 2 AWG

Determining Unbalanced and Maximum Unbalanced Current for 1φ Neutral Conductors 77
1. 15 A
   30 – 15 = 15 A
2. 30 A
3. 60 A
   125 – 65 = 60 A
4. 125 A

Determining Unbalanced and Maximum Unbalanced Load for 3φ Wye Neutral Conductors 78
1. 34.6 A
2. 80 A
3. 69.3 A
4. 100 A
5. 52.9 A

Determining Neutral Demand 79
1. 456 A
2. 516 A

Chapter 7 — Conductor Ampacity and Protective Devices

Determining Allowable Ampacity 88
1. 130 A
2. 135 A
3. 85 A

Determining Allowable Ampacity Using Correction Factors 89
1. 114.4 A
2. 110.7 A
3. 74.8 A

Determining Ampacity Using Current-Carrying Conductors 90
1. 104 A
2. 135 A
3. 42.5 A

Determining Conductor Size for 1φ Residential Services 91
1. No. 1/0 AWG
2. 350 kcmil

Determining Ampacity Using Temperature and Conduit-Fill Derating 92
1. 91.52 A

Determining Conductor Size Using Terminal Temperature Rating 93
1. No. 8 AWG
2. 1/0 AWG

Determining Current for Continuous-Load Conductors 94
1. 80 A
2. 42.5 A

Determining Busbar Ampacity 94
1. 750 A
2. 2800 A

Determining Overcurrent Protection for Devices Rated 800 A or Less 95
1. 400 A

Determining Overcurrent Protection for Devices Rated over 800 A 96
1. 1500 A

Determining Overcurrent Protection Size for Small Conductors 97
1. 15 A

Determining Standard Overcurrent Protection Size 98
1. No

Determining Overcurrent Protection Size for Noncontinuous Loads 100
1. 60 A

Determining Overcurrent Protection Size for Continuous Loads 100
1. 160 A
2. 50 A
3. 40 A

Determining Overcurrent Protection Size for Noncontinuous and Continuous Loads 101
1. 60 A
2. 40 A

Chapter 8 — Voltage Drop

Determining Voltage Drop Using Ohm's Law 107
1. 10.44 V

Determining Voltage Drop for 1φ Circuits 108
1. 3.75 V
2. 9.67 V

Determining Conductor Size for 1φ Circuits 109
1. No. 6 AWG
2. No. 1 AWG

Determining Maximum Load on a 1φ Circuit 110
1. 14.9 A

Determining Maximum Distance between a 1φ Power Source and Load 111
1. 121′

Determining Voltage Drop for 3φ Circuits 111
1. 11 V

Determining Conductor Size for 3φ Circuits 113
1. No. 10 AWG

Determining Maximum Load on a 3φ Circuit 113
1. 34.3 A

Determining Maximum Distance between a 3φ Power Source and Load 114
1. 121.58′

Determining Power Loss for 1φ and 3φ Circuits 116
1. 45 W
2. 997 W

Determining Voltage Drop Calculations for Feeders and Branch Circuits 117
1. 1/0

Chapter 9—Motors

Determining Motor Full-Load Current 123
1. 30.8 A
2. 124 A
3. 222 A

Determining Conductor Size for Continuous-Duty Motors 124
1. 2/0 AWG

Determining Conductor Size for Other Than Continuous-Duty Motors 124
1. No. 1 THW

Determining Motor Overload Protection Size 126
1. 18.75 A
2. 37.5 A
3. 34.5 A

Determining the Higher-Size Overload 127
1. 21 A
2. 42 A

Determining Integral Thermal Protector Size 128
1. 91 A

Determining Branch/Short-Circuit Ground Fault Protection 129
1. 60 A
2. 350 A

Determining Feeder Conductor Size for Motors of the Same Voltage 129
1. 440 A
2. 700 kcmil

Determining Short-Circuit Ground Fault Protection 132
1. 400 A

Determining Volt-Amp Rating for Single-Phase (1φ) AC and DC Motors 132
1. 2760 VA

Determining Volt-Amp Rating for Three-Phase (3φ) AC Motors 132
1. 6055 VA
2. 6055 VA

Determining Synchronous Speed 133
1. 200 RPM

Determining Actual Shaft Speed 134
1. 1140 RPM

Determining Motor Torque 134
1. 52.52 ft-lb

Determining Horsepower 136
1. 10.76 HP

Determining Motor Nameplate Current 136
1. 16.8 A

Determining Locked-Rotor Current Ratings 140
1. 540 LRC

Determining VFD to Motor Conductor Size 140
1. No. 8 THW

Determining Bypass Conductor Size 141
1. No. 10 THW

Chapter 10—Conductor-Fill, Box-Fill, and Pull Box and Junction Box Sizing

Determining Conductor Fill for Same-Size Conductors 149
1. 1¼"
2. 2½"

Determining Fill in Conduits Using Table 4 149
1. 0.829 sq in.
2. 2.742 sq in.

Determining the Cross-Sectional Area of Conductors 150
1. 0.0211 sq in.
2. 0.0556 sq in.

Determining the Cross-Sectional Area of Compact Conductors 151
1. 0.3421 sq in.

Determining Conduit Size for Conductors of Different Sizes 151
1. 1¼"

Determining Conduit Nipple Fill for Conductors of Various Sizes 152
1. 0.519 sq in.
2. 1"

Determining the Maximum Number of Same-Size Conductors in a Wireway 154
1. 17

Determining Wireway Size 154
1. 4" × 4"

Determining Box Size and Fill for Same-Size Conductors 158
1. 4" × 1½"
2. 7

Determining Box Size and Fill for Different-Size Conductors 160
1. 23.5 cu in.

Determining Box Size with Plaster Rings and Raised Covers 160
1. 2⅛"

Determining Box Size for Straight Pulls—1000 Volts and Under 161
1. 28"

Determining Box Size for Angle Pulls—1000 Volts and Under 161
1. 21" × 29.5"

Determining Box Size for U-Pulls—1000 Volts and Under 162
1. 24.5"
2. 21"

Determining Depth of Boxes with Removable Covers 162
1. 5"

Determining Distance between Conduits Containing the Same Conductors—1000 Volts and Under 162
1. 21"

Determining Box Size for Straight Pulls—Over 1000 Volts 163
1. 66"

Determining Box Size for Angle Pulls and U-Pulls—Over 1000 Volts 164
1. 57"

Determining Distance between Conduits in Boxes—Over 1000 Volts 165
1. 46.8"

Chapter 11—Dwellings

Determining Lighting Load for One-Family Dwellings—Standard Calculation Method ..172
1. 3 VA/sq ft
2. 9000 VA
3. 9900 VA

Determining the Minimum Number of Branch Circuits for One-Family Dwellings—Standard Calculation Method ...174
1. 5
2. 7

Determining General Lighting Load for One-Family Dwellings—Standard Calculation Method 174
1. 6675 VA
2. 6570 VA

Determining Fixed-Appliance Load for One-Family Dwellings—Standard Calculation Method 175
1. 7500 VA

Determining Dryer Load for One-Family Dwellings—Standard Calculation Method 176
1. 10,000 W
2. 29,250 W
3. 3500 W

Determining Cooking Equipment Load for One-Family Dwellings—Standard Calculation Method ...177
1. 17 kW

Determining Cooking-Equipment Load for One-Family Dwellings—Table 220.55, Note 1 178
1. 23,000 W

Determining Cooking Equipment Load for One-Family Dwellings—Table 220.55, Note 2 178
1. 17,850 W

Determining Cooking-Equipment Load for One-Family Dwellings—Table 220.55, Note 3 180
1. 5250 W
2. 13,600 W
3. 25 kW Col. C is smaller than 26,175 W

Determining Cooking-Equipment Branch-Circuit Load for One-Family Dwellings—Table 220.55, Note 4 181
1. 8000 W
2. 5000 W
3. 8800 W

Determining Cooking-Equipment Load for One-Family Dwellings—Table 220.55, Note 5 181
1. 8400 W
2. 24,150 W
3. 17,025 W

Determining Heating or A/C Loads for One-Family Dwellings—Standard Calculation Method 181
1. 10 kW

Determining Largest Motor Load for One-Family Dwellings 181
1. 375 VA

Determining Service Conductor Size for One-Family Dwellings 183
1. 3/0 AWG

Determining Heating and A/C Load for One-Family Dwellings—Optional Calculation Method 184
1. 8.5 kW

Determining the General Load for One-Family Dwellings—Optional Calculation Method ...185
1. 20,920 W

Determining Service Conductor Size for One-Family Dwellings—Optional Calculation Method 185
1. 2/0 AWG

Determining the General Lighting Load for Multifamily Dwellings—Standard Calculation Method 187
1. 214,950 W

Determining Fixed-Appliance Load for Multifamily Dwellings—Standard Calculation Method 187
1. 67,500 W

Determining Dryer Load for Multifamily Dwellings—Standard Calculation Method 188
1. 47,250 W

Determining Cooking-Equipment Load for Multifamily Dwellings—Standard Calculation Method 189
1. 45 kW

Determining Heating or A/C Load for Multifamily Dwellings—Standard Calculation Method 190
1. 240 kW

Determining Service Size For Multifamily Dwellings—Optional Calculation Method ...192
1. 1600 A

Determining Service Demand for Farms 193
1. 165 A

Determining Service Size for Mobile and Manufactured Homes 194
1. 50 A

Determining Service Size for Mobile Home Parks 196
1. 690 A
2. 18 A

Practice Question Answers **291**

Chapter 12 – Nondwelling Occupancies

Determining Continuous Lighting Loads 202
1. 46,875 VA

Determining Lighting Loads with Demand Factors 203
1. 39,000 VA

Determining the Required Amount of Lighting Circuits ...203
1. 39 Circuits

Determining Show Window Lighting Load 204
1. 7500 VA

Determining Track Lighting Load 205
1. 2812.5 VA

Determining Receptacle Load – Nondwelling 206
1. 11,750 VA

Determining Receptacle Load for Multioutlet Assemblies – Nondwelling 206
1. 1260 VA
2. 6300 VA

Determining Commercial Cooking Equipment Load 207
1. 30,400 VA

Determining the Load for a School Using the Optional School Calculation Method 210
1. 183,750 VA

Determining the Load – Optional Restaurant Calculation 210
1. 260 kVA

Determining Service Demand for Farm Loads 211
1. 33,685 VA

Appendix

CONTENTS

Answer Sheet, Practice Exam 14-1	294
Answer Sheet, Practice Exam 14-2	295
Answer Sheet, Practice Exam 14-3	296
Answer Sheet, Practice Exam 14-4	297
Recommended NEC® Sections to Tab	298
Standard Calculation: One-Family Dwelling	299
Standard Calculation: Multifamily Dwelling	300
Optional Calculation: One-Family Dwelling	301
Optional Calculation: Multifamily Dwelling	302
Standard Calculation: Mobile and Manufactured Homes	303
Cables	304
Raceways	304
Ohm's Law and Power Formula	304
Common Electrical Insulations	304
Sine Waves	304
AC/DC Formulas	305
Horsepower Formulas	305
Voltage Drop Formulas—1ϕ, 3ϕ	305
Powers of 10	306
Units of Energy	306
Units of Power	306
Standard Sizes of Fuses and CBs	306
Voltage Conversion	306
Electrical Symbols	307
Abbreviations	311
Hazardous Locations	315
Motor Calculation Form	316

Practice Exam 14-1

Answer Sheet

Name _____   Exam _____

Answer	NEC® Reference	Answer	NEC® Reference	Answer	NEC® Reference	Answer	NEC® Reference
1. ___ ___		26. ___ ___		51. ___ ___		76. ___ ___	
2. ___ ___		27. ___ ___		52. ___ ___		77. ___ ___	
3. ___ ___		28. ___ ___		53. ___ ___		78. ___ ___	
4. ___ ___		29. ___ ___		54. ___ ___		79. ___ ___	
5. ___ ___		30. ___ ___		55. ___ ___		80. ___ ___	
6. ___ ___		31. ___ ___		56. ___ ___		81. ___ ___	
7. ___ ___		32. ___ ___		57. ___ ___		82. ___ ___	
8. ___ ___		33. ___ ___		58. ___ ___		83. ___ ___	
9. ___ ___		34. ___ ___		59. ___ ___		84. ___ ___	
10. ___ ___		35. ___ ___		60. ___ ___		85. ___ ___	
11. ___ ___		36. ___ ___		61. ___ ___		86. ___ ___	
12. ___ ___		37. ___ ___		62. ___ ___		87. ___ ___	
13. ___ ___		38. ___ ___		63. ___ ___		88. ___ ___	
14. ___ ___		39. ___ ___		64. ___ ___		89. ___ ___	
15. ___ ___		40. ___ ___		65. ___ ___		90. ___ ___	
16. ___ ___		41. ___ ___		66. ___ ___		91. ___ ___	
17. ___ ___		42. ___ ___		67. ___ ___		92. ___ ___	
18. ___ ___		43. ___ ___		68. ___ ___		93. ___ ___	
19. ___ ___		44. ___ ___		69. ___ ___		94. ___ ___	
20. ___ ___		45. ___ ___		70. ___ ___		95. ___ ___	
21. ___ ___		46. ___ ___		71. ___ ___		96. ___ ___	
22. ___ ___		47. ___ ___		72. ___ ___		97. ___ ___	
23. ___ ___		48. ___ ___		73. ___ ___		98. ___ ___	
24. ___ ___		49. ___ ___		74. ___ ___		99. ___ ___	
25. ___ ___		50. ___ ___		75. ___ ___		100. ___ ___	

Practice Exam 14-2

Answer Sheet

Name _____   Exam _____

Answer	NEC® Reference	Answer	NEC® Reference	Answer	NEC® Reference	Answer	NEC® Reference
1. ___ ___		26. ___ ___		51. ___ ___		76. ___ ___	
2. ___ ___		27. ___ ___		52. ___ ___		77. ___ ___	
3. ___ ___		28. ___ ___		53. ___ ___		78. ___ ___	
4. ___ ___		29. ___ ___		54. ___ ___		79. ___ ___	
5. ___ ___		30. ___ ___		55. ___ ___		80. ___ ___	
6. ___ ___		31. ___ ___		56. ___ ___		81. ___ ___	
7. ___ ___		32. ___ ___		57. ___ ___		82. ___ ___	
8. ___ ___		33. ___ ___		58. ___ ___		83. ___ ___	
9. ___ ___		34. ___ ___		59. ___ ___		84. ___ ___	
10. ___ ___		35. ___ ___		60. ___ ___		85. ___ ___	
11. ___ ___		36. ___ ___		61. ___ ___		86. ___ ___	
12. ___ ___		37. ___ ___		62. ___ ___		87. ___ ___	
13. ___ ___		38. ___ ___		63. ___ ___		88. ___ ___	
14. ___ ___		39. ___ ___		64. ___ ___		89. ___ ___	
15. ___ ___		40. ___ ___		65. ___ ___		90. ___ ___	
16. ___ ___		41. ___ ___		66. ___ ___		91. ___ ___	
17. ___ ___		42. ___ ___		67. ___ ___		92. ___ ___	
18. ___ ___		43. ___ ___		68. ___ ___		93. ___ ___	
19. ___ ___		44. ___ ___		69. ___ ___		94. ___ ___	
20. ___ ___		45. ___ ___		70. ___ ___		95. ___ ___	
21. ___ ___		46. ___ ___		71. ___ ___		96. ___ ___	
22. ___ ___		47. ___ ___		72. ___ ___		97. ___ ___	
23. ___ ___		48. ___ ___		73. ___ ___		98. ___ ___	
24. ___ ___		49. ___ ___		74. ___ ___		99. ___ ___	
25. ___ ___		50. ___ ___		75. ___ ___		100. ___ ___	

Practice Exam 14-3

Answer Sheet

Name _____ Exam _____

Answer	NEC® Reference	Answer	NEC® Reference	Answer	NEC® Reference	Answer	NEC® Reference
1. ___ ___		26. ___ ___		51. ___ ___		76. ___ ___	
2. ___ ___		27. ___ ___		52. ___ ___		77. ___ ___	
3. ___ ___		28. ___ ___		53. ___ ___		78. ___ ___	
4. ___ ___		29. ___ ___		54. ___ ___		79. ___ ___	
5. ___ ___		30. ___ ___		55. ___ ___		80. ___ ___	
6. ___ ___		31. ___ ___		56. ___ ___		81. ___ ___	
7. ___ ___		32. ___ ___		57. ___ ___		82. ___ ___	
8. ___ ___		33. ___ ___		58. ___ ___		83. ___ ___	
9. ___ ___		34. ___ ___		59. ___ ___		84. ___ ___	
10. ___ ___		35. ___ ___		60. ___ ___		85. ___ ___	
11. ___ ___		36. ___ ___		61. ___ ___		86. ___ ___	
12. ___ ___		37. ___ ___		62. ___ ___		87. ___ ___	
13. ___ ___		38. ___ ___		63. ___ ___		88. ___ ___	
14. ___ ___		39. ___ ___		64. ___ ___		89. ___ ___	
15. ___ ___		40. ___ ___		65. ___ ___		90. ___ ___	
16. ___ ___		41. ___ ___		66. ___ ___		91. ___ ___	
17. ___ ___		42. ___ ___		67. ___ ___		92. ___ ___	
18. ___ ___		43. ___ ___		68. ___ ___		93. ___ ___	
19. ___ ___		44. ___ ___		69. ___ ___		94. ___ ___	
20. ___ ___		45. ___ ___		70. ___ ___		95. ___ ___	
21. ___ ___		46. ___ ___		71. ___ ___		96. ___ ___	
22. ___ ___		47. ___ ___		72. ___ ___		97. ___ ___	
23. ___ ___		48. ___ ___		73. ___ ___		98. ___ ___	
24. ___ ___		49. ___ ___		74. ___ ___		99. ___ ___	
25. ___ ___		50. ___ ___		75. ___ ___		100. ___ ___	

Practice Exam 14-4

Answer Sheet

Name _____   Exam _____

Answer	NEC® Reference	Answer	NEC® Reference	Answer	NEC® Reference	Answer	NEC® Reference
1. ___ ___		26. ___ ___		51. ___ ___		76. ___ ___	
2. ___ ___		27. ___ ___		52. ___ ___		77. ___ ___	
3. ___ ___		28. ___ ___		53. ___ ___		78. ___ ___	
4. ___ ___		29. ___ ___		54. ___ ___		79. ___ ___	
5. ___ ___		30. ___ ___		55. ___ ___		80. ___ ___	
6. ___ ___		31. ___ ___		56. ___ ___		81. ___ ___	
7. ___ ___		32. ___ ___		57. ___ ___		82. ___ ___	
8. ___ ___		33. ___ ___		58. ___ ___		83. ___ ___	
9. ___ ___		34. ___ ___		59. ___ ___		84. ___ ___	
10. ___ ___		35. ___ ___		60. ___ ___		85. ___ ___	
11. ___ ___		36. ___ ___		61. ___ ___		86. ___ ___	
12. ___ ___		37. ___ ___		62. ___ ___		87. ___ ___	
13. ___ ___		38. ___ ___		63. ___ ___		88. ___ ___	
14. ___ ___		39. ___ ___		64. ___ ___		89. ___ ___	
15. ___ ___		40. ___ ___		65. ___ ___		90. ___ ___	
16. ___ ___		41. ___ ___		66. ___ ___		91. ___ ___	
17. ___ ___		42. ___ ___		67. ___ ___		92. ___ ___	
18. ___ ___		43. ___ ___		68. ___ ___		93. ___ ___	
19. ___ ___		44. ___ ___		69. ___ ___		94. ___ ___	
20. ___ ___		45. ___ ___		70. ___ ___		95. ___ ___	
21. ___ ___		46. ___ ___		71. ___ ___		96. ___ ___	
22. ___ ___		47. ___ ___		72. ___ ___		97. ___ ___	
23. ___ ___		48. ___ ___		73. ___ ___		98. ___ ___	
24. ___ ___		49. ___ ___		74. ___ ___		99. ___ ___	
25. ___ ___		50. ___ ___		75. ___ ___		100. ___ ___	

RECOMMENDED NEC® SECTIONS TO TAB

ARTICLE/SECTION		ARTICLE/SECTION		ARTICLE/SECTION	
90	Introduction	358	Electrical Metallic Tubing: Type EMT	630	Electric Welders
100	Definitions			645	Information Technology Equipment
110.26	Spaces About Electrical Equipment	376	Metal Wireways		
		392	Cable Trays	680	Swimming Pools, Fountains, and Similar Installations
210	Branch Circuits	400	Flexible Cords and Cables		
220	Branch-Circuit, Feeder, and Service Calculations	404	Switches	695	Fire Pumps
		406	Receptacles, Cord Connectors, and Attachment Plugs (Caps)	700	Emergency Systems
220.55	Calculations			702	Optional Standby Systems
230	Services	408	Switchboards and Panelboards	725	Class 1, Class 2, and Class 3 Remote-Control, Signaling, and Power-Limited Circuits.
240	Overcurrent Protection	410	Luminaires, Lampholders, and Lamps		
250	Grounding and Bonding				
250.66	Size of Alternating-Current Grounding Electrode Conductor	422	Appliances	760	Fire Alarm Systems
		430	Motors, Motor Circuits, and Controllers	800	Communications Circuits
250.122	Size of Equipment Grounding Conductors			820	Community Antenna Television and Radio Distribution Systems
		430.22	Single Motor		
300	Wiring Methods	430.32	Continuous-Duty Motors	Chapter 9	Table 1 Percent of Cross Section of Conduit and Tubing for Conductors
300.5	Underground Installations	430.52	Rating or Setting for Individual Motor Circuit		
310.15(B)(16)	Table				
		430.250	Table	Chapter 9	Table 4 Dimensions and Percent Area of Conduit and Tubing
312	Cabinets, Cutout Boxes, and Meter Socket Enclosures	450	Transformers and Transformer Vaults (Including Secondary Ties)		
314.16	Number of Conductors in Outlet, Device, and Junction Boxes, and Conduit Bodies	500	Class I Locations	Chapter 9	Table 5 Dimensions of Insulated Conductors and Fixture Wires
		502	Class II Locations		
314.28	Pull and Junction Boxes and Conduit Bodies	511	Commercial Garages, Repair, and Storage	Chapter 9	Table 8 Conductor Properties
330	Metal-Clad Cable: Type MC	514	Motor Fuel Dispensing Facilities	Annex C	Conduit and Tubing Fill Tables for Conductors and Fixture Wires of the Same Size
334	Nonmetallic-Sheathed Cable: Types NM, NMC, and NMS	518	Assembly Occupancies		
		550	Mobile Homes, Manufactured Homes, and Mobile Home Parks	Annex D	Examples
344	Rigid Metal Conduit: Type RMC				
348	Flexible Metal Conduit: Type FMC	590	Temporary Installations		
		600	Electric Signs and Outline Lighting		

STANDARD CALCULATION: ONE-FAMILY DWELLING

1. GENERAL LIGHTING: *Table 220.12*

_____ sq ft × 3 VA = _____ VA

Small appliances: *220.52(A)*

_____ VA × _____ circuits = _____ VA

Laundry *220.52(B)*

_____ VA × 1 = _____ VA
_____ VA

©2017 by American Technical Publishers, Inc. This form may be reproduced for instructional use only. It shall not be reproduced and sold.

Applying Demand Factors: *Table 220.42.*

First 3000 VA × 100% = 3000 VA

Next _____ VA × 35% = _____ VA   PHASES   NEUTRAL

Remaining _____ VA × 25% = _____ VA

Total _____ VA   _____ VA   _____ VA

2. FIXED APPLIANCES: *220.53*

Dishwasher = _____ VA
Disposer = _____ VA
Compactor = _____ VA
Water Heater = _____ VA
_____ = _____ VA
_____ = _____ VA
_____ = _____ VA

(120 V Loads × 75%)

Total _____ VA × 75% = _____ VA   _____ VA   _____ VA

3. DRYER: *220.54; Table 220.54*

_____ VA × _____ % = _____ VA   _____ VA × 70% = _____ VA

4. COOKING EQUIPMENT: *Table 220.55; Notes*

Col A _____ VA × _____ % = _____ VA
Col B _____ VA × _____ % = _____ VA
Col C _____ VA × _____ % = _____ VA
Total _____ VA   _____ VA × 70% = _____ VA

5. HEATING or AC: *220.60*

Heating unit = _____ VA × 100% = _____ VA
A/C unit = _____ VA × 100% = _____ VA
Heat pump = _____ VA × 100% = _____ VA
Largest Load _____ VA   _____ VA   _____ VA

6. LARGEST MOTOR: *220.14(C)*

φ _____ VA × 25% = _____ VA   _____ VA

N _____ VA × 25% = _____ VA   _____ VA

1φ service: PHASES $I = \dfrac{_____ \text{ VA}}{\text{V}} = _____$ A

             PHASES $I = \dfrac{_____ \text{ VA}}{\text{V}} = _____$ A   [____] VA   [____] VA

220.61(B) First 200 A × 100% = 200 A

Remaining _____ A × 70% = _____ A

Total _____ A

STANDARD CALCULATION: MULTIFAMILY DWELLING

1. GENERAL LIGHTING: *Table 220.12*
  _____ sq ft × 3 VA × _____ units = _____ VA
  _____ sq ft × 3 VA × _____ units = _____ VA
  _____ sq ft × 3 VA × _____ units = _____ VA
  Small appliances: *220.52(A)*
  _____ VA × _____ circuits × _____ units = _____ VA
  Laundry *220.52(B)*
  _____ VA × 1 × _____ units = _____ VA
  _____ VA

  ©2017 by American Technical Publishers, Inc.
  This form may be reproduced for instructional use only.
  It shall not be reproduced and sold.

  Applying Demand Factors: *Table 220.42.*
  First 3000 VA × 100% = 3000 VA
  Next _____ VA × 35% = _____ VA   PHASES   NEUTRAL
  Remaining _____ VA × 25% = _____ VA
  Total   _____ VA   _____ VA   _____ VA

2. FIXED APPLIANCES: *220.53*
  Dishwaher = _____ VA
  Disposer = _____ VA
  Compactor = _____ VA
  Water Heater = _____ VA
  _____ = _____ VA
  _____ = _____ VA
  _____ = _____ VA
                                                                    (Four or more 120 V Loads × 75%)
  Total   _____ VA × 10 units × 75% = _____ VA   _____ VA   _____ VA

3. DRYER: *220.54; Table 220.54*
  _____ VA × _____ units × _____ % = _____ VA   _____ VA × 70% = _____ VA

4. COOKING EQUIPMENT: *Table 220.55; Notes*
  Col A _____ units = _____ VA × _____ % = _____ VA
  Col B _____ VA × _____ units × _____ % = _____ VA
  Col C _____ VA × _____ units × _____ % = _____ VA
  Total   _____ VA   _____ VA × 70% = _____ VA

5. HEATING or AC: *220.60*
  Heating unit = _____ VA × 100% × _____ units = _____ VA
  A/C unit = _____ VA × 100% × _____ units = _____ VA
  Heat pump = _____ VA × 100% × _____ units = _____ VA
  Largest Load   _____ VA   _____ VA   _____ VA

6. LARGEST MOTOR: *220.14(C)*
  φ _____ VA × 25% = _____ VA   _____ VA
  N _____ VA × 25% = _____ VA   _____ VA

1φ service: PHASES $I = \dfrac{\text{_____ VA}}{\text{V}} = $ _____ A

              PHASES $I = \dfrac{\text{_____ VA}}{\text{V}} = $ _____ A   [_____] VA   [_____] VA

220.61(B) First 200 A × 100% =   200 A
Remaining _____ A × 70% =   _____ A
Total   _____ A

OPTIONAL CALCULATION: ONE-FAMILY DWELLING

1. HEATING or A/C: *220.82(C)(1–6)*

Heating units (3 or less) = _____ VA × 65% = _____ VA
Heating units (4 or more) = _____ VA × 40% = _____ VA
A/C unit = _____ VA × 100% = _____ VA
Heat pump = _____ VA × 100% = _____ VA
Largest Load _____ VA
Total _____ VA

©2017 by American Technical Publishers, Inc.
*This form may be reproduced for instructional use only.
It shall not be reproduced and sold.*

PHASES

_____ VA

2. GENERAL LOADS: *220.82(B)(1 – 4)*

General lighting: *280.82(B)(1)*
_____ sq ft × 3 VA   _____ VA

Small appliance and laundry loads: *220.82(B)(2)*
_____ VA × _____ circuits =   _____ VA

Special Loads: *220.82(B)(3 –4)*
Dishwasher = _____ VA
Disposer = _____ VA
Compactor = _____ VA
Water Heater = _____ VA
_____ = _____ VA
_____ = _____ VA
_____ = _____ VA
_____ = _____ VA
_____ = _____ VA
_____ VA
                                                Total _____ VA

Applying Demand Factors: *120.82(B)*
First 10,000 VA × 100% =   10,000 VA
Remaining _____ VA × 40% =   _____ VA
Total   _____ VA   _____ VA

NEUTRAL (Loads from Standard Calculation)

1. General Lighting = _____ VA
2. Fixed appliances = _____ VA
3. Dryer = _____ VA
4. Cooking equipment = _____ VA
5. Heating or A/C = _____ VA
6. Largest motor = _____ VA

Total [_____] VA

1ϕ service: PHASES $I = \dfrac{_____\ VA}{V} = _____ A$

PHASES $I = \dfrac{_____\ VA}{V} = _____ A$

[_____] VA

OPTIONAL CALCULATION: MULTIFAMILY DWELLING

1. HEATING or A/C: *220.82(C)(1–6)*

Heating unit = _____ VA × 100% × _____ units = _____ VA
A/C unit = _____ VA × 100% = _____ units = _____ VA
Heat pump = _____ VA × 100% = _____ units = _____ VA
Largest Load _____ VA   PHASES
Total   _____ VA

©2017 by American Technical Publishers, Inc.
This form may be reproduced for instructional use only.
It shall not be reproduced and sold.

2. GENERAL LOADS: *220.82(B)(1 – 4)*

General lighting: *280.82(B)(1)*

_____ sq ft × 3 VA × _____ units =   _____ VA
_____ sq ft × 3 VA × _____ units =
_____ sq ft × 3 VA × _____ units =   _____ VA
_____ sq ft × 3 VA × _____ units =

Small appliance and laundry loads: *220.82(B)(2)*

_____ VA × _____ circuits × _____ units =

Special Loads: *220.82(B)(3 –4)*

Dishwasher =   _____ VA
Disposer =   _____ VA
Compactor =   _____ VA
Water Heater =   _____ VA
_____ =   _____ VA
_____ =   _____ VA
_____ =   _____ VA
_____ =   _____ VA
_____ =   _____ VA
Total   _____ VA × _____ units = _____ VA _____ VA

                                Total Connected Load _____ VA

Applying Demand Factors: *Table 220.84*
_____ VA × _____ % = _____ VA   _____ VA

NEUTRAL (Loads from Standard Calculation)

1. General Lighting =   _____ VA
2. Fixed appliances =   _____ VA
3. Dryer =   _____ VA
4. Cooking equipment =   _____ VA
5. Heating or A/C =   _____ VA
6. Largest motor =   _____ VA

Total   _____ VA

1φ service: PHASES $I = \dfrac{VA}{V} = $ _____ A   **3φ service:** $I = \dfrac{VA}{V \times \sqrt{3}} = $ _____ A

                  NEUTRAL $I = \dfrac{VA}{V} = $ _____ A $I = \dfrac{VA}{V \times \sqrt{3}} = $ _____ A

220.84; First 200 A × 100% =   200 A
Remaining _____ A × 70%   _____ A
                                      _____ A

STANDARD CALCULATION: MOBILE AND MANUFACTURED HOMES

1. GENERAL LIGHTING: *Table 550.18(A)(1)*
_____ sq ft × 3 VA = _____ VA
Small appliances: *550.18(A)(2)*
_____ VA × _____ circuits = _____ VA
Laundry *550.18(A)(2)*
_____ VA × 1 = _____ VA
_____ VA

©2017 by American Technical Publishers, Inc.
*This form may be reproduced for instructional use only.
It shall not be reproduced and sold.*

Applying Demand Factors: *Part 550.18(A)(5).*
First 3000 VA × 100% = 3000 VA
Next _____ VA × 35% = _____ VA   PHASE A   PHASE B
Remaining _____ VA × 25% = _____ VA
Total   _____ VA ÷ _____ V = _____ A _____ A

2. Motors and Heaters: *550.18(B)(2)*
Heating unit = _____ VA × 100% = _____ VA
A/C unit = _____ VA × 100% = _____ VA
Largest Load _____ _____ VA ÷ _____ V = _____ A _____ A
Blower Motor = _____ VA × 100% = _____ VA ÷ _____ V = _____ A _____ A

3. LARGEST MOTOR: *550.18(B)(3)*
φ _____ VA × 25% = _____ VA ÷ _____ V = _____ A _____ A

4. FIXED APPLIANCES: *550.18(B)(4)*
Dishwasher = _____ VA
Disposer =   _____ VA
Compactor =   _____ VA
Water Heater = _____ VA
_____ = _____ VA
_____ = _____ VA
_____ = _____ VA
Total   _____ VA × _____ % _____ VA ÷ _____ V = _____ A _____ A

5. RANGE: *550.18(B)(5)*
Range = _____ VA × _____ % = _____ VA ÷ _____ V = _____ A _____ A

[_____] A [_____] A

CABLES

AC	Armored Cable
BX	Tradename for AC
FCC	Flat Conductor Cable
IGS	Integrated Gas Spacer Cable
MC	Metal-Clad Cable
MI	Mineral-Insulated, Metal Sheathed Cable
MV	Medium Voltage
NM	Nonmetallic-Sheathed Cable (dry)
NMC	Nonmetallic-Sheathed Cable (dry or damp)
NMS	Nonmetallic-Sheathed Cable (dry)
SE	Service-Entrance Cable
TC	Tray Cable
UF	Underground Feeder Cable
USE	Underground Service-Entrance Cable

RACEWAYS

EMT	Electrical Metallic Tubing
ENT	Electrical Nonmetallic Tubing
FMC	Flexible Metal Conduit
FMT	Flexible Metallic Tubing
IMC	Intermediate Metal Conduit
LFMC	Liquidtight Flexible Metal Conduit
LFNC	Liquidtight Flexible Nonmetallic Conduit
RMC	Rigid Metal Conduit
RNC	Rigid Nonmetallic Conduit

OHM'S LAW AND POWER FORMULA

Ohm's law and power formula wheel showing relationships:
$E = R \times I$, $E = \dfrac{P}{I}$, $E = \sqrt{P \times R}$, $E = \dfrac{E^2}{R}$ ... center variables P, E, I, R; $R \times I^2$, $P = E \times I$, $E \times I$, $\sqrt{\dfrac{P}{R}}$, $I = \dfrac{P}{E}$, $I = \dfrac{E}{R}$, $\dfrac{P}{E}$, $\dfrac{E}{R}$, $\dfrac{E^2}{P}$, $\dfrac{P}{I^2}$, $R = \dfrac{E}{I}$

VALUES IN INNER CIRCLE ARE EQUAL TO VALUES IN CORRESPONDING OUTER CIRCLE

COMMON ELECTRICAL INSULATIONS

60°C / 140°F	75°C / 167°F	90°C / 194°F
TW	FEPW	TBS
UF	RH	SA
	RHW	SIS
	THHW	FEP
	THW	FEPB
	THWN	MI
	XHHW	RHH
	USE	RHW-2
	ZW	THHN
		THHW
		THW-2
		THWN-2
		USE-2
		XHH
		XHHW
		XHHW-2
		ZW-2

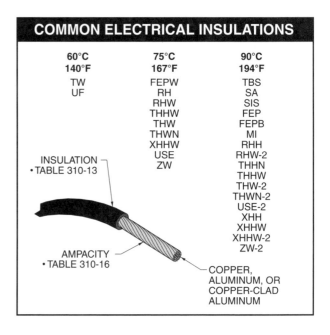

INSULATION • TABLE 310-13
AMPACITY • TABLE 310-16
COPPER, ALUMINUM, OR COPPER-CLAD ALUMINUM

SINE WAVES

Frequency	Period	Peak-to-Peak Value
$f = \dfrac{1}{T}$ where f = frequency (in hertz) 1 = constant T = period (in seconds)	$T = \dfrac{1}{f}$ where T = period (in seconds) 1 = constant f = frequency (in hertz)	$V_{p\text{-}p} = 2 \times V_{max}$ where 2 = constant $V_{p\text{-}p}$ = peak-to-peak value V_{max} = peak value

Average Value	rms Value
$V_{avg} = V_{max} \times 0.637$ where V_{avg} = average value (in volts) V_{max} = peak value (in volts) 0.637 = constant	$V_{avg} = V_{max} \times 0.707$ where V_{rms} = rms value (in volts) V_{max} = peak value (in volts) 0.707 = constant

AC/DC FORMULAS

To Find	DC	AC		
		1φ, 115 or 220 V	1φ, 208, 230, or 240 V	3φ—All Voltages
I, HP known	$\dfrac{HP \times 746}{E \times Eff}$	$\dfrac{HP \times 746}{E \times Eff \times PF}$	$\dfrac{HP \times 746}{E \times Eff \times PF}$	$\dfrac{HP \times 746}{1.73 \times E \times Eff \times PF}$
I, kW known	$\dfrac{kW \times 1000}{E}$	$\dfrac{kW \times 1000}{E \times PF}$	$\dfrac{kW \times 1000}{E \times PF}$	$\dfrac{kW \times 1000}{1.73 \times E \times PF}$
I, kVA known		$\dfrac{kW \times 1000}{E}$	$\dfrac{kW \times 1000}{E}$	$\dfrac{kVA \times 1000}{1.763 \times E}$
kW	$\dfrac{I \times E}{1000}$	$\dfrac{I \times E \times PF}{1000}$	$\dfrac{I \times E \times PF}{1000}$	$\dfrac{I \times E \times 1.73 \times PF}{1000}$
kVA		$\dfrac{I \times E}{1000}$	$\dfrac{I \times E}{1000}$	$\dfrac{I \times E \times 1.73}{1000}$
HP (output)	$\dfrac{I \times E \times Eff}{746}$	$\dfrac{I \times E \times Eff \times PF}{746}$	$\dfrac{I \times E \times Eff \times PF}{746}$	$\dfrac{I \times E \times 1.73 \times Eff \times PF}{746}$

Note: *Eff* = efficiency

HORSEPOWER FORMULAS

To Find	Use Formula	Example		
		Given	Find	Solution
HP	$HP = \dfrac{I \times E \times Eff}{746}$	240 V, 20 A, 85% Eff	HP	$HP = \dfrac{I \times E \times Eff}{746}$ $HP = \dfrac{20 \text{ A} \times 240 \text{ V} \times 85\%}{746}$ $HP = \mathbf{5.5}$
I	$I = \dfrac{HP \times 746}{E \times Eff \times PF}$	10 HP, 240 V, 90% Eff, 88% PF	I	$I = \dfrac{HP \times 746}{E \times Eff \times PF}$ $I = \dfrac{10 \text{ HP} \times 746}{240 \text{ V} \times 90\% \times 88\%}$ $I = \mathbf{39 \text{ A}}$

VOLTAGE DROP FORMULAS—1φ, 3φ

Phase	To Find	Use Formula	Example		
			Given	Find	Solution
1φ	VD	$VD = \dfrac{2 \times R \times L \times I}{1000}$	240 V, 40 A, 60′ L, 0.764 R	VD	$VD = \dfrac{2 \times R \times L \times I}{1000}$ $VD = \dfrac{2 \times 0.764 \times 60 \times 40}{1000}$ $VD = \mathbf{3.67 \text{ V}}$
3φ	VD	$VD = \dfrac{2 \times R \times L \times I}{1000} \times 0.866$	208 V, 110 A, 75′ L, 0.194 R, 0.866 multiplier	VD	$VD = \dfrac{2 \times R \times L \times I}{1000} \times 0.866$ $VD = \dfrac{2 \times 0.194 \times 75 \times 110}{1000} \times 0.866$ $VD = \mathbf{2.77 \text{ V}}$

Note: $\dfrac{\sqrt{3}}{2} = 0.866$

POWERS OF 10

1×10^4	= 10,000	= $10 \times 10 \times 10 \times 10$	Read ten to the fourth power
1×10^3	= 1000	= $10 \times 10 \times 10$	Read ten to the third power or ten cubed
1×10^2	= 100	= 10×10	Read ten to the second power or ten squared
1×10^1	= 10	= 10	Read ten to the first power
1×10^0	= 1	= 1	Read ten to the zero power
1×10^{-1}	= 0.1	= 1/10	Read ten to the minus first power
1×10^{-2}	= 0.01	= $1/(10 \times 10)$ or 1/100	Read ten to the minus second power
1×10^{-3}	= 0.001	= $1/(10 \times 10 \times 10)$ or 1/1000	Read ten to the minus third power
1×10^{-4}	= 0.0001	= $1/(10 \times 10 \times 10 \times 10)$ or 1/10,000	Read ten to the minus fourth power

UNITS OF ENERGY

Energy	Btu	ft-lb	J	kcal	kWh
British thermal unit	1	777.9	1.056	0.252	2.930×10^{-4}
Foot-pound	1.285×10^{-3}	1	1.356	3.240×10^{-4}	3.766×10^{-7}
Joule	9.481×10^{-4}	0.7376	1	2.390×10^{-4}	2.778×10^{-7}
Kilocalorie	3.968	3.086	4.184	1	1.163×10^{-3}
Kilowatt-hour	3.413	2.655×10^6	3.6×10^6	860.2	1

UNITS OF POWER

Power	W	ft lb/s	HP	kW
Watt	1	0.7376	$.341 \times 10^{-3}$	0.001
Foot-pound/sec	1.356	1	$.818 \times 10^{-3}$	1.356×10^{-3}
Horsepower	745.7	550	1	0.7457
Kilowatt	1000	736.6	1.341	1

STANDARD SIZES OF FUSES AND CBs

NEC® 240.6(A) lists standard ampere ratings of fuses and fixed-trip CBs as follows:
15, 20, 25, 30, 35, 40, 45,
50, 60, 70, 80, 90, 100, 110,
125, 150, 175, 200, 225,
250, 300, 350, 400, 450,
500, 600, 700, 800,
1000, 1200, 1600,
2000, 2500, 3000, 4000, 5000, 6000

VOLTAGE CONVERSIONS

To Convert	To	Multiply By
rms	Average	0.9
rms	Peak	1.414
Average	rms	1.111
Average	Peak	1.567
Peak	rms	0.707
Peak	Average	0.637
Peak	Peak-to-Peak	2

ELECTRICAL SYMBOLS...

Lighting Outlets

Symbol Name	
OUTLET BOX AND INCANDESCENT LIGHTING FIXTURE	─◯─ ◯ (CEILING / WALL)
INCANDESCENT TRACK LIGHTING	[ooo] [ooo]
BLANKED OUTLET	(B) (B)
DROP CORD	(D)
EXIT LIGHT AND OUTLET BOX. SHADED AREAS DENOTE FACES.	⊗ ⊗
OUTDOOR POLE-MOUNTED FIXTURES	◯─•─◯
JUNCTION BOX	(J) (J)
LAMPHOLDER WITH PULL SWITCH	(L)$_{PS}$ (L)$_{PS}$
MULTIPLE FLOODLIGHT ASSEMBLY	▽▽▽
EMERGENCY BATTERY PACK WITH CHARGER	[B]
INDIVIDUAL FLUORESCENT FIXTURE	[□] [□]
OUTLET BOX AND FLUORESCENT LIGHTING TRACK FIXTURE	├──┤
CONTINUOUS FLUORESCENT FIXTURE	[□──□]
SURFACE-MOUNTED FLUORESCENT FIXTURE UNDERFLOOR DUCT AND	[□] ─[□]

Panelboards

FLUSH-MOUNTED PANELBOARD AND CABINET	▬▬▬
SURFACE-MOUNTED PANELBOARD AND CABINET	▬▬▬

Convenience Outlets

SINGLE RECEPTACLE OUTLET	─⊖
DUPLEX RECEPTACLE OUTLET—120 V	─⊖
TRIPLEX RECEPTACLE OUTLET—240 V	─⊕
SPLIT-WIRED DUPLEX RECEPTACLE OUTLET	─⊖
SPLIT-WIRED TRIPLEX RECEPTACLE OUTLET	─⊕
SINGLE SPECIAL-PURPOSE RECEPTACLE OUTLET	─△
DUPLEX SPECIAL-PURPOSE RECEPTACLE OUTLET	─△
RANGE OUTLET	─⊖ R
SPECIAL-PURPOSE CONNECTION	─▲ DW
CLOSED-CIRCUIT TELEVISION CAMERA	▭◁
CLOCK HANGER RECEPTACLE	─(C)
FAN HANGER RECEPTACLE	(F)
FLOOR SINGLE RECEPTACLE OUTLET	⊖
FLOOR DUPLEX RECEPTACLE OUTLET	⊖
FLOOR SPECIAL-PURPOSE OUTLET	△
JUNCTION BOX FOR TRIPLE, DOUBLE, OR SINGLE DUCT SYSTEM AS INDICATED BY NUMBER OF PARALLEL LINES	▯

Busducts and Wireways

SERVICE, FEEDER, OR PLUG-IN BUSWAY	[B] [B] [B]
CABLE THROUGH LADDER OR CHANNEL	[C] [C] [C]
WIREWAY	[W] [W] [W]

Switch Outlets

SINGLE-POLE SWITCH	S
DOUBLE-POLE SWITCH	S$_2$
THREE-WAY SWITCH	S$_3$
FOUR-WAY SWITCH	S$_4$
AUTOMATIC DOOR SWITCH	S$_D$
KEY-OPERATED SWITCH	S$_K$
CIRCUIT BREAKER	S$_{CB}$
WEATHERPROOF CIRCUIT BREAKER	S$_{WCB}$
DIMMER	S$_{DM}$
REMOTE CONTROL SWITCH	S$_{RC}$
WEATHERPROOF SWITCH	S$_{WP}$
FUSED SWITCH	S$_F$
WEATHERPROOF FUSED SWITCH	S$_{WF}$
TIME SWITCH	S$_T$
CEILING PULL SWITCH	Ⓢ
SWITCH AND SINGLE RECEPTACLE	─⊖$_S$
SWITCH AND DOUBLE RECEPTACLE	─⊖$_S$
A STANDARD SYMBOL WITH AN ADDED LOWERCASE SUBSCRIPT LETTER IS USED TO DESIGNATE A VARIATION IN STANDARD EQUIPMENT	◯$_{a,b}$ ─⊖$_{a,b}$ S$_{a,b}$

...ELECTRICAL SYMBOLS...

CONTACTS								OVERLOAD RELAYS	
INSTANT OPERATING				TIMED CONTACTS—CONTACT ACTION RETARDED AFTER COIL IS:				THERMAL	MAGNETIC
WITH BLOWOUT		WITHOUT BLOWOUT		ENERGIZED		DE-ENERGIZED			
NO	NC	NO	NC	NOTC	NCTO	NOTO	NCTC		

SUPPLEMENTARY CONTACT SYMBOLS

SPST NO		SPST NC		SPDT		TERMS
SINGLE BREAK	DOUBLE BREAK	SINGLE BREAK	DOUBLE BREAK	SINGLE BREAK	DOUBLE BREAK	SPST SINGLE-POLE, SINGLE-THROW
DPST, 2NO		DPST, 2NC		DPDT		SPDT SINGLE-POLE, DOUBLE-THROW
SINGLE BREAK	DOUBLE BREAK	SINGLE BREAK	DOUBLE BREAK	SINGLE BREAK	DOUBLE BREAK	DPST DOUBLE-POLE, SINGLE-THROW
						DPDT DOUBLE-POLE, DOUBLE-THROW
						NO NORMALLY OPEN
						NC NORMALLY CLOSED

METER (INSTRUMENT)

INDICATE TYPE BY LETTER	TO INDICATE FUNCTION OF METER OR INSTRUMENT, PLACE SPECIFIED LETTER OR LETTERS WITHIN SYMBOL.			
	AM or A	AMMETER	VA	VOLTMETER
	AH	AMPERE HOUR	VAR	VARMETER
	µA	MICROAMMETER	VARH	VARHOUR METER
	mA	MILLAMMETER	W	WATTMETER
	PF	POWER FACTOR	WH	WATTHOUR METER
	V	VOLTMETER		

PILOT LIGHTS

INDICATE COLOR BY LETTER	
NON PUSH-TO-TEST	PUSH-TO-TEST

INDUCTORS

IRON CORE
AIR CORE

COILS

DUAL-VOLTAGE MAGNET COILS		BLOWOUT COIL
HIGH-VOLTAGE	LOW-VOLTAGE	
LINK	LINKS	

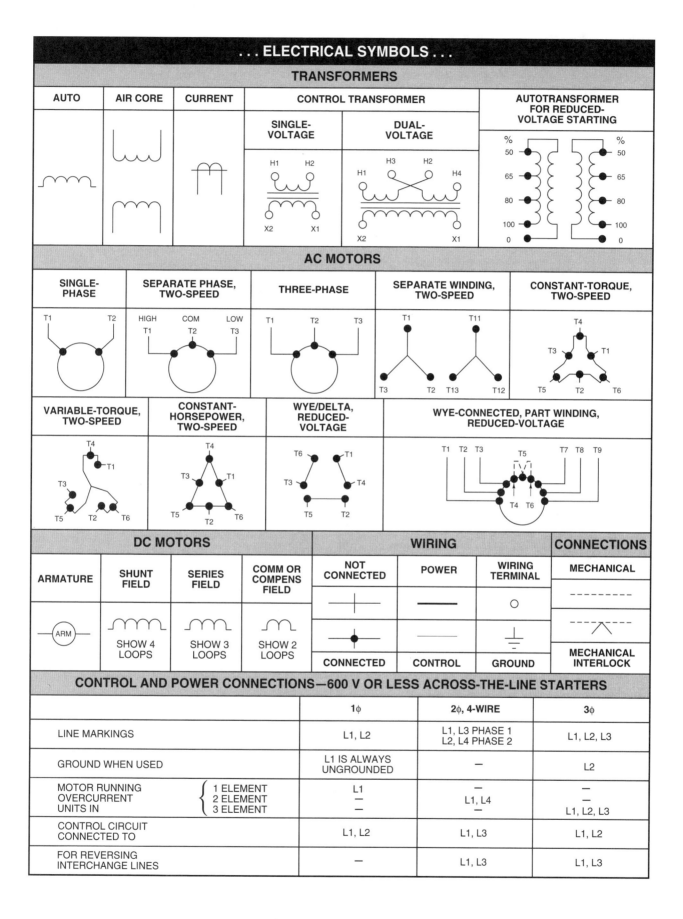

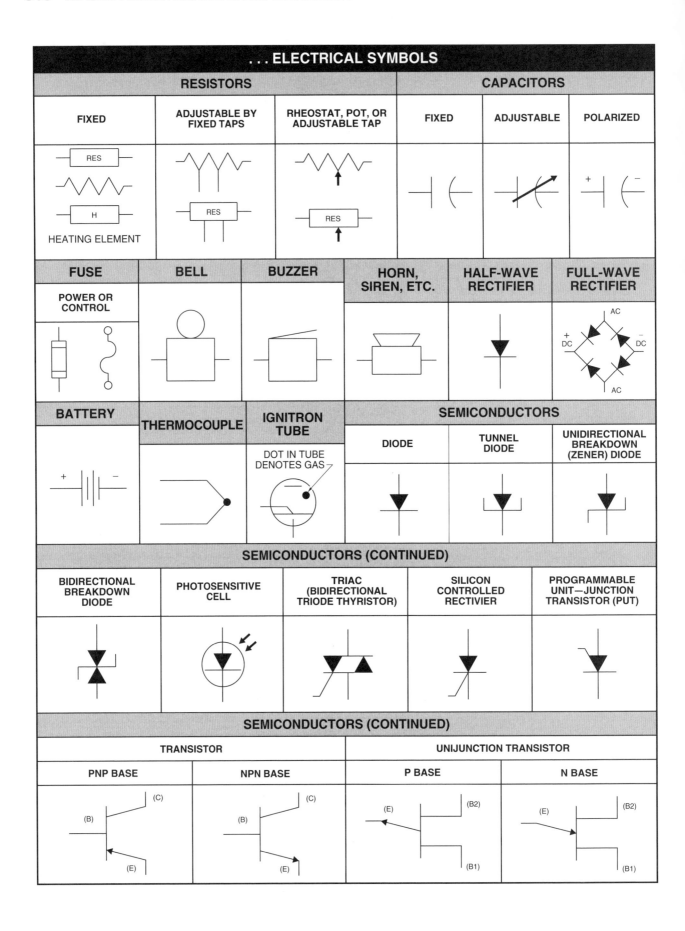

ABBREVIATIONS...

A

abbreviation	ABBR
acoustic	ACST
acoustical tile	ACT. or AT.
adhesive	ADD. or ADH
adjustable	ADJ
adjustable-trip circuit breaker	ATCB
aggregate	AGGR
aileron	AIL
air conditioner	A/C
air handler	A/H
air tight	AT
alarm	ALM
alloy	ALY
alternating current	AC
aluminum	Al
ambient	AMB
Ambulatory Health Care Center	AHCC
American National Standards Institute	ANSI
American Wire Gauge	AWG
ammeter	A or AM
ampere	A or AMP
ampere interrupting rating	AIR.
amps	A
anchor bolt	AB
anode	A
antenna	ANT.
apartment	APT
appliance	APPL
approved	APPD or APVD
approximate	APPROX
approximately	APPROX
architectural	ARCH.
architecture	ARCH.
area	A
area drain	AD
armature	A or ARM.
asphalt	ASPH
asphalt tile	AT.
as required	AR
Assured Equipment Grounding Program	AEGP
astragal	A
Authority Having Jurisdiction	AHJ
automatic	AU or AUTO
automatic sprinkler	AS
auxiliary	AUX
avenue	AVE
azimuth	AZ

B

basement	BSMT
bathroom	B
bathtub	BT
battery (electric)	BAT.
beam	BM
bearing	BRG
bearing plate	BPL or BRG PL
bedroom	BR
benchmark	BM
black	BK
block	BLK
blocking	BLKG
blue	BL
board	BD
board foot	BF
bonding jumper	BJ
boulevard	BLVD
brake relay	BR
brass	BRS
brick	BRK
bridge	BRDG
bronze	BRZ
brown	BR
building	BL or BLDG
building line	BL
built-up roofing	BUR
bypass	BYP

C

cabinet	CAB.
cable	CA
Cable Antenna Television	CATV
cantilever	CANTIL
capacitor	CAP.
cased opening	CO
casement	CSMT
casing	CSG
cast iron	CI
cast-iron pipe	CIP
cast steel	CS
cast stone	CS or CST
catch basin	CB
cathode	K
ceiling	CLG
cellar	CEL
Celsius	°C
cement	CEM
cement floor	CF
center	CTR
centerline	CL
center-to-center	C to C
centigrade	C
central processing unit	CPU
ceramic	CER
ceramic tile	CT
ceramic-tile floor	CTF
channel	CHAN
chapter	CH
chimney	CHM
circuit	CIR or CKT
circuit breaker	CB
circuit interrupter	CI
circular mils	CM
cleanout	CO
clockwise	CW
closet	CLO
coarse	CRS
coated	CTD
coaxial	COAX
Code-Making Panel	CMP
cold air	CA
cold water	CW
column	COL
compacted	COMP
concrete	CONC
concrete block	CCB or CONC BLK
concrete floor	CCF
concrete pipe	CP
condenser	COND
conductor resistivity	K
conduit	C or CND
construction joint	CJ
continuous	CONT
contour	CTR
control	CONT
control joint	CJ or CLJ
control relay	CR
control relay master	CRM
copper	Cu
corner	COR
cornice	COR
corrugated	CORR
counterclockwise	CCW
counter electromotive force	CEMF
county	CO
cubic	CU
cubic foot	CU FT
cubic foot per minute	CFM
cubic foot per second	CFS
cubic inch	CU IN.
cubic yard	CU YD
current	I
current transformer	CT
cut out	CO
cycles per second	CPS

D

damper	DMPR
dampproofing	DP
dead load	DL
decibel	DB
deck	DK
demolition	DML
depth	DP
detail	DET or DTL
diagonal	DIAG
diagram	DIAG
diameter	D or DIA
dimension	DIM.
dimmer	DMR
dining room	DR
diode	DIO
direct current	DC
disconnect switch	DS
dishwasher	DW
distribution	DISTR
distribution panel	DPNL
division	DIV
door	DR
dormer	DRM
double-acting	DBL ACT
double hung window	DHW
double-pole	DP
double-pole double-throw	DPDT
double-pole double-throw switch	DPDT SW
double-pole single-throw	DPST
double-pole single-throw switch	DPST SW
double-pole switch	DP SW
double strength glass	DSG
double-throw	DT
down	D or DN
downspout	DS
drain	DR
drain tile	DT
drawing	DWG
drinking fountain	DF
drum switch	DS
dryer	D
drywall	DW
duplex	DX
dust tight	DT
dutch door	DD
duty cycle	DTY CY
dynamic braking contactor or relay	DB

E

each	EA
east	E
efficiency	Eff
ejector pump	EP
electric	ELEC

Appendix 311

...ABBREVIATIONS...

electrical	ELEC	frequency	FREQ	inch	IN.
electrical metallic tubing	EMT	front view	FV	inch per second	IPS
electric panel	EP	full-load amps	FLA	inch-pound	IN. LB
electromechanical	ELMCH	full-load current	FLC	infrared	IR
electromotive force	EMF	full-load torque	FLT	inside diameter	ID
electronic	ELEK	furnace	FUR.	instantaneous overload	IOL
elevation	EL	furring	FUR.	instantaneous-trip circuit breaker	ITB
elevator	ELEV	fuse	FU	insulation	INSUL
enamel	ENAM	fuse block	FB	integrated circuit	IC
entrance	ENTR	fuse box	FUBX	interior	INT or INTR
equipment	EQPT	fuse holder	FUHLR	interlock	INTLK
equipment bonding jumper	EBJ	fusible	FSBL	intermediate	INT
equipment grounding conductor	EGC	future	FUT	intermediate metal conduit	IMC
equivalent	EQUIV			International Electrotechnical Commission	IEC
estimate	EST	**G**		interrupt	INT
excavate	EXC			inverse time breaker	ITB
exception	Ex.	gallon per hour	GPH	inverse-time circuit breaker	ITCB
exhaust	EXH	gallon per minute	GPM	iron	I
existing	EXIST. or EXST	galvanized	GALV	iron pipe	IP
expanded metal	EM	garage	GAR	isolated ground	IG
expansion joint	EXP JT	gas	G		
explosionproof	EP	gate	G	**J**	
exterior	EXT	gauge	GA		
exterior grade	EXT GR	generator	GEN	jamb	JB or JMB
		glass	GL or GLS	joint	JT
F		glass block	GLB	joist	J or JST
		glaze	GLZ	junction	JCT
face brick	FB	gold	Au	junction box	JB
Fahrenheit	°F	grade	GR		
fast	F	grade line	GL	**K**	
field	F	gravel	GVL		
figure	FIG.	gray	GY	key way	KWY
fine print note	FPN	green	G or GR	kick plate	KPL
finish	FNSH	gross vehicle weight	GVW	kiln-dried	KD
finish all over	FAO	ground	G, GND, GRD, or GRND	kilo (1000)	k
finished floor	FNSH FL	grounded (outlet)	G	1000 circular mils	kcmil
finished grade	FG or FIN GR	ground-fault circuit interrupter	GFCI	1000'	kFT
finish one side	F1S	ground fault protection of equipment	GFPE	kilovolt amps	kVA
finish two sides	F2S	grounding electrode conductor	GEC	kilowatt	kW
firebrick	FBCK	grounding electrode system	GES	kilowatt-hour	kWh
fire door	FDR	gypsum	GYP	kitchen	K or KIT.
fire extinguisher	FEXT	gypsum sheathing board	GSB	knife switch	KN SW
fire hydrant	FHY			knockout	KO
fireplace	FP	**H**			
fireproof	FPRF			**L**	
fire wall	FW	hand-off-auto	HOA		
fixed window	FX WDW	handrail	HNDRL	lamp	LT
fixture	FXTR	hardboard	HBD	lath	LTH
flammable	FLMB	hardware	HDW	laundry	LAU
flashing	FL	hardwood	HDWD	laundry tray	LT
flat	FL	hazardous	HAZ	lavatory	LAV
flexible metallic conduit	FMC	header	HDR	left	L
float switch	FS	heating	HTG	left hand	LH
floor	FL	heating, air-conditioning, refrigeration	HACR	less-flammable, liquid-insulated	LFLI
floor drain	FD	heating, ventilating, and air conditioning	HVAC	library	LBRY or LIB
flooring	FLG or FLR	heavy-duty	HD	light	LT
floor line	FL	hertz	Hz	lighting	LTG
flow switch	FLS	highway	HWY	lighting panel	LP
fluorescent	FLUOR or FLUR	hollow core	HC	lights	LTS
flush	FL	hollow metal door	HMD	limit switch	LS
flush mount	FLMT	horizontal	HOR	line	L
footing	FTG	horsepower	HP	linoleum	LINO or LINOL
foot per minute	FPM	hose bibb	HB	lintel	LNTL
foot per second	FPS	hot water	HW	live load	LL
foot switch	FTS	hot water heater	HWH	living room	LR
forward	F or FWD	hours	HRS	load	L
foundation	FDN	hydraulic	HYDR	location	LOC
four-pole	4P			locked-rotor ampacity	LRA
four-pole double-throw switch	4PDT SW	**I**		locked-rotor current	LRC
four-pole single-throw switch	4PST SW			louver	LVR or LV
four-pole switch	4PSW	immersion detection circuit interrupter	IDCI	lumber	LBR
frame	FR				

...ABBREVIATIONS...

M

magnetic brake	MB
main	MN
main bonding jumper	MBJ
main control center	MCC
manhole	MH
manual	MAN., MN, or MNL
manufacturer	MFR
marble	MRB
masonry	MSNRY
masonry opening	MO
material	MATL or MTL
maximum	MAX
maximum working pressure	MWP
mechanical	MECH
medicine cabinet	MC
medium	MED
memory	MEM
metal	MET or MTL
metal anchor	MA
metal door	METD
metal flashing	METF
metal jalousie	METJ
metal lath and plaster	MLP
metal threshold	MT
mezzanine	MEZZ
miles per gallon	MPG
miles per hour	MPH
minimum	MIN
mirror	MIR
miscellaneous	MISC
molding	MLDG
monolithic	ML
mortar	MOR
motor	M, MOT, or MTR
motor branch-circuit, short-circuit, ground-fault	MBCSCGF
motor circuit switch	MCS
motor control center	MCC
motor starter	M
motor switch	MS
mounted	MTD

N

nameplate	NPL
National Electrical Code®	NEC®
National Electrical Manufacturers Association	NEMA
National Electrical Safety Code	NESC
National Fire Protection Association	NFPA
negative	NEG
net weight	NTWT
neutral	N or NEUT
nominal	NOM
nonadjustable-trip circuit breaker	NATCB
non-time delay fuse	NTDF
normally closed	NC
normally open	NO
north	N
nosing	NOS
not to scale	NTS
number	NO.

O

Occupational Safety and Health Administration	OSHA
ohmmeter	OHM.
on center	OC
opaque	OPA
opening	OPNG
open web joist	OJ or OWJ
orange	O
ounces per inch	OZ/IN.
outlet	OUT.
outside diameter	OD
overall	OA
overcurrent	OC
overcurrent protection device	OCPD
overhang	OVHG
overhead	OH.
overload	OL
overload relay	OL

P

paint	PNT
painted	PTD
pair	PR
panel	PNL
pantry	PAN.
parallel	PRL
peak-to-peak	P-P
perpendicular	PERP
personal computer	PC
phase	PH
piece	PC
piling	PLG
pillar	PLR
pilot light	PL
piping	PP
pitch	P
plank	PLK
plaster	PLAS
plastic	PLSTC
plate	PL
plate glass	PLGL
plugging switch	PLS
plumbing	PLMB or PLBG
plywood	PLYWD
pneumatic	PNEU
point of beginning	POB
pole	P
polyvinyl chloride	PVC
porcelain	PORC
porch	P
positive	POS
pound(s)	LB
pounds per feet	LB-FT
pounds per inch	LB-IN.
pounds per square foot	PSF
pounds per square inch	PSI
power	P or PWR
power consumed	P
power factor	PF
precast	PRCST
prefabricated	PFB or PREFAB
prefinished	PFN
pressure switch	PS
primary switch	PRI
property line	PL
pull box	PB
pull switch	PS
pull-up torque	PUT
pushbutton	PB

Q

quadrant	QDRNT
quantity	QTY
quarry tile	QT
quarry-tile roof	QTR
quart	QT
quick-acting	QA

R

radius	R
raintight	RT
random	RDM or RNDM
range	R or RNG
receipt of comments	ROC
receipt of proposals	ROP
receptacle	RECEPT or RCPT
recess	REC
recessed	REC
rectifier	REC
red	R
reference	REF
refrigerator	REF
register	REG or RGTR
reinforce	RE
reinforced concrete	RC
reinforcing steel	RST
reinforcing steel bar	REBAR
required	REQD
resistance	R
resistor	R or RES
return	RTN
reverse	R or REV
reverse-acting	RACT
revision	REV
revolutions per minute	RPM
revolutions per second	RPS
rheostat	RH
ribbed	RIB
right	R
right hand	RH
rigid	RGD
riser	R
road	RD
roll roofing	RR
roof	RF
roof drain	RD
roofing	RFG
room	R or RM
root mean square	RMS
rotor	RTR
rough	RGH
rough opening	RO
rough sawn	RS

S

safety switch	SSW
sanitary	SAN
scale	SC
schedule	SCH or SCHED
screen	SCR
scuttle	S
secondary	SEC
section	SECT
selector switch	SS
series	S
service	SERV
service entrance	SE
service factor	SF
sewer	SEW.
shake	SHK
sheathing	SHTHG
sheet	SH or SHT
sheet metal	SM
shelf and rod	SH & RD

...ABBREVIATIONS

shelving	SHELV	**T**		**V**	
shingle	SHGL	telephone	TEL	valley	VAL
shower	SH	television	TV	valve	V
shutter	SHTR	temperature	TEMP	vapor seal	VS
siding	SDG	tempered	TEMP	vaportight	VT
silicon controlled rectifier	SCR	terazzo	TER	vent	V
sill cock	SC	terminal	T or TERM.	ventilation	VENT.
silver	Ag	terminal board	TB	vent pipe	VP
single-phase	1PH	terra cotta	TC	vent stack	VS
single-pole	SP	thermal	THRM	vertical	V or VERT
single-pole circuit breaker	SPCB	thermally protected	TP	vinyl tile	VTILE or VT
single-pole double-throw	SPDT	thermostat	THERMO	violet	V
single-pole double-throw switch	SPDT SW	thermostat switch	THS	volt	V
single-pole single-throw	SPST	three-phase	3PH	voltage	E or V
single-pole single-throw switch	SPST SW	three-pole	3P	voltage drop	VD
single-pole switch	SP SW	three-pole double-throw	3PDT	volt amps	VA
single strength glass	SSG	three-pole single-throw	3PST	volts	V
sink	S or SK	three-way	3WAY	volts alternating current	VAC
skylight	SLT	three-wire	3W	volts direct current	VDC
slate	S, SL, or SLT	threshold	TH	volume	VOL
sliding door	SLD	time	T		
slope	SLP	time delay	TD	**W**	
slow	S	time-delay fuse	TDF		
smoke detector	SD	time delay relay	TR	wainscot	WAIN
socket	SOC	toilet	T	walk-in closet	WIC
soffit	SF	tongue-and-groove	T & G	warm air	WA
soil pipe	SP	torque	T	washing machine	WM
solenoid	SOL	transformer	T, TRANS, or XFMR	waste pipe	WP
solid core	SC	transformer, primary side	H	waste stack	WS
south	S	transformer, secondary side	X	water	WTR
spare	SP	tread	TR	water closet	WC
specification	SPEC	triple-pole double-throw	3PDT	water heater	WH
splash block	SB	triple-pole double-throw switch	3PDT SW	water meter	WM
square	SQ	triple-pole single-throw	3PST	waterproof	WP
square foot	SQ FT	triple-pole single-throw switch	3PST SW	watt(s)	W
square inch	SQ IN.	triple-pole switch	3P SW	weatherproof	WP
square yard	SQ YD	truss	TR	welded	WLD
stack	STK	two-phase	2PH	welded wire fabric	WWF
stainless steel	SST	two-pole	DP	west	W
stairs	ST	two-pole double-throw	DPDT	white	W
standard	STD	two-pole single-throw	DPST	wide flange	WF
standpipe	SP	typical	TYP	wire gauge	WG
starter	START or STR			wire mesh	WM
steel	STL	**U**		with	W/
stone	STN			without	W/O
storage	STOR	unclamp	UCL	wood	WD
street	ST	underground	UGND	wrought iron	WI
structural glass	SG	underground feeder	UF		
sump pump	SP	undervoltage	UV	**Y**	
supply	SPLY	Underwriters Laboratories Inc.	UL		
surface four sides	S4S	unexcavated	UNEXC	yellow	Y
surface one side	S1S	unfinished	UNFIN		
switch	S or SW	up	U		
switched disconnect	SWD	utility room	U RM		

NEC® HAZARDOUS LOCATIONS

Hazardous Location—A location where there is an increased risk of fire or explosion due to the presence of flammable gases, vapors, liquids, combustible dusts, or easily-ignitable fibers or flyings.

Location—A position or site.

Flammable—Capable of being easily ignited and of burning quickly.

Gas—A fluid (such as air) that has no independent shape or volume but tends to expand indefinitely.

Vapor—A substance in the gaseous state as distinguished from the solid or liquid state.

Liquid—A fluid (such as water) that has no independent shape but has a definite volume. A liquid does not expand indefinitely and is only slightly compressible.

Combustible—Capable of burning.

Ignitable—Capable of being set on fire.

Fiber—A thread or piece of material.

Flyings—Small particles of material.

Dust—Fine particles of matter.

Classes	Likelihood that a flammable or combustible concentration is present
I	Sufficient quantities of flammable gases and vapors present in air to cause an explosion or ignite hazardous materials
II	Sufficient quantities of combustible dust are present in air to cause an explosion or ignite hazardous materials
III	Easily ignitable fibers or flyings are present in air, but not in a sufficient quantity to cause an explosion of ignite hazardous materials.

Divisions	Location containing hazardous substances
1	Hazardous location in which hazardous substance is normally present in air in sufficient quantities to cause an explosion or ignite hazardous materials
2	Hazardous location in which hazardous substance is not normally present in air in sufficient quantities to cause an explosion or ignite hazardous materials

Groups	Atmosphere containing flammable gases or vapors or combustible dust	
Class I	Class II	Class III
A	E	
B	F	none
C	G	
D		

DIVISION I EXAMPLES

Class I:

Spray booth interiors

Areas adjacent to spraying or painting operations using volatile flammable solvents

Open tanks or vats of volatile flammable liquids

Drying or evaporation rooms for flammable vents

Areas where fats and oil extraction equipment using flammable solvents are operated

Cleaning and dyeing plant rooms that use flammable liquids that do not contain adequate ventilation

Refrigeration or freezer interiors that store flammable materials

All other locations where sufficient ignitable quantities of flammable gases or vapors are likely to occur during routine operations

Class II:

Grain and grain products

Pulverized sugar and cocoa

Dried egg and milk powders

Pulverized spices

Starch and pastes

Potato and wood flour

Oil meal from beans and seeds

Dried hay

Any other organic material that may produce combustible dusts during their use or handling

Class III:

Portions of rayon, cotton, or other textile mills

Manufacturing and processing plants for combustible fibers, cotton gins, and cotton seed mills

Flax processing plants

Clothing manufacturing plants

Woodworking plants

Other establishments involving similar hazardous processes or conditions

MOTOR CALCULATION FORM

	Motor 1	Motor 2	Motor 3
	HP ☐ V ☐ φ ☐	HP ☐ V ☐ φ ☐	HP ☐ V ☐ φ ☐
FLC-Table 430.247, 248, 249, & 250	☐ A	☐ A	☐ A
Branch Circuit Conductors 430.22, FLC × 1.25	☐ × 1.25 = ☐ A	☐ × 1.25 = ☐ A	☐ × 1.25 = ☐ A
Conductor Size 310.16	☐ AWG	☐ AWG	☐ AWG
Overloads 430.32 Nameplate Amps	Nameplate Amps = ☐ A	Nameplate Amps = ☐ A	Nameplate Amps = ☐ A
Standard = Np × 1.25 or 1.15 based on SF or Temp Rise. 430.32.A.1	☐ × ☐ = ☐ A	☐ × ☐ = ☐ A	☐ × ☐ = ☐ A
Maximum = Np × 1.40 or 1.30 based on SF or Temp Rise. 430.32.C	☐ × ☐ = ☐ A	☐ × ☐ = ☐ A	☐ × ☐ = ☐ A
Short Circuit Ground Fault Protection Device 430.52 FLC × Table 430.52	☐ × ☐ = ☐ A	☐ × ☐ = ☐ A	☐ × ☐ = ☐ A
Up to standard size if answer not standard size fuse or breaker 240.6	☐ A	☐ A	☐ A
Feeder Conductor Largest FLC × 1.25 + FLC of other motors 430.24	☐ × 1.25 = ☐ + ☐ + ☐ = ☐ A		
Conductor Size 310.16(B)(16)	☐ AWG		
Feeder Protection 430.62 Largest branch circuit OCPD + FLC of other motors	☐ + ☐ + ☐ = ☐ A		
Down to standard size if answer not standard size fuse or breaker 240.6	☐ A		

Glossary

A

adjustable-trip circuit breaker (ATCB): A circuit breaker with a trip setting that can be changed by adjusting the amp setpoint, trip-time characteristics, or both within a particular range.

ampacity: The maximum amount of current a conductor can carry continuously without exceeding its temperature rating.

angle pull: A condition where conductors are pulled into a box and routed to a raceway entering an adjacent wall in the box.

architect's seal: An embossed stamp or insignia affixed to the construction drawing that signifies that a licensed professional architect has reviewed and approved the project drawings.

armature: The movable coil of wire in a generator that rotates through a magnetic field.

atom: The smallest particle to which an element can be reduced and still maintain the properties of the element.

B

battery: An electrical energy storage device made up of cell plates placed together in a series or parallel configuration to produce a voltage.

bonding: The process of establishing continuity and conductivity through a connection.

bonding jumper: A conductor that electrically connects two or more metal parts.

box fill: The total volume of all conductors, devices, and fittings in a box.

branch-circuit conductor: A conductor that makes up the portion of an electrical circuit between the circuit breaker or fuse and the loads connected in the circuit.

buck-boost transformer: A type of transformer designed to buck (lower) or boost (raise) line voltage.

busbar: A copper or aluminum bar, usually rectangular in shape, that conducts electricity within an electrical device.

C

capacitance (C): The ability of a component or circuit to store energy in the form of an electrical charge.

capacitor: An electrical device designed to store electrical energy by means of an electrostatic field.

cartridge fuse: A fuse constructed of a metallic link or links that are designed to open at predetermined current levels to protect circuit conductors and equipment.

circuit breaker (CB): A reusable overcurrent protective device that can be reset after it trips due to a short or a ground fault in the circuit.

combination circuit: A circuit that has both series and parallel connections.

conductor: A material that has low electrical resistance and permits electrons to move through it easily.

conductor fill: The maximum percentage of cross-sectional area that can be occupied by conductors inside a raceway.

current (I): The flow of electrons moving past a point in a circuit.

current transformer (CT): A transformer that is used to isolate an ammeter to prevent the hazards caused by connecting the meter to current.

D

drawing note: A note on a drawing that contains information specific to the drawing on that particular page or sheet.

drawing scale: A system of drawing representation in which drawing elements are proportional to actual elements.

E

effective voltage (V_{EFF}): The voltage value of a sine wave that produces the same amount of heat in a pure resistive circuit as DC of the same value.

elevation view: An orthographic drawing showing vertical planes of a building.

equipment bonding jumper (EBJ): A conductor that connects two or more parts of the equipment grounding conductor.

equipment grounding conductor (EGC): An electrical conductor that provides a low-impedance path between electrical equipment and enclosures and the system-grounded conductor and grounding electrode conductor.

F

feeder conductor: A conductor that carries more than one load at a time.

field winding: A magnet used to produce the magnetic field in a generator.

fixed load: A load where the current does not fluctuate.

floor plan: An orthographic drawing of a building as though cutting planes were made through it horizontally.

frequency (f): The speed (cycles per second) at which an armature spins in a magnetic field.

G

generator: A machine that converts mechanical energy into electrical energy by means of electromagnetic induction.

grounded conductor: A conductor that has been intentionally grounded to earth.

grounding: The connection of all exposed non-current-carrying metal parts to the earth.

grounding electrode conductor (GEC): A conductor that connects the grounding electrode(s) to the system-grounded conductor and/or the equipment grounding conductor.

H

handhole: A box, large enough to reach into but not large enough for a person to get into, that is placed in the ground and used for installing, operating, or maintaining equipment or wires.

horsepower: A unit of power equal to 746 W or 33,000 lb–ft per minute (550 lb–ft per second).

I

impedance (Z): The total opposition of any combination of resistance, inductive reactance, and capacitive reactance offered to the flow of an AC circuit.

inductance (L): A property that opposes the change of current in an AC circuit.

inductive circuit: A circuit in which current lags voltage. The greater the inductance is in a circuit, the larger the phase shift will be.

inductive reactance (X_L): The opposition of an inductor to alternating current.

inductor: An electrical device designed to store electrical energy by means of a magnetic field.

inverse-time circuit breaker (ITCB): A circuit breaker with an intentional delay between the time when the fault or overload is sensed and the time when the circuit breaker operates.

J

junction box: A box in which splices, taps, or terminations are made.

L

lampholder: A device that functions as a support for a lamp for the purpose of making electrical contact with the lamp.

luminaire schedule: A schedule that supplies specific information regarding the lighting distribution systems.

M

main bonding jumper (MBJ): The connection at the service equipment that bonds together the equipment grounding conductor, grounded conductor, and grounding electrode conductor.

matter: Anything that has mass and occupies space.

mechanical equipment circuit schedule: A schedule that provides specific information about the loads supplied by the distribution system.

motor full-load current (FLC): The current required by a motor to produce full-load torque at the motor's rated speed.

motor-volt amp (VA) rating: The amount of power used to produce horsepower output.

multiwire circuit: A circuit that consists of two or more ungrounded conductors sharing a common grounded neutral conductor.

N

neutral conductor: A current-carrying conductor that is intentionally grounded and connected to the neutral point of the system.

nipple: A conduit raceway that is 24″ or shorter in length.

noncoincidental load: A load that is not normally on at the same time as another load.

O

one-line riser drawing: An electrical drawing that shows the arrangement of the electrical service and main distribution equipment in block form.

P

panelboard schedule: A schedule that supplies detailed information about individual panelboards.

parallel circuit: A circuit that has two or more components connected so that there is more than one path for current to flow.

parallel conductors: Two or more conductors that are electrically connected at both ends to form a single conductor.

peak voltage (V_p): The maximum voltage of either a positive or negative alternation.

plot plan: An orthographic drawing that shows the location and orientation of a structure on a lot and the size of the lot.

plug fuse: A fuse that uses a metallic strip that melts when a predetermined amount of current flows through it.

potential transformer: A precision two-winding transformer that is used to step down high voltage to allow safe voltage measurement. Also known as an instrument transformer.

power factor (PF): The electrical efficiency of a circuit.

print: A detailed plan of a building drawn orthographically using a conventional representation of lines and includes abbreviations and symbols.

pull box: A box used as a point to route electrical conductors into a raceway system.

R

reactive power: The circulating current in a winding such as a motor or transformer.

resistance (R): The opposition to current flow.

resonance: A condition where the inductive reactance equals capacitive reactance.

S

schedule: A detailed list on a print that provides information about building components such as doors, windows, and electrical and mechanical equipment.

series circuit: A circuit that has two or more components connected so that there is only one current path.

service factor (SF): A number that represents the percentage of extra demand that can be placed on a motor for short time intervals without damaging the motor.

short circuit: Any circuit in which current takes a shortcut around the normal path of current flow.

show window lighting: Lighting used to enhance the window of a store that displays merchandise for sale.

sine wave: A symmetrical waveform that shows how AC varies over time.

single phasing: The operation of a motor that is designed to operate on three phases but is only operating on two phases because one phase is lost.

slip ring: A metallic ring connected to the ends of the armature that is used to connect the induced voltage to the brushes.

specifications: Sheets of paper, gathered together into a pamphlet (or into several volumes for a large building) covering a number of subjects in detail.

starting torque: The torque required to start a motor with a load applied.

straight pull: A condition where conductors are pulled into a box and routed to a raceway on the opposite wall of the box.

symbol: A graphic representation of an object.

synchronous speed: The speed in which the magnetic field travels around the windings of a motor.

system: Any separately derived power source such as a transformer, standby generator, fuel cell, or other power production system.

T

temperature rise (TR): The amount of heat a motor generates without destroying the motor insulation.

title block: An area on a print used to provide information about the print.

torque: The amount of twisting power that the shaft of a motor delivers.

track lighting: Lighting used to illuminate an area using adjustable spotlight fixtures that are fastened to a current-carrying metal track.

true power: The actual work being done by the load using electricity.

U

ungrounded system: A system that does not have a winding grounded at the supply transformer.

unity circuit: A pure inductive circuit that is 90° out-of-phase lagging or a pure capacitive circuit that is 90° out-of-phase leading.

U-pull: A condition in which conductors are pulled into a box and routed to another raceway entering the same wall of the box.

V

variable-frequency drive (VFD): An electronic unit designed to control the speed of a motor using solid-state components.

variable load: A load where the current fluctuates.

voltage (E): The amount of electrical pressure in a circuit.

voltage drop: The voltage that is lost due to the resistance of conductors.

W

watt: A unit of measure equal to the power produced by a current of 1 A across a potential difference of 1 V.

wireway: A sheet metal or nonmetallic enclosure with a cover that opens to provide access to the conductors inside.

Index

Page numbers in italic refer to figures

A

AC (alternating current), 10
AC generators, 22, 22–23, 23
actual shaft speed calculations, 133–134, *134*
adjustable-trip circuit breakers (ATCBs), 98
air conditioning loads
   multifamily dwellings, 189–190, *190*
   nondwelling occupancies, 206, 208
   one-family dwellings, 181, *182*, 183–184, *184*
allowable conductor ampacity, *88*, 88
alternating current (AC), 10
alternating current (AC) generators, 22, 22–23, 23
alternations, 23, *24*, 26
ambient temperature along conductors, 89
American wire gauge (AWG), 9, *9*
ammeters, 16–17, *17*
ampacity
   allowable conductors, *88*, 88
   and ambient temperature, 89
   busbars, *94*, 94
   derating factors, 92
   ratings for small conductors, 96, *97*
   and terminal temperature ratings, 92–93, *93*
amp-hours, 10
angle pulls, *160*, *161*, 161, *164*, 164
answers, 5
apparent power, *31*, 32
appliances, 208
apprentice electricians, 2
architect's scales, *221*, 221
architect's seals, 219, *220*
arc welders, *211*, 211
armatures, 22, 22–23, *23*, *24*
ATCBs (adjustable-trip circuit breakers), 98
atoms, *8*, 8, *9*
autotransformers, 55–56, *56*
auxiliary gutters, 92
auxiliary wireways, 92
AWG (American wire gauge), 9, *9*

B

balancing motor loads, 130, *131*
bare conductors, 151
batteries, 10

bonding, 66
bonding jumpers, *72*, 72–73, *73*, *74*
box fill, 155–158, *156*, *157*, *159*
branch-circuit conductors, 123–125
branch circuit protection, motors, 128–129
branch circuits
   one-family dwellings, 173
   overcurrent protection, 99–100, *100*
   sizing conductors, 93–94
   voltage drop
      and feeders, 106, *116*, 116–117, *117*
      single-phase (1ϕ), *107*, 107–111, *108*, *109*, *110*, *111*
      three-phase (3ϕ), 111–114, *112*, *114*, *115*
brushes, 22, *22*, *23*
buck-boost transformers, 61, *62*
busbars, *94*, 94
bypass devices, variable-frequency drive, 140–141

C

capacitance, 28, *30*
capacitive reactance, 28
capacitors, 28, *28*, 212
cards, electrician's, 2
cartridge fuses, 99, *100*
cells, battery, 10
circuit breakers (CBs)
   adjustable-trip, 98, 317
   defined, 99
   instantaneous trip, 128
   interrupting capacity of, 101
   inverse-time, *97*, 97–98, *98*, 128, 318
   for transformers, *57*, 57
circuit breaks, 36, 39
circuits. *See also* single-phase (1ϕ) circuits; three-phase (3ϕ) circuits; branch circuits
   breaks, 36, 39
   capacitive, *30*, 30
   combination, *41*, 41–43, *42*
   inductive, *30*, 30
   loads of, 106, 209
   multiwire, *43*, 43–44, *44*
   parallel, *38*, 38–40
   power loss of, 114–116, *115*
circuits (*continued*)

series, *36*, 36–38
testing of, 16–18
voltage drop of, 106
circular mils, 106, 108, 113
clamp-on ammeters, *17*, 17
closed delta transformers, 53–54, *54, 61*, 61
coated conductors, 107
combination circuits, *41*, 41–43, *42*
compact conductors, 150, *151*
conductor fill
　defined, *148*, 148
　for nipples, 152, *153*
　sizing conduit, 151, *152*
　tables, 148–151, *149, 150, 151*
conductors, 8
　allowable ampacity of, *88*, 88, 96, *97*
　bare, 151
　and boxes, *156*, 156–158, *157, 159*
　coated, 107
　compact, 150, *151*
　current-carrying, 86, 90, *91*
　dimensions of, *150*, 150, *151*
　feeder, 91
　for general wiring, *86*, 86–87
　grounded, 66–67, *67*
　in gutters, 92
　identification of, 87
　installing in raceways, *155*, 155
　insulated, *86*, 86–87, *88*
　materials for, 9, *106*, 106
　neutral as current-carrying, 90, *91*
　overcurrent protection for, 96–97, *97*
　parallel, 86, *87*
　properties of, 106–107
　rated from 2001 V to 35,000 V, 92
　service, *182*, 182–183, 185
　service entrance, 91
　sizing of
　　for branch circuits, 93–94, *94*
　　for motors, 123–125, *124, 125*
　　for 1φ circuits, 108–109, *109*
　　for 3φ circuits, *112*, 112–113
　strand numbers of, 106
　types of, *86*
　uncoated, 107
　for variable-frequency drives, 140
　in wireways, 92, *153*, 153–154, *154*
conduit fill, 89, 92
conduits
　and ambient temperature, 89–90, *90*
　distance between, 162, *163*, 164–165, *165*
　properties of, 149, *150*
　sizing of, 151, *152*
connected loads, 202
connections, motor, 136–137
Construction Specifications Institute (CSI), 219
continuous-duty motors, 123–124, *124*
continuous lighting loads, *202*, 202

continuous loads, 93–94, 100–101
control circuit conductors, 138
cooking equipment
　commercial, 207, *208, 209*, 209
　multifamily dwellings, 188–189, *189*
　one-family dwellings, 177–181, *178, 179*
copper atoms, 8
covers, pull box, 162
CSI MasterFormat, 219, *220*
CTs (current transformers), *56*, 56
current
　calculating using Ohm's law, *12*, 12
　calculating using power formula, *15*, 15
　defined, *9*, 9
　measuring with meters, 16–17, *17*
　and parallel circuit calculations, 39
　relationships, 11
　and series circuit calculations, 37
current-carrying conductors, 86, 90, *91*
current ratios, 49–50
current transformers (CTs), *56*, 56
cycles, 25

D

DC (direct current), *9, 10*, 10
delta connections, 53–55, *54, 55, 61*, 61
delta neutral conductors, *78*, 78
delta-wye connections, 58–59, *59*
demand factors, 203, 209–210
demand loads, 193
derating factors, 92
derating neutral conductors, 79
dielectric, *28*, 28
direct current (DC), *9, 10*, 10
discharging capacitors, 28
display window lighting. *See* show window lighting
drawing notes, 221
drawings, 218, *219*, 320
drawing scales, *221*, 221
dry cells, 10
dryer loads, 175–177, *176, 188*, 188
dual-element fuses, 128
duplex receptacles, *205*

E

EBJs (equipment bonding jumpers), *73*, 73, 318
effective voltage, *26*, 26
EGCs (equipment grounding conductors), 69–71, *70, 71*, 318
electrical heating appliances, 212
electrician's cards, 2
electrons, 8, *9*, 9
electrostatic charge, 28
elevation views, 216–217, *218*

ELI the ICE man, 30
equipment bonding jumpers (EBJs), *73*, 73, 318
equipment grounding conductors (EGCs), 69–71, *70*, *71*, 318
exams, 2–5, *3*, *6*

F

farm dwellings, 192–193
farm loads, 211
feeder circuits, *116*, 116–117, *117*
feeder conductors, 91, 129–132, *130*
field windings, *22*, 22
fixed appliances, 175, *176*, 187, *188*, 208
fixed loads, 212
fixed multioutlet assemblies, 206, *207*, 209
FLC (full-load current), 122–123, *123*
floor plans, 216, *217*
flux, 22, *23*
formula wheels, *15*, 15
frequencies, 25
full-load current (FLC), 122–123, *123*
fuses
    cartridge, 99, *100*
    dual-element (time-delay), 128
    nontime-delay, 128
    plug, 98, *99*
    standard ratings for, *97*, 97
    for transformers, *57*, 57

G

GECs (grounding electrode conductors), 69, 74–75, *75*, 318
generators, *22*, 22–23, *23*, 212
grounded conductors, 66–67, *67*, 318
ground-fault current paths, 66
ground-fault detection systems, 79
ground fault protection, motors, 128–129, 131–132
grounding, *66*, 66, *70*
grounding electrode conductors (GECs), 69, 74–75, *75*, 318
grounding electrodes, 69, *70*
gutters, 92

H

handholes, 163
heating loads
    commercial buildings, 206
    multifamily dwellings, 189–190, *190*
    one-family dwellings, 181, *182*, 183–184, *184*
    optional school calculation method, 208
high-intensity discharge (HID) lighting, 203, *204*
high legs, 61
high-resistance ground systems (HRGSs), *80*, 80
horsepower calculations, motors, *135*, 135–136

I

impedance, 29
index, NEC, 4
inductance, *27*, 27
inductive circuits, *30*, 30
inductive reactance, 27
inductors, 27
in-phase circuits, 30, *31*
instantaneous trip breakers, 128
instrument (potential) transformers, 56
insulated conductors, *86*, 86–87, *88*
insulation classifications, *88*, 88
insulators, 8, *28*, 28
integral thermal protectors, 128
intermittent-duty motors, 124
interrupting capacity, 101
inverse-time circuit breakers (ITCBs), *97*, 97–98, *98*, 128
iron core transformers, 51

J

journeyman electricians, 2
junction boxes
    1000 volts and under, *160*, 160–163, *161*, *162*
    defined, *160*, 160
    depth of, *162*, 162
    over 1000 volts, 163–165, *164*, *165*

K

kilovolt-amps (kVA), 51, 132–133

L

lampholders, 205
laundry circuits, *174*, 174
laundry loads, 193–194, *195*
lighting
    circuits, 203
    loads, *202*, 202–205, 208
    for manufactured homes, 193–194, *195*
    for mobile homes, 193–194, *195*
    for multifamily dwellings, 186–187, *187*
    for nondwelling occupancies, 203, *204*
    for one-family dwellings, 172–174, *173*, *174*, *175*
    show window, 203–204, *204*
    track, 204–205, *205*
    transformers for, 54, *55*
lighting circuits, 203
lighting transformers, 54, *55*
loads
    maximum distance from power source, 110–111, *111*, 114, *115*
    maximum placed on circuits, 109–110, *110*, 113, *114*
    overcurrent protection for, 100–101

load-side bonding jumpers, 73
locked-rotor current (LRC), 138–140, *139*
luminaire schedules, 222, *223*

M

magnetic lines of force, 22, *23*
magnetic poles, 25
main bonding jumpers (MBJs), *68*, 68, *70,* 318
manufactured homes, 193–194, *195*
master electricians, 2
master electrician's card, *4*
master electrician's exams, 2–4, *3*
MasterFormat. *See* CSI MasterFormat
matter, 8
maximum distance from power source to loads, 110–111, *111,* 114, *115*
maximum loads on circuits, 109–110, *110,* 113, *114*
maximum-size motor overloads, *127*
MBJs (main bonding jumpers), *68*, 68, *70,* 318
mechanical equipment circuit schedules, 222
megohmmeters, *18*, 18
meters, *16,* 16–18, *17, 18*
mobile homes, *192,* 193–195, *195*
motors
 branch short-circuit calculations, 128–129
 calculation forms, *122,* 122
 conductor sizing, 123–125, *124, 125*
 connections, 136–137, *137, 138*
 control circuit conductors, 138
 feeder conductor sizing, 129–132, *130*
 full-load current (FLC), 122–123, *123*
 ground-fault protection calculations, 128–129
 horsepower calculations, *135,* 135–136
 locked-rotor current ratings, 138–140, *139*
 for multifamily dwellings, 190
 nameplate current calculations, 136
 for nondwelling occupancies, 208
 for one-family dwellings, 181, *182*
 overload calculations, 126–128, *127*
 speed calculations, 133–134, *134*
 torque calculations, 134
 voltage drop, 137
 volt-amp (VA) calculations, 132–133
multifamily dwellings
 optional calculation method, 190–192, *191*
 standard calculation method, *186,* 186–190
multioutlet assemblies, 206, *207,* 209
multiplier resistors, 16
multiwire circuits, *43,* 43–44, *44*

N

nameplate current, 136
National Electrical Code (NEC), 4, *5*

national testing organizations, 4
NEC index, 4
negative alternations, 23, *26,* 26
neutral conductors, 76–79, *77, 78,* 90, *91*
neutral points, *77,* 77
neutrons, 8
nipples, 152, *153*
noncoincidental loads, 181, 319
noncontinuous loads, 100–101
nondwelling occupancy lighting, 203, *204*
non-motor-operated appliances, 212
nontime-delay fuses, 128–129
notes, drawing, 221
nuclei, *8,* 8

O

OCPDs (overcurrent protective devices), 126–129, *127,* 138
ohmmeters, *17,* 17
ohms, 9
Ohm's law
 calculating voltage drop, *107,* 107
 determining current, *12,* 12
 determining resistance, *12,* 12
 determining voltage, *11,* 11
 overview, 10, *11*
 and parallel circuits, 39–40
 and power formula, *15,* 15
 and series circuits, 37–38
one-family dwellings
 optional calculation method, *183,* 183–185, *184, 185*
 standard calculation method, *172,* 172–183
  air conditioning loads, 181, *182,* 183–184, *184*
  branch circuit requirements, 173–174
  cooking equipment, 177–181, *178, 179*
  dryer loads, 175–177, *176*
  fixed appliances, 175, *176*
  general loads, 184, *185*
  heating loads, 181, *182,* 183–184, *184*
  laundry circuits, *174,* 174
  lighting, 172–174, *173, 174, 175*
  motors, 181, *182*
  sizing service conductors, *182,* 182–183
  small-appliances, *174,* 174
one-line riser drawings, 218, *219*
open delta transformers, *55,* 55, 61
open neutral (multiwire circuits), *44,* 44
optional calculation method
 multifamily dwellings, 190–192, *191*
 one-family dwellings, *183,* 183–185, *184, 185*
 schools, 208
orthographic projections, 216
outlines, exam, 2, *3*
out-of-phase circuits, 30, *31*
output voltage per degree of rotation, *22, 23, 24*
overcurrent protection
 for branch circuits, 99–101

common devices for, 94, *95*
for continuous loads, 100–101
for devices rated 800 A or less, 95, *96*
for devices rated over 800 A, 95–96
interrupting capacity, 101
for noncontinuous loads, 100–101
for small conductors, 96–97, *97*
standard ratings, *97*, 97–98
for transformers, 56–57, *57*
types of, 98–99, *99*, *100*
overcurrent protective devices (OCPDs), 126–129, *127*, 138

P

pacing, exam, 4
panelboard schedules, 222, 319
parallel circuits, *38*, 38–40
parallel conductors, 66–67, 86, *87*
parallel connections, 10
part-winding motors, 124–125
peak voltage, 26
periodic-duty motors, 124
periods, 25
phase angles, *31*
phase coils, 58, *59*
phase converters, 212
phase lines, 58, *59*
phase shift, 30
plans, 216, *217*, *218*
plaster rings, 158–160, *159*
plot plans, 216, *218*
plug fuses, 98, *99*
points of beginning, 216
poles, 25
positive alternation, 23, *26*, 26
potential transformers, 56
power, 37–38, 40
power factors, 29–30
power formulas, *13*, 13–15, *14*, *15*
power loss, *51*, 51, 114–116, *115*
power sources, 110–111, *111*, 114, *115*
power transformers, 54, *55*
practice tests, 5–6
preparing for the exam, 2–5, *3*, *6*
primary windings, transformer, *48*, 48
prime movers, 23
prints, *216*, 216–218
protons, 8
pull boxes, 160–165
defined, *160*, 160
depth of, *162*, 162
1000 volts and under, *160*, 160–163, *161*, *162*, *163*
over 1000 volts, 163–165, *164*, *165*
Pythagorean theorem, 32

R

raceways, 89, *90*, 151
raised covers, 158–160, *159*, *160*
ratios, transformer, *48*, 48–50, *49*
reactive power, *31*, 31–32
receptacle loads, *205*, 205–208, *207*, *209*
receptacles, *205*, 205
refrigeration equipment, 212
removable covers, pull box, 162
requirements, exam, 4
resistance
calculating using Ohm's law, *12*, 12
combination circuit calculations, 41–43, *42*
defined, 9
measuring with meters, *17*, 17–18, *18*
parallel circuit calculations, 40
relationships, 11
series circuits calculations, 37
resistance welders, 211
resonance, 30
restaurant calculation method, 210
riser drawings, 218, *219*
root-mean-square (RMS) voltage, 26
rotation, armature, *22*, 22–23, *23*, *24*

S

scales, *221*, 221
schedules, 222, *223*, 319
school calculation method, 208–209, *209*
seals, architect's, 219, *220*
secondary windings, transformers, *48*, 48
series circuits, *36*, 36–38
series connections, 10
service conductors, *182*, 182–183, 185
service entrance conductors, 91
service factors (SFs), 126–127, *127*
service loads, 193–194, *195*
shaft speed calculations, 133–134, *134*
shells, electron, *8*, 8
short-circuit ground fault protection, 128–129, 131–132
short circuits, 36, 39
short-time duty motors, 124, *125*
show window lighting, 203–204, *204*
signs, 205
sine waves, *22*, 22, *23*, *24*, *31*
single non-motor-operated appliances, 212
single-phase (1ϕ) circuits
branch circuit calculations, *107*, 107–111, *108*, *109*, *110*, *111*
maximum distance from power sources to loads, 110–111, *111*
neutral conductors, *77*, 77
single-phase (1ϕ) generators, *22*, 22
single-phase (1ϕ) motors
connections, *137*, 137
feeders, 130, *131*
volt-amp (VA) calculations, 132

single-phase (1φ) transformers, 52, *60*, 60
single phasing, 126
single receptacles, *205*
skin effect, *27*, 27
slip rings, *22*, 22, *23*
small-appliances, *174*, 174, 193–194, *195*
small motors, 125–126
space heating equipment, 212
specifications, 218–219, *220*
standard calculation method
    manufactured homes, 193–194, *195*
    mobile homes, 193–194, *195*
    multifamily dwellings, *186*, 186–190
    one-family dwellings, *172*, 172–183
        air conditioning loads, 181, *182*, 183–184, *184*
        branch circuit requirements, 173–174
        cooking equipment, 177–181, *178*, *179*
        dryer loads, 175–177, *176*
        fixed appliances, 175, *176*
        general loads, 184, *185*
        heating loads, 181, *182*, 183–184, *184*
        laundry circuits, *174*, 174
        lighting, 172–174, *173*, *174*, *175*
        motors, 181, *182*
        sizing service conductors, *182*, 182–183
        small-appliances, *174*, 174
starting torque, 134
stepping up/down voltage, 55–56, *56*
straight pulls, *160*, 160, 163, *164*
strands per conductor, 106
study tips, 4–5
supply-side bonding jumpers, *72*, 72–73
symbols, 222, *223*
synchronous speeds, 133
system bonding jumpers, 73, *74*
system-grounded conductors, 75–76, *76*
system grounding electrode conductors, 74–75, *75*
systems, 73

T

tabbing the NEC, 4
tap connections, 59
temperature derating factors, 92
temperature rise (TR), 126
terminal temperature ratings, 92–93, *93*
testing circuits, 16–18
thermal protectors, 128
three-phase (3φ) circuits
    branch circuit calculations, 111–114, *112*, *114*, *115*
    neutral conductors, *78*, 78–79, *79*
three-phase (3φ) generators, *23*, 23
three-phase (3φ) motors
    connections, 137, *138*
    feeders, 130, *131*
    volt-amp calculations, 132

three-phase (3φ) transformers
    connections, 60–61, *61*, *62*
    delta-wye calculations, 58–59, *59*
    sizing, 52–55, *53*, *54*, *55*
    volt-amp ratings, *50*, 50
time-delay (dual-element) fuses, 128
timing, 4
tips, exam preparation, 5
title blocks, 219, *220,* 320
torque, 134
track lighting, 204–205, *205*
training, 2
transformers
    autotransformers, 55–56, *56*
    buck-boost, 61, *62*
    closed delta, 53–54, *54*, *61*, 61
    connections of, 58–61, *60*, *61*, *62*
    current, *56*, 56
    current ratios for, 49–50
    delta connections of, 58–59, *59*, *61*, 61
    delta-wye calculations for, 58–59, *59*
    efficiency of, *51*, 51
    open delta, 52, *55*, 55, 61
    overcurrent protection of, 56–57, *57*
    potential (instrument), 56
    ratings of, *50*, 50–51
    single-phase (1φ). *See separate main entry*
    sizing of, 52–55, *53*, *54*, *55*
    three-phase (3φ). *See separate main entry*
    turns ratios for, *48*, 48
    voltage ratios for, 48–49, *49*
    wye connections of, 52–53, *53*, 60, *61*
true power, 29, *31*, 32
turns ratios, *48*, 48

U

unbalanced current, 76, 77, *78*
unbalanced loads, 76
uncoated conductors, 107
ungrounded systems, 79
unity circuits, 30
U-pulls, 161–162, *162*

V

valance shells, *8*, 8
variable-frequency drives (VFDs), 140
variable loads, 212
varying-duty motors, 124
VFDs (variable-frequency drives), 140
voltage
    and AC generators, *22*, 22–23, *23*, 24
    calculating using Ohm's law, *11*, 11
    calculating using power formula, *14*, 14

defined, 10
effective voltage, *26*, 26
measuring with meters, *16*, 16
parallel circuit calculations, 39
peak voltage, 26
relationships, 11
series circuits calculations, 37
stepping up/down, 55–56, *56*
voltage drop
  calculations
    for feeder and branch circuits combined, *116*, 116–117, *117*
    for 1ϕ circuits, *107*, 107–111, *108*, *109*, *110*, *111*
    power loss, solving for, 114–116, *115*
    for 3ϕ circuits, 111–114, *112*, *114*, *115*
    using Ohm's law, *107*, 107
    using voltage drop formula, *108*, 108, 111, *112*
  conductor properties of, 106–107
  defined, 106
  factors of, *106*, 106
  for feeder and branch circuits combined, *116*, 116–117, *117*
  formulas for, *108*, 108, 111, *112*
  of motors, 137
  NEC recommendations for, 106
  reducing, 106
voltage output per degree of rotation, *22*, *23*, *24*
voltage ratios, 48–49, *49*
volt-amp ratings, *50*, 50–51
voltmeters, *16*, 16
volts (V), 10

W

water heaters, 212
watt-hour meters, 18
watts, 32, 135
welding equipment, *211*, 211
wet cells, 10
wild legs, 61
windings, 55–56, *56*
wire coils, 22
wires, *9*, 9
wireways, 92, 152–154, *153*, *154*, *155*
workbooks, 5–6, *6*
wye connections, transformers, 52–53, *53*, 60, *61*
wye neutral conductors, *79*, 79
wye-start, delta-run motors, 124–125